Modelling Parasite Transmission and Control

ADVANCES IN EXPERIMENTAL MEDICINE AND BIOLOGY

Modelling Parasite Transmission and Control

Edited by

Edwin Michael, PhD

Department of Infectious Disease Epidemiology, Imperial College, London, UK

Robert C. Spear, PhD

Center for Occupational and Environmental Health, School of Public Health, University of California, Berkeley, California, USA

Springer Science+Business Media, LLC

Landes Bioscience

Springer Science+Business Media, LLC
Landes Bioscience

Printed in the USA.

Springer Science+Business Media, LLC, 233 Spring Street, New York, New York 10013, USA
http://www.springer.com

Please address all inquiries to the publishers:
Landes Bioscience, 1002 West Avenue, Austin, Texas 78701, USA
Phone: 512/ 637 6050; FAX: 512/ 637 6079
http://www.landesbioscience.com

The chapters in this book are available in the Madame Curie Bioscience Database.
http://www.landesbioscience.com/curie

Modelling Parasite Transmission and Control, edited by Edwin Michael and Robert C. Spear. Landes Bioscience / Springer Science+Business Media, LLC dual imprint / Springer series: Advances in Experimental Medicine and Biology.

ISBN: 978-1-4419-6063-4

Library of Congress Cataloging-in-Publication Data

Modelling parasite transmission and control / edited by Edwin Michael, Robert C. Spear.
p. ; cm. -- (Advances in experimental medicine and biology ; 673)
Includes bibliographical references and index.
ISBN 978-1-4419-6063-4
1. Communicable diseases--Transmission--Mathematical models. I. Michael, Edwin. II. Spear, Robert C., 1939- III. Series: Advances in experimental medicine and biology ; v. 673.
[DNLM: 1. Parasites--pathogenicity. 2. Disease Vectors. 3. Models, Theoretical. W1 AD559 v.673 2010 / QZ 85 M689 2010]
RA639.M627 2010
362.196'9--dc22

2010001263

DEDICATION

To our immediate families who have lived with this project, Shirin, Priya, and Patty and to numerous colleagues for their forbearance as well as their support.

PREFACE

Modelling parasite transmission has made enormous strides since the seminal models of Ross for describing malaria transmission developed during the early 1900s (Ross, 1911). McDonald's use of the early malaria models to show that killing adult mosquitoes would be particularly effective in reducing infection transmission was a major advance in demonstrating the usefulness of theoretical analysis and population dynamics modelling in particular for guiding parasite control programmes, and since then parasite transmission models have also been used to guide the onchocerciasis control programme in Africa (Habbema et al, 1992), as well as for investigating best strategies for controlling a host of other parasites, including tuberculosis, trachoma and lately helminth infections, such as schistosomiasis and filariasis (Chan et al, 1995; Laing et al, 2007; Michael et al, 2004). The importance of this work is highlighted by greater understanding of threshold phenomena in transmission dynamics leading to the concept that natural "breakpoints" occur below which parasite systems will go extinct to the roles that worm mating behaviour and infection aggregation can play in both helminth transmission and control (Hairston, 1962; Macdonald, 1961, 1965). The emerging trend from this work is thus the increasing use of understanding parasite transmission dynamics via the construction and analysis of mathematical models for use in guiding the development of informed parasite control strategies, so much so that this twin objective, viz improving understanding of parasite transmission dynamics and applying models to guide parasite control, has almost become a *de facto* goal of most recent work in parasite transmission modelling.

Another growing theme in parasite transmission modelling is the evolution of modelling techniques and conceptual frameworks, from the phenomenological investigation of key factors governing transmission dynamics at large population scales, based on the generalized Anderson and May models synthesized during the 1980s and early 1990s, to the more detailed treatment of localized transmission of parasites at smaller population and spatial scales. Mathematically, this has meant that modelling techniques have increasingly moved from the application of simple dynamic differential equations-based analytic approaches to the use of simulation and other computationally intensive methods that focus on fitting and calibrating models to local data. This has raised debate about the roles of simplified a priori-derived mechanistic models versus more complex but a posteriori models derived from data for generalizing understanding of transmission processes, and therefore how best to use models to further the theory of parasite transmission and accordingly its use for designing useful control options. This includes at its core the issue of how best to reconcile and model the impact of ecological and spatial scales as well as the effective addressing of other key, possibly scale-dependent heterogeneities of increasing complexity, such as impacts of host immunity, genetic diversity, multi-species interactions, spatial heterogeneity and between and within community connections, and external forcing factors linked to environmental variation and change.

The increasing recognition of the public health importance of parasites has led to a more recent flowering of interest to tackle and resolve these modelling issues, culminating in the convening of both the Dahlem Conference on the population biology of infectious diseases held in 1981 and the follow up Issac Newton Institute programme on modelling infectious diseases held during 1993. These meetings summarized the then contemporary state of parasite ecology and epidemiology (Anderson and May, 1982; Grenfell and Dobson, 1995), but also underscored gaps and future research needs. This book is an attempt to present the progress made since these meetings in terms of both the conceptual and practical advances made in addressing key gaps in parasite transmission modelling (focus here restricted to the major microbial and helminth parasites), the new understanding these have yielded regarding the transmission and control of these parasites, and the future challenges they portend for population dynamics modelling.

We have organized the material in the book into two major sections, the first presenting the state of the art in models aimed at capturing complex or detailed aspects of transmission dynamics beginning with a review of the evolution of modelling malaria transmission. This first chapter, by Smith and Ruktanonchai, describes the process by which the simplifying assumptions underlying early modelling efforts were relaxed and models made more realistic (and generally more complex) over time as they became informed by new information from observational studies of vector biology and epidemiology. A key highlight is the need to consider the spatial scale in mosquito biting behaviour and pattern if more robust models of malaria transmission in real populations are to be developed. The second chapter, by Michael and Gambhir, focuses on the implications for disease transmission and control of differences in density-dependent processes regulating larval infection in the two major mosquito vectors transmitting filariasis. The authors demonstrate how a careful consideration and modeling of the occurrence and operation of multiple density-dependent processes acting on different stages of the parasite life cycle will be required to explain differential system persistence, transitions, and resilience to perturbations in different host-vector-parasite systems. In a similar spirit, but in a more general case, is the chapter by Pugliese (Chapter 3) on the dynamics of multiple parasite species infecting a host population. This might be characterized as a phenomenological investigation focusing on general questions of competition or co-existence between parasite species. Given the prevalence of co-infections commonly observed in the field, this chapter presents a foundation from which studies of specific co-infections might be explored using models.

Chapters 4 and 5 by Gaff and Schaefer and by Cornell exemplify current advances in modeling tick-borne disease transmission and helminths, respectively, but are also of particular interest because of the modeling approaches utilized. Gaff and Schaefer use metapopulation models, which have been common in population biology for many years, because tick population dynamics are very sensitive to local environmental conditions. In addition, patch connectivity is shown to be of considerable importance, specifically with respect to control strategies. The notion of connectivity resurfaces in a similar context in the chapter by Seto and Carlton in Part II.

The chapter by Cornell is similar to that of Pugliese in its phenomenological orientation, and stems from the observation that in the case of helminths "the processes affecting parasite demographics are density dependent—i.e., depend nonlinearly on the number of worms in the host—such as the risk that a female cannot find a mate, or the modulation of infection rate by the host's immunity." Such questions are naturally addressed through the use of stochastic models as shown by the author. Although this chapter mainly focuses on simulation tools useful in the study of such models, Cornell also makes the case for the continued utilization of analytically tractable models in search of general principles not easily discernable from simulation studies.

In common with the metapopulation chapter of Gaff and Schaefer, Chapters 6 and 7 by Remais and by Spear and Hubbard represent a transition from phenomenological studies at an unspecified scale to a local scale where the issue is to specify effective and sustainable local control tactics. Remais (Chapter 6) describes a variant of the Anderson-May model of schistosome transmission which is adjoined to a submodel of uninfected snail density as a function of temperature and rainfall. To the extent possible, the model was parameterized to utilize both site-specific and site-invariant data for model calibration and subsequent use in short-term forecasting. Spear and Hubbard (Chapter 7) elaborate on the parameter estimation issues and residual parameter uncertainty central to model use at this scale and propose an approach tailored to deal with the heterogeneous nature of local data typically available for this purpose.

Part II of the book serves to highlight the current use of transmission models in the planning, monitoring and evaluation of parasite control programmes, and begins with Chapter 8 by Buckee and Gupta, who review advances made in using mathematical models of transmission for the design of malaria control programmes. The chapter shows how ultimately the design of appropriate control programmes will rely on better understanding of the complex dynamical interactions between vector ecology, generation and impact of pathogen diversity on host immunity and development of drug resistance, and the pathology of the disease in humans, which is only now beginning to be studied as a result of advances particularly in pathogen genomics. In a similar vein, White and Garnett (Chapter 9) highlight the advances made in modeling TB to date but also underscore how further progress in predicting the effects of control will crucially require gaining a better knowledge of the natural history and processes involved in disease transmission, particularly rates of change in latency, and impacts of exogenous reinfection, active disease and recovery from TB disease. They suggest that a key role of models in this regard is to pinpoint existing gaps in knowledge, and in conjunction with developments in diagnostics and genetic analysis, help guide empirical research to resolve such gaps.

One study that has begun to model the impact of control programmes on both infection dynamics and disease pathogenesis (blinding sequelae) is that described by Ghambir and colleagues for trachoma in Chapter 10. A novel aspect of their modelling framework, with implications for other diseases, is the combination of a compartmental SI contact type module to model microparasite infection dynamics with a module for capturing disease prevalence based on quantifying the number of prior infections (or infection "burden") experienced by individuals in a population.

The last two chapters in this section address the roles of parasite transmission models for decision-making in disease control. In Chapter 11, Michael and Ghambir describes the role that models can play in resolving the management questions regarding community level interventions for lymphatic filariasis elimination, including evaluating thresholds to meet various intervention endpoints, optimising control parameters for different endemic conditions, addressing the impacts of vector and spatial heterogeneities, and developing and applying rational monitoring and evaluation protocols for assessing programme success. The results indicate that instituting an adaptive management approach to parasite control, focusing on supporting the development of locally appropriate and ecologically resilient control strategies will be crucial to resolving the uncertainties surrounding parasite population dynamics and the effectiveness of presently proposed intervention options. Seto and Carlton (Chapter 12), by contrast, emphasize how site-specific epidemiologic and disease control options should and can be incorporated into modelling frameworks to support routine decision-making by public health officials in endemic regions, focusing on the model developed by their group for *Schistosoma japonicum* transmission in China. They demonstrate in particular how their model structure can be used to consider the spatial interconnectivity of local populations, which has profound implications for deciding not only the spatial scale of control but also the potential of the parasite to invade, persist and spread as a result of changes in host, vector and pathogen migratory patterns.

We have chosen to separate the final two chapters of the book into an Epilogue because they almost certainly point to important future applications or developments in the mathematical modelling of infectious and parasitic diseases. There are immediate challenges in addressing the implications of climate change on the extent and intensity of environmentally mediated infectious diseases in general and parasitic diseases in particular. In Chapter 13, Parham and Michael present a simple model of malaria transmission sensitive to changes in rainfall and temperature variables. Their example makes a convincing case that climate-driven transmission models will be crucial to understanding the rate at which *P. falciparum* and *P. vivax* may either infect, expand into, or go extinct in populations as local environmental conditions change. The chapter prompts two more general points, the first being that mechanistically-based models will be centrally important tools in assessing the impact of various climate change scenarios on disease transmission because, unlike statistical models, they possess at least some credible basis for extrapolating beyond the range of current climatological experience. The second key point is that mechanistic models present a framework for assessing the relative impact of, for example, agricultural adaptation to climate change versus the narrow impact of temperature and rainfall effects in an otherwise unchanging environment.

The second dimension of future developments that is nearly upon us relates to the impact of genomics on the modelling of disease transmission and control. It is not difficult to foresee modest changes to current modelling structures to take advantage of elements of this vast new source of data. But it is also highly probable that altogether new approaches will emerge. Indeed, the utilization of genetic data relating to parasites, hosts, or vectors is only now beginning to occur in ecological or epidemiological studies of parasitic disease transmission. In Chapter 14, Tibayrenc argues for "an integrated approach to the genetic epidemiology and population genomics of Chagas disease." Just as mathematical modelling of disease transmission originally grew from foundational studies of epidemiology and vector biology, it seems likely that we are at the beginning of a new class of modelling studies that will take advantage of the molecular epidemiology of disease transmission. While Tibayrenc argues "that Chagas disease potentially constitutes a paradigm model for the integrated genetic epidemiology and integrated population genomic approaches to understanding of disease transmission," we suggest that his argument also points to new opportunities and challenges for the next generation of mathematical modelling studies for clarifying parasite transmission dynamics in general, as highlighted by the work of Buckee and Gupta (Chapter 8) in using information from sequence data for modelling the impact of parasite diversity on host immunity and development of resistance in malaria.

It is clear that many fascinating problems still remain to be addressed in parasite transmission modelling, from better understanding of transmission processes and natural history of infection to investigating the impact of ecological and spatial scales, climate change, host immunity and social behaviour, parasite-host evolutionary dynamics and parasite community ecology on parasite transmission. This book captures some of the advances made in recent years and provides indications of ways forward for addressing these questions by shedding light on developments in conceptual frameworks and modelling tools as well as the emergence of new data forms for aiding model construction, testing and analysis. Another important advance has been the parallel development of robust computationally-intensive statistical methods to allow model testing and parameterization by aiding the fitting of models to complex data. This is an exciting area of work, which we believe will broaden the scope of mathematical modelling in investigating parasite transmission processes. In particular, we expect this advance will now allow modellers to begin the successful development and analysis of mechanistically-rich models of parasite transmission that will facilitate better integration of the variety of mechanisms increasingly recognized as important in simultaneously affecting transmission, including abiotic processes, trophic and evolutionary interactions, movement in space, and behaviour and even physiology of the individual. We foresee a continuing bright future for using mathematical modelling to clarify parasite transmission dynamics and address problems related to effective parasite control. Ultimately, through this improved application of models to research and management, we expect that parasite control would be an achievable goal bringing benefits to a vast number of our fellow human beings.

Edwin Michael, PhD
Department of Infectious Disease Epidemiology
Imperial College
London, UK

Robert C. Spear, PhD
Center for Occupational and Environmental Health
School of Public Health
University of California
Berkeley, California, USA

References

1. Ross R. Some quantitative studies in epidemiology. Nature 1911; 87:466-467.
2. Habbema JDF, Alley ES, Plaisier AP et al. Epidemiological modelling for onchocerciasis control. Parasit Today 1992; 8:99-103.
3. Chan MS, Guyatt HL, Bundy DAP et al. The development of an age structured model of schistosomiasis transmission dynamics and control and its validation for Schistosoma mansoni. Epidemiol Infect 1995; 115:325-344.
4. Liang S, Seto E, Remais J et al. Environmental effects on transmission and control of parasitic diseases exemplified by schistosomiasis in Western China. Proc Natl Acad Sci USA 2007; 104(17):7110-7115
5. Spear RC, Hubbard A, Liang S et al. The use of disease transmission models for public health decision-making: towards an approach for designing intervention srategies for schistosomiasis japonicum. Environ Hlth Perspectives 2002; 110:907-915.
6. Michael E, Malecela-Lazaro MN, Simonsen PE et al. Mathematical modelling and the control of lymphatic filariasis. Lancet Infect Dis 2004; 4:223-234.
7. Hairston NG. Population ecology and epidemiological problems. In: Wolstenholme GEW, O'Connor M, eds. Bilharziasis. CIBA Foundation Symposium on Bilharziasis. London: Churchill, 1962:36-62.
8. Macdonald G. Epidemiologic models in studies of vector-borne diseases. Public Hlth Rep 1961; 76:753-764.
9. Macdonald G. The dynamics of helminth infections, with special reference to schistosomes. Trans R Soc Trop Med Hyg 1965; 59:489-506.
10. Anderson RM, May RM, eds. Infectious Diseases of Humans: Dynamics and Control. Oxford: Oxford University Press, 1991.
11. Anderson RM, May RM, eds. Population Biology of Infectious Diseases. Dahlem Konference. Berlin: Springer Verlag, 1982.
12. Grenfell BT, Dobson AP, eds. Ecology of Infectious Diseases in Natural Populations. Publications of the Newton Institute. Cambridge: Cambridge University Press, 1995.

ABOUT THE EDITORS...

EDWIN MICHAEL is currently a Senior Lecturer in infectious disease epidemiology at Imperial College London, UK, with a research focus on modelling the transmission and control of tropical parasitic and infectious diseases. His main interest lies in developing a system dynamics approach to gaining a better understanding of parasite transmission, immunology, genetics and economics, in order to develop integrated mathematical models of pathogen transmission as a tool for aiding the rational design, monitoring and evaluation of large-scale intervention programmes, ranging from vector control, chemotherapy to vaccinations. He has worked extensively in Africa (primarily East Africa), India, Vietnam and Papua New Guinea, particularly over the past decade (in partnership with international (WHO, World Bank) and national institutions), in translating research on disease population biology, spatial dynamics and public health decision-making for developing reliable model-based spatial decision support tools to aid the design, surveillance and evaluation of ecologically resilient and sustainable intervention programmes against parasitic diseases of major public health importance in developing countries. His current interest is in extending this work to developing integrated ecological, economic and social systems approaches for investigating interactions between climate change, ecosystem dynamics and the socio-ecology of disease transmission in vulnerable communities.

ABOUT THE EDITORS...

ROBERT C. SPEAR is an engineer by training, having received the BS and MS degrees in Engineering Science and Mechanical Engineering, respectively, from the University of California at Berkeley and the PhD degree in Control Engineering from Cambridge University in 1968. After several years in the aerospace industry his interests turned to environmental issues and he returned to Berkeley in 1970 to take up a post-doctoral position in this field in the School of Public Health. He was appointed to a faculty position in 1971 and is now Professor of the Graduate School at Berkeley.

His research interests focus on the assessment and quantification of human exposures to toxic and hazardous agents in the environment. His early work concerned the exposure of agricultural workers to pesticides. In more recent years his work has concerned applications of mathematical and statistical techniques in the assessment and control of exposures to both chemical and biological agents. For the past 15 years his work has been increasingly focused on determinants of the prevalence and control of the parasitic disease schistosomiasis in the mountainous regions of Sichuan Province in southwestern China.

PARTICIPANTS

María-Gloria Basáñez
Department of Infectious Disease Epidemiology
Imperial College London
London
UK

Isobel M. Blake
Department of Infectious Disease Epidemiology
Imperial College London
London
UK

Caroline O. Buckee
Department of Zoology
University of Oxford
Oxford
UK

Elizabeth J. Carlton
School of Public Health
University of California
Berkeley, California
USA

Stephen J. Cornell
Institute of Integrative and Comparative Biology
University of Leeds
Leeds
UK

Holly Gaff
Community and Environmental Health
College of Health Sciences
Old Dominion University
Norfolk, Virginia
USA

Manoj Gambhir
Department of Infectious Disease Epidemiology
Imperial College London
London
UK

Geoff P. Garnett
MRC Centre for Outbreak Analysis and Modeling
Department of Infectious Disease Epidemiology
Imperial College London
London
UK

Nicholas C. Grassly
Department of Infectious Disease Epidemiology
Imperial College London
London
UK

Sunetra Gupta
Department of Zoology
University of Oxford
Oxford
UK

A. Hubbard
Center for Occupational
and Environmental Health
School of Public Health
University of California
Berkeley, California
USA

Edwin Michael
Department of Infectious Disease
Epidemiology
Imperial College London
London
UK

Paul E. Parham
Grantham Institute for Climate Change
Department of Infectious Disease
Epidemiology
Imperial College London
London
UK

Andrea Pugliese
Dipartimento di Matematica
Università di Trento
Povo
Italy

Justin Remais
Rollins School of Public Health
Emory University
Atlanta, Georgia
USA

Nick Ruktanonchai
Department of Zoology and Emerging
Pathogens Institute
University of Florida
Gainesville, Florida
USA

Elsa Schaefer
Department of Mathematics
Marymount University
Arlington, Virginia
USA

Edmund Y.W. Seto
School of Public Health
University of California
Berkeley, California
USA

David L. Smith
Department of Zoology
and Emerging Pathogens Institute
University of Florida
Gainesville, Florida
USA

Robert C. Spear
Center for Occupational
and Environmental Health
School of Public Health
University of California
Berkeley, California
USA

Michel Tibayrenc
IRD Representative Office
French Embassy
Bangkok
Thailand

Peter J. White
MRC Centre for Outbreak Analysis
and Modeling
Imperial College London
and
Modelling and Economics Unit
Health Protection Agency Centre
for Infections
London
UK

CONTENTS

PART 1. MODELLING PARASITE TRANSMISSION

PART 2. APPLICATION OF MODELS TO PARASITE CONTROL

ACKNOWLEDGEMENTS

We would like to thank all the authors for their enthusiasm and positive response to our invitations to contribute chapters for this book. Their diligence in producing work in a form that is engaging, thought-provoking and scientifically comprehensive is especially appreciated. We would also like to thank Celeste Carlton at Landes Bioscience for her constant encouragement and patience.

Chapter 1

Progress in Modelling Malaria Transmission

David L. Smith* and Nick Ruktanonchai

Abstract

Transmission of human malaria is a complicated dynamic process that involves populations of humans, parasites, and vectors. The first mathematical models of malaria are now more than a century old, and they are still a useful conceptual synthetic description of transmission, but they fail in some important ways. To address some of those failures, malaria transmission models have now been extended to consider malaria immunity, superinfection, and heterogeneous biting, among other factors. These extensions of the basic theory often arise from field studies in a single place, but tests of the theory come comparing standard measures of malaria taken from many places across the transmission spectrum. Several good models now exist that describe these basic patterns across the spectrum from low to high endemicity. The future of malaria modeling will involve applying these models to make decisions about real systems and finding new ways to test the underlying causes of the patterns.

Modelling Malaria Transmission, a Historical Introduction

Ronald Ross demonstrated that mosquitoes transmit malaria parasites.[1] In the years that followed, malaria control programs focused on reducing mosquito densities, largely through larval vector control. Eleven years after Ross's seminal discovery, he wrote about malaria transmission in Report on the Prevention of Malaria in Mauritius:[2]

> The infection rate varies, not only in neighbouring places, but even in the same place from year to year.... What is the cause of these variations?
>
> This question is so important as regards both the general theory of malaria and the subject of prevention that we must endeavour to obtain clear ideas about it by careful reasoning (pp. 30).

In the subsequent pages, Ross made clear that "careful reasoning" meant a quantitative description of malaria transmission and proceeded to describe the first mathematical model of malaria.[2] Waite provided a longer description of the model and detailed analysis.[3] He noted that the malaria rate, now called the parasite rate or the prevalence of malaria, was constantly changing due to new infections, recovery, human migration, births and deaths. He considered migration to be too complicated to treat sensibly in a simple model, so like Ross he avoided the distraction and considered the malaria rate in a closed population. He described other aspects of the model in the following way:

*Corresponding Author: David L. Smith—Department of Zoology and Emerging Pathogens Institute, University of Florida, Gainesville, Florida, USA.
Email: smitdave@gmail.com

Modelling Parasite Transmission and Control, edited by Edwin Michael and Robert C. Spear.

> I shall assume ... that both the human population and the anophelines are uniformly distributed over the area in question; that persons of all ages, whether they have previously suffered from malaria or not, are equally liable to infection; that the number of anophelines per unit of the population remains approximately constant during the period under review; and further that infected and non-infected persons are equally exposed to the risk of being bitten (pp. 422).

Waite was forthright about the limitations of his model; he considered the conclusions to be "tentative and suggestive," but he hoped they would throw light on the factors that affected the malaria rate for "specialists who were considering remedies for the disease." He pointed out that all of the simplifying assumptions of his model should be reconsidered and refined. Over the past century, knowledge about the parasite, vector and humans and their interactions has improved substantially and the terminology and technology has changed. Malaria models have expanded to consider most aspects of transmission, but the malaria-modelling pioneers laid out the basic malaria modelling agenda. From the beginning, malaria modelling has been motivated by malaria prevention. Models were developed to provide some useful general guidance provided the limitations were considered and soon there was progress in moving beyond the limiting assumptions of the simple models.

The next seventy years saw a part of the agenda fulfilled. Ross published a second mathematical model in 1911 in the second edition of The Prevention of Malaria[4] and again in an article to Nature;[5] the model retained the same simplifying assumptions, but it was formulated as a set of ordinary differential equations. This model was analyzed in great detail by Alfred J. Lotka a year after its publication[6] and in a series of five papers, expanded on some of the earlier themes,[7-11] notably the delay caused by sporogony.[10] George Macdonald also considered the importance of this delay for control, noting that only old mosquitoes are infectious, so killing adult mosquitoes to shorten lifespan would be especially effective at reducing transmission.[43] Macdonald's analysis underestimated the potential benefits of killing adults since, by the assumptions of his model, the density of adult mosquitoes would also be reduced and the depletion of adults could potentially reduce recruitment of adults from larval habitat.[12] Macdonald's analysis also had some historical significance because it helped explain why DDT had been so effective at controlling malaria in early trials. This analysis helped to launch the Global Malaria Eradication Programme, operationalized as time-limited, intensive indoor residual spraying with DDT. Controversy over the programmatic implementation of malaria control during the Global Malaria Eradication Program and the failure to eradicate malaria continues, but the successes of malaria control during GMEP should also be remembered. This simple elaboration on the model fulfilled the mission of providing useful advice to people who do malaria control: few would argue against the great control potential of reducing the mosquito lifespan and that it should be pursued as an option if it is operationally feasible.

An important simplifying assumption from the first models was that untreated malaria infections in humans would clear at a constant per-capita rate. This is equivalent to assuming that the waiting time to clear an infection is exponentially distributed and that it would not differ in a person reinfected with one or multiple different parasites of the same species, or with multiple parasite species. In 1915, McKendrick first considered the problem of multiple infections and relapses in persons that have been infected "once, twice, thrice, etc. by the disease in question.[13]" In 1947, Walton revisited the question of multiple infections and provided a formula to describe the distribution of the number of infections that would be carried in a population.[14] Macdonald also considered the question of multiple infections,[15] but as others have described in some detail, his mathematical formulation does not match the formal description in the paper: the model was supposed to describe infections that clear independently, but it actually described infections that queue up and clear sequentially.[16,17] Dietz, et al derived a simple formula for the waiting time to clear parasites when infections are complex, when the system is at the steady state and when new infections arrive at a constant rate, and clear independently at a constant per-capita rate.[16,18] The model that was developed for a large-scale malaria control trial in Garki, Nigeria is now called the Garki Model, which also considered some aspects of immunity. Dynamical equations describing changes in the distribution of the number of infections were described and analyzed by Bailey[19] and Dietz has written a synthetic review that considers all of these issues in some detail.[20,21]

By 1980 many of the assumptions of the simple models had thus been challenged, but the ones describing transmission remained fairly unchanged. Models up to that point had assumed that populations were closed (i.e., no migration), that human population sizes were essentially infinite and that biting was evenly distributed over space and time. Is biting even over space and time? How far do parasites move in a generation? How far do mosquitoes disperse after completing 4-6 feeding cycles, the time required to complete sporogony? How many humans live within the radius of parasite or mosquito movement? Is biting within those distances randomly or proportionally distributed or is it poorly mixed? These questions are, perhaps, most relevant if the goal of malaria control is to understand how malaria control programs work or if the goal is to contain the spread of antimalarial drug resistance. There is substantial evidence that some people are bitten more than others based on body size,[22,23] body odors,[24] proximity to larval habitat[25] and potential alternate hosts,[54] housing design,[26,27] preferential biting of infectious hosts[28] and other factors. Recent theory has progressively relaxed the assumptions of ideal mixing to include heterogeneous biting,[29-31] preferential biting of infectious hosts,[32] patch-based models of transmission on heterogeneous landscapes,[33] mosquito oviposition behavior[34] and finite host populations.[29] These are increasingly drawing the connection between reasonably simple descriptions of heterogeneous biting in closed populations (i.e., without migration) and "realistic" descriptions of transmission that focus on the details of mosquito behavior. In this chapter, we will review theoretical and empirical studies that have moved beyond the simple assumptions that were made by Waite and Ross.

Complexity, Parsimony and Robust Descriptions of Transmission

Why should malaria be modelled and what sort of model is worth making? Again, Ross provides one of the best answers in the introduction to a seminal paper about the use of mathematics in epidemiology, in which he points out that the study of epidemics is fundamentally concerned with numbers.[35] He also notes that different Climates display striking differences in the seasonal patterns of cases and in their frequency and he asks what accounts for these differences. "Why, indeed, should epidemics occur at all?" He proposes that there are two ways to answer this question that he called the a priori and a posteriori approaches. A posteriori approaches were synonymous with statistics; "we commence with observed statistics, endeavour to fit analytical laws to them and so work backwards to the underlying cause." To work a priori, "we assume a knowledge of the causes, construct our differential equations on that supposition, follow-up the logical consequences and finally test the calculated results by comparing them with the observed statistics." Ross did not believe that it would be possible to answer his general question by relying on one approach or the other. He notes that many papers had investigated epidemics a posteriori, but few had done so a priori. He then proceeds to develop a purely theoretical approach to the "Theory of Happenings" and calls his theory "a priori pathometry". The name has not caught on, but the paper established some of the foundations of epidemiology and provided rigor to ideas that are so basic in epidemiology, we almost forget they exist.

It is also worth asking what sort of model is worth making. On the subject of model building, George P. Box, a statistician, famously reminds us that "all models are wrong, but some are useful." It is not enough, however, to merely know that models are wrong—"the practical question is how wrong do they have to be to not be useful?"[36] Few would deny that the Ross-Macdonald model has been conceptually useful, but the model does not describe malaria well enough under field conditions to serve as an adequate quantitative guide for most purposes.[18,37]

Since there are many uses for models, before building a model, it is worth thinking about how the model would be used. Models serve many purposes, including conceptual understanding, prediction, statistical inference, strategic planning, persuasion and decision-making. It may not be possible to describe how a model will be used in advance, so it is often best to build a range of different models and a solid basis for understanding how the models correspond and how well they serve different purposes.

One of the most important functions of a model is to structure thought about quantitative problems.[38] In this context, it is important to recognize the distinction between "theory" and "models." In science, theory refers to a set of deliberately oversimplified models that have been extensively tested and that are consistent with existing evidence collected from observations and experiments. Developing and testing models form the core activities of scientists, but the word "model" applies equally well to animal models, in vitro models, conceptual models, statistical models and mathematical models. Likewise, mathematical models can have many purposes as part of basic and applied scientific processes and each model should be suited to its purpose. Theory is the unifying framework of science, but it is rarely codified into a single, grand unified model. No model can serve all purposes equally well, because models inevitably make tradeoffs between generality, realism and precision; to be useful, a model must be robust.[39] As an example, a malaria model can strive to be "realistic," in the sense that the model is increasingly representational, but as the model becomes more representational, it comes to describe one particular location at a certain point in time. Moreover, the more realistic a model becomes, the greater the data needs; however, the availability of data has been a limiting factor. In modelling malaria, it is useful to conduct studies that build a broad range of models and that explicitly consider the question of robustness. Some property of a model is robust if the predictions hold over a wide range of assumptions and models. When models fail to agree, we would say that a finding is not robust and careful analysis would provide guidelines about when the simpler assumptions breakdown. It may not be possible to describe malaria under every circumstance, but it might be useful to find "rules of thumb" and simple functional forms that describe malaria transmission in different contexts and at a range of spatial scales—what approximations are necessary to strike the right balance to have a model that is both simple and generally useful? It may not be possible to build a single model that would work in every place, but it may not be wise to try and build a different model for every place. A practical way forward is to consider adequate models for malaria transmission by context, including the biting habits of the vector and their bionomics and the transmission intensity.

Transmission Intensity and Its Estimations

Before reviewing mathematical models of malaria transmission, it is worth describing how malaria is measured in different contexts.[40] What malariometric indices have been developed to measure malaria and its intensity? One of the earliest indices used was the spleen rate, or the proportion of a population with a palpably enlarged spleen. The spleen rate is reasonably noninvasive, easy to measure and it is closely associated with malaria. Other causes of splenomegaly are reasonably rare, making a high spleen rate a good indicator of ongoing malaria transmission. After parasites could be reliably identified by microscopy, the parasite rate became a more commonly used index of malaria transmission intensity, a measure that addresses the proportion of humans with parasites in their blood.[41] A third method of measuring malaria examines the prevalence of some marker of previous infection that is present in blood serum—a measure of the proportion of humans currently with antibodies that developed in response to malaria infection.[42] Typically, seroprevalence rises with age and gives a robust measure of previous infection compared with age. The spleen rate, the parasite rate and seroprevalence all measure the fraction of the population with particular conditions at some point in time. Measures of prevalence are useful in malaria if they consider something that changes slowly—immunological responses wane slowly, untreated infections and splenomegaly last for several months.

An alternative way to measure malaria in humans is to ask how often a person gets infected. Measuring incidence (i.e., counting the number of events) is different from measures of prevalence. Prevalence measures the proportion of the population in some given state, while incidence measures the number of events—how many new infections occurred in a population over some fixed period of time. In models, the per-capita incidence rate is called the force of infection or the hazard rate for infection. Ross called this quantity the happenings rate, h.[35] It is difficult to measure the force of infection, however, since it is impossible to observe infections directly and since many of them never present with clinical symptoms. Despite these difficulties, force of infection can be measured by tracking a cohort of uninfected people until they become infected. The average waiting time to infection is the inverse of the force of infection, 1/h. Since infants are born uninfected, the waiting time to the first infection, called the infant conversion rate, is a measure of the force of infection. So, too, is a measure of the rise in seroprevalence or the parasite rate with age.

Clinical malaria is defined as fever with parasites. The definition of clinical malaria counts anyone who is infected with malaria and gets a fever from any cause, such as a viral or bacterial infection, thus, it is not always known how many of those fevers were due to malaria. An alternative index of transmission intensity would be the prevalence of clinical malaria, the population with clinical malaria at any point in time. Fevers are short compared with infection response, so clinical malaria is less common than infection with malaria. Fevers have multiple causes, moreover, so measures of disease prevalence tend to be a less specific as an index of malaria. Splenomegaly, by comparison is also a measure of disease prevalence, but it is generally more informative because the duration of fever is comparatively short and the incidence of splenomegaly from other causes is lower. Because the prevalence of clinical malaria is not reliable, clinical incidence is often measured by counting the number new cases of clinical malaria that occurred in a population over a fixed period of time. This measure is generally standardized to a fixed population and time, i.e., 200 cases of clinical malaria (or malaria-attributable clinical episodes), per thousand people, per year.

The spleen rate, the parasite rate, seroprevalence, the force of infection and clinical incidence all measure malaria transmission intensity on the human population, but none of them are perfect indices. Some infections are treated and clear before an immune response develops; this and waning immune responses can affect the accuracy of seroprevalence. Treated infections can clear rapidly and untreated infections also clear after some time, so the parasite rate is a measure of recent infection, modulo drug treatment to achieve a cure. Parasite rates tend to rise with age in young children and they decline with age throughout adolescence and adulthood, so crude parasite rate are a poor index of transmission intensity. The parasite rate in older children (aged greater than 2 but less than 10), called the standard parasite rate does provide a reliable, if imperfect, index of transmission intensity.[43] Although complete immunity to infection may not occur in malaria, the incidence of clinical malaria does decline with age.[44] Clinical incidence displays age-specific patterns that differ by endemicity and this makes it an unreliable index of transmission intensity.

Transmission intensity can also be measured by counting infections in the invertebrate host and entomological indices of transmission intensity can and should play a major role in defining interventions. The human biting rate is defined as the number of bites by vectors, per human, per day. The entomological inoculation rate (EIR) is the product of the human biting rate and the sporozoite rate, the proportion of mosquitoes that have sporozoites in their salivary glands. Another way to characterize EIR is the expected number of infectious bites, per person, per day (dEIR) or per year (aEIR). Just as there is a force of infection for humans, there is also a force of infection for mosquitoes. There are two ways to characterize transmission of the parasites from humans to mosquitoes—what fraction of bites on humans infects a mosquito and what fraction of bites on an infected (and presumably infectious) human infects a mosquito? The former measure is called the net infectivity of the human reservoir and the latter is called the efficiency of transmission (denoted c).

HBR and EIR are measures of actual transmission intensity, but there is another class of indices that measure potential transmission intensity. One of these measures is the stability index (S): how many human bloodmeals does a vector take over its lifetime? The stability index is a composite measure, the product of three quantities: the mosquito lifespan (denoted 1/g; the probability a mosquito survives one day is denoted $p = e^{-g}$), the length of a feeding cycle (denoted 1/f; the time elapsed between two bloodmeals) and the number of human bloodmeals as a proportion of all bloodmeals (Q). The stability index is the product ($S = fQ/g$). The parasite requires several days to complete sporogony (denoted n), and the probability that a mosquito survives long enough for the parasite to complete sporogony (e^{-gn}) also serves as a measure of potential transmission. Additionally, a measure of potential transmission is the number of vectors per human (m), which can help measure potential transmission. This is directly related to the rate that new adult vectors are recruited into the mosquito population, scaled by human population density (denoted l and m = l/g). Vectorial capacity (V) is defined to be the number of infectious bites that would eventually arise from all the mosquitoes that bite a single human on a single day. It is related to the other indices by the formula ($V = l\, S^2\, e^{-gn}$).

The most useful index of potential transmission is the basic reproductive number (denoted R_0), defined as the expected number of mosquitoes that would arise from a single infected mosquito after one complete generation of the parasite. It is calculated by summing all of the infectious bites that arise from the mosquitoes that are infected by a single person (i.e., vectorial capacity) over the entire duration of a simple, untreated infection. If parasites clear from humans at the per-capita rate r, then the average waiting time to clear an infection is 1/r. Not all infectious bites actually produce an infection, only a fraction do (denoted b). It is defined for a population with no immunity and no malaria control. The formula for the basic reproductive number is related to vectorial capacity:

$$R_0 = bcV/r$$

This number forms a foundation for understanding malaria transmission, but its derivation relies on the simplifying assumptions made by Ross of an ideal, well-mixed population.

This list of malariometric indices is not comprehensive, but it does give a conceptual overview of the points in the life cycle where the intensity of malaria transmission can be measured. In malaria epidemiology, it is important to measure these quantities for many reasons. These malariometric indices provide a critical test of the models (for example, see ref. 45). How well do the models capture the variability in these indices and their relations to one another across the range of transmission intensity? The Ross-Macdonald model proposes that there is a functional relationship between these indices, but it does not incorporate immunity or a range of other factors that are known to be important. The Garki model does incorporate some aspects of superinfection immunity, but it does not correctly predict what would happen if malaria transmission were abruptly interrupted. Indeed, the frontier in malaria modelling today is to fulfil Ross's vision of creating a model based on a set of a priori assumptions that is capable of explaining all of the patterns and much of this involves retreading old ground.

Preferential Biting and Uneven Exposure

The first major departure from ideal population mixing was motivated by studies from sexually transmitted diseases. The number of sexual partners and sexual activity in a population of humans are clearly uneven; many people are not sexually active, some are absolutely monogamous, some people have multiple partners and a few people in the population are highly sexually active. The idea is intuitive to most of us who suspect that everyone else is having more sex than us. If one thinks of sex as a risky behavior, it serves as a useful metaphor for malaria exposure.

It would be surprising if exposure to malaria were evenly distributed among all the humans in a population, but how large is the variance in biting heterogeneity and how is it described? One way is to characterize the proportion of bites received by a small fraction of humans. It has been proposed that 20% of the population gets 80% of the bites. Mosquito behavior and ecology affect the variability in exposure to malaria and the issue is especially complicated because malaria is transmitted by dozens of anopheline species with different biting preferences, biting habits, larval habitat and bionomics. Heterogeneous biting may also depend on other contextual details, such as the distribution of larval habitats and humans and the availability of alternative human hosts. Despite the variability, most studies of malaria have tended to focus on vectors that bite indoors and at night.

Muirhead-Thomson's classical study of biting counted bites on family members outside their huts and concluded that adults were bitten more than children and men were bitten more than women.[22] Muirhead-Thomson also observed that infants behaved differently from older children and adults—the infants were more likely to shake off the mosquitoes, even while asleep. Other studies caught blood fed mosquitoes, analyzed bloodmeals and identified the host they had fed on. These studies found that the proportion of bites that a person received was roughly proportional to the proportion of skin in the household.[23] Another curious pattern was that, relative to body size, children younger than 18 months old were bitten at a much higher rate than their siblings who were older than 18 months. However, Spencer's study of 8 pairs of mothers and infants sleeping side by side found that the infants were never bitten, even though the mothers were bitten frequently.[62] Pregnant women account for a larger portion of bites.[63] These studies and many others like them (see Port and Boreham,[23] for a review of the early studies) have found different and contradictory patterns depending on where the samples were taken and depending on whether the study evaluated landing catches or did bloodmeal analysis. The robust pattern from all these studies has been that larger individuals were bitten more than smaller individuals.

Other studies have demonstrated that mosquitoes prefer to feed on individuals who carry gametocytes, potentially through certain symptoms such as increased skin temperature and CO_2 expiration.[55] Mosquitoes with sporozoites often spend significantly more time probing blood vessels and are more likely to attempt to resume a bloodmeal if the feeding has been interrupted.[56] This altered feeding behaviour also contributes to a higher mortality of infected mosquitoes, probably due to increased feeding activity, which increases likelihood of mortality from host defensive behaviour.[57] Early studies suggested that multiple feeding was rare, but these studies tend to sample mosquitoes at random. Multiple feeding may be more important than it seems if it is more frequent among infectious mosquitoes.

These studies have described the proportion of bites received by individuals who live within the same household, but they don't address the questions of how many bites are received by a household. One way to count household heterogeneity is to use insecticides to knockdown all the vectors within a household; the distribution of vectors per household is extremely heterogeneous (for example, see Fig. 2a in ref. 46). This raises the question of what accounts for the huge variability in exposure among households.

The choice of a bloodmeal host is determined by the mosquito search algorithm and chance; mosquitoes follow a set of rules, but host choice is probably opportunistic. Mosquitoes begin their search for a bloodmeal host from the larval habitat where they deposited their eggs. As mosquitoes tack across the wind, odor plumes cause mosquitoes to change direction and follow the plume upwind. As they approach the neighborhood of a potential host, different olfactory cues, CO_2, movement and body temperature draw mosquitoes to a particular household or host.[58] Mosquito behavior at the point of contact probably depends on host defensive behavior, personal protection and the mosquito's physiological state: a mosquito that has few energy resources in reserve would be willing to take more risks than one with plentiful resources.

It follows that one would expect to find enormous variability in household risk that will be related to a large set of factors. The easiest one to characterize is proximity to larval habitat; indeed, many studies have found that proximity to larval habitat was an important risk factor for infection or clinical malaria.[59] Other factors include the odors that emanate from a particular host—including the "limberger cheese" smell.[24] The type of house is an important factor in determining household risk—some kinds of houses and housing designs are easier for mosquitoes to find and enter.[26,27,47,48]

It is also important to note that malaria control can be a cause of heterogeneous biting.[29] If some people in a house use an insecticide-treated bednet (ITN), the mosquitoes may be deflected onto those who do not. ITNs can have an area repelling effect and this would deflect some of the mosquitoes from houses that use ITNs onto those who do not. Since ITNs also kill some mosquitoes, delay feeding and redirect mosquitoes onto alternative hosts that would have fed on a human,[61] the average level of risk may be reduced.[49]

All of this discussion has focused on biting that occurs indoors and at night, but vectors bite in the forest in some areas and in the city in others. In different contexts, the cause of heterogeneous biting can will be related to the factors that predispose people to be at risk.

Although it is often useful to try and explain the variance in exposure to malaria, the total variance in exposure is, itself, an interesting quantity for transmission. When biting is heterogeneous, it means that biting is more intense on a small portion of the population and less intense on the rest. This makes it easier for malaria to invade a population when rare, essentially because person who is bitten most frequently is most likely to become infected and most likely to infect a mosquito. Thus, malaria would be able to persist at lower levels of vectorial capacity; if a^2 is the coefficient of variation of biting exposure, then $R_0 = bcV(1 + a)/r$.[30,31,50] On the other hand, at some higher level of vectorial capacity, the average level of risk would be reduced in the population. This leads to a different relationship between EIR and PR.[45]

Immunity and the Infectious Reservoir

The flip side of human exposure is to ask how often vectors become infected after biting a human. Another way to ask the question is which humans infect most of the vectors. Immunity to malaria is not like immunity to measles—it does not fully protect against infection or disease and it may have poor immune memory.[51] Thus, herd immunity does not tend to develop. In highly endemic areas, immunity reduces parasite densities in older children and adults and it reduces the frequency and severity of disease.[52] Gametocyte densities do decline and adults also naturally develop transmission blocking immunity, where human antibodies in the bloodmeal prevent infection of the mosquitoes.[53] The infectivity of humans thus declines, but direct observations demonstrate that the adult humans in highly endemic areas remain infectious to mosquitoes.[54]

The Garki model described some of these effects of immunity by considering semi-immune individuals that would gain and lose infections; non-immune infected individuals would pass into the semi-immune individuals at some low rate, reflecting the idea that immunity is acquired after exposure to malaria.[18] The Garki model assumed that lower gametocyte densities in semi-immune individuals would render them uninfectious, despite being infected.

After Garki, a new trend started in malaria modelling. The Garki Model was the motivation for considering malaria as a familiar compartmental model, called an SIRS model. The infective classes are typically interpreted as susceptible (S), infected and infectious (I) and recovered and immune (R). The trick was to relabel all "semi-immune" individuals as being recovered and immune because they were not infectious, whether or not they were infected.[16,55-57] The strength of the Garki and SIRS models were that they provided a simple conceptual way of explaining why the PR declines in older age classes, when the relationship between PR and age was far from the steady state. PR tends to rise with age in infants to a plateau in young children. At approximately age ten, it starts to decline. It then falls by about a third of the maximum by age 20 and by two-thirds of the maximum late in life.[41] Both the Garki model and the SIRS model produce curves with this basic shape and a simple version of this model has been used to describe shifts in the age at infection when ITNs are deployed,[58] although the models predict that the age when PR first begins to decline would rise sharply at low PR, a trend that was not observed in analysis of data.[41] The failing of the SIRS models is that it often invites misleading comparisons with acute, immunizing childhood diseases such as measles. In malaria, immunity to severe malaria is acquired rapidly and usually early in life,[59] but infection continues throughout life.

Following in the SIRS tradition, the first epidemiological model concluded that immunity would not affect the evolution of resistance.[60] A subsequent model considered the dynamics of infection in semi-immunes, like the Garki model and showed that immunity would, indeed, affect the evolution of resistance.[61] The different predictions arose because the simplifying assumption about immunity in the SIRS model forced a kind of rigidity—a person was either immune or not. When that stiffness was relaxed, the quantitative effects of immunity could play out in the models in interesting ways. Despite being less infective, they were also less likely to have clinical malaria. Since the demand for antimalarial drugs was lower, they would form a natural refuge for drug-sensitive parasites.

One of the great failures of malaria modelling has been its neglect of disease as an important epidemiological state that is distinct from infection. Indeed, there is a sharp distinction between infection and disease in malaria. Clinical malaria is important for estimating burden, for understanding the need for antimalarial drugs and their potential to control malaria and for understanding drug pressure and the evolution of antimalarial drug resistance. Most models assume that blood stage infection is a good measure of the infectious reservoir; this may not be true in places with good health systems where most clinical malaria is treated and cured. In those places, gametocytemia could persist in people who have no other signs of infection. Clinical disease and gametocytemia as important epidemiological states apart from asexual parasitemia remain largely unexplored by mathematical models (but see ref. 61). Incorporating clinical disease, the gain and loss of immunity and infectivity and identifying where these factors are important are important questions in malaria modelling.

Malaria Transmission in Real Populations

The assumption that populations are infinite and well-mixed are useful for formulating theory based on ordinary differential equations, but to apply these to real populations, it is necessary to tie the model populations at a particular place and time. For such populations, the distribution of human populations, local density and the flight distances of mosquitoes matter. Other aspects of mosquito behavior also affect the validity of the basic models. Mosquitoes can senesce, feed multiple times per life-cycle, postpone oviposition (called gonotrophic disassociation), rest in warm humid refugia in otherwise inhospitable environments, feed on nectar and engage in other behaviors that are not accounted for by the simple formulas. Understanding the factors that explain malaria transmission may require some understanding of these factors, but it is probably not necessary to describe every detail of mosquito behaviour in a mathematical model.

As a starting point, it is worth asking how the simple models get things wrong. Models that consider finite human populations and their spatial distribution provide a starting point for thinking about transmission in a more general setting. In a very large, well-mixed population, every infectious bite will land on a different person. In a small population with proportional mixing, the same individuals who make it easier for malaria to persist absorb multiple infections and this will tend to reduce the risk for everyone else and slow down the invasion.[29] Bailey's approach to space reformulated partial differential equation models of malaria to consider diffusive movement of mosquitoes and hosts.[19] Later work considered diffusive movement of mosquitoes on heterogeneous landscapes by partitioning space into patches and showed that the distributions of larval habitat and humans form a template for mosquito aggregation and hence, for transmission.[33,34] These extensions of the basic theory beg the question of whether it is necessary to consider local transmission in an explicitly spatial context. How far do mosquitoes move from the time when they become infected and when they become infectious? How bad are models of proportional mixing (i.e., heterogeneous biting) as an approximation to poor mixing (i.e., locally small populations)?[62]

These ideas are particularly important for control. How far do the protective effects of ITNs extend beyond the populations where they are used?[64] How wide does a malaria control zone have to be to minimize the chances of transmission by mosquitoes?[64] How close do cows or other alternative hosts have to be to provide a zooprophylactic effect?[65]

At larger spatial scales, the movement of infected mosquitoes is not nearly as important as the movement of infected humans. Some infected mosquitoes do move long distances in the cargo holds of airplanes and ships[63,64] and one such event sparked a major epidemic in Brazil.[65-67] Perhaps the most striking example of the global spread of malaria has been the spread of antimalarial drug resistance throughout southeast Asia and South America from a few points of origin and from southeast Asia into Africa. Human movement among areas that have the potential for malaria transmission is hard to describe, but it is an important aspect of malaria spread. This calls for embedding existing models of malaria transmission into a broader metapopulation context as a way of understanding how malaria parasites persist in their vector and human host populations.

Conclusion

Mathematical models are nothing more than a tool for thinking carefully and quantitatively about malaria and they may be the only way for thinking rigorously about complex quantitative aspects of biological systems. As a scientific activity, they often engage in a different way of doing science and they provide a natural bridge to the applied sciences and public health. Models begin from a set of assumptions and work forward—if the model is correct, then the predictions of the model should be reflected in real world phenomena. This methodological difference often makes the logic of modelling seem backwards to other scientists. Ross's vision was that mathematical modelling would provide a complementary methodology to more conventional statistical approaches. There is, however, an enormous gap between the kinds of predictions that models make and the kind and amount of data that is collected in the field. This is often a failure of modellers to engage with field biologists, but it is also a failure of field biologists to recognize how collaboration with modellers could improve their study designs to account for the real complexity of malaria. The most notable exception of this was the Garki model that was field-tested during the Garki Project. It could be argued, however, that the Garki model was the first time that the models were field tested, more than six decades after the first malaria model. In response to a new call for malaria eradication, there are many questions that arise in planning, monitoring and evaluating malaria control programs that could be informed by modelling. Models could be used in the process, at least in theory, but to be useful, models and modellers must engage with data and many different kinds of scientists to repeat the aims of the Garki project many times, although probably not on the same scale. By iteratively working from a priori to a posteriori and back, we can anticipate a race to the finish. Will we first eradicate malaria or understand it? It may not be possible to do one without the other.

References

1. Ross R. On some peculiar pigmented cells found in two mosquitoes fed on malarial blood. BMJ 1897; 18:1786-1788.
2. Ross R. Report on the prevention of malaria in mauritius. London: Waterlow and Sons Limited, 1908.
3. Waite H. Mosquitoes and malaria. A study of the relation between the number of mosquitoes in a locality and the malaria rate. Biometrika 1910; 7(4):421-436.
4. Ross R. The prevention of malaria. Second ed. London: John Murray, 1911.
5. Ross R. Some quantitative studies in epidemiology. Nature 1911; 87:466-467.
6. Lotka AJ. Quantitative studies in epidemiology. Nature 1912; 88(2206):497-498.
7. Lotka AJ. Contributions to the analysis of malaria epidemiology. I. General part. Amer J Hyg 1923; 3(Suppl 1):1-36.
8. Lotka AJ. Contributions to the analysis of malaria epidemiology. II. General part (continued). Comparison of two formulae given by Sir Ronald Ross. Amer J Hyg 1923; 3(Suppl 1):38-54.
9. Lotka AJ. Contributions to the analysis of malaria epidemiology. III. Numerical part. Amer J Hyg 1923; 3(Suppl 1):55-95.
10. Lotka AJ, Sharpe FR. Contributions to the analysis of malaria epidemiology. IV. Incubation lag. Amer J Hyg 1923; 3 (Suppl 1):96-112.
11. Lotka AJ. Contributions to the analysis of malaria epidemiology. V. Summary Amer J Hyg 1923; 3(Suppl 1):113-121.
12. Smith DL, McKenzie FE. Statics and dynamics of malaria infection in Anopheles mosquitoes. Malar J 2004; 3:13.
13. McKendrick AG. The epidemiological significance of repeated infections and relapses. Indian J Med Res 1915; 3:266-270.
14. Walton GA. On the control of malaria in Freetown, Sierra Leone. I. Plasmodium falciparum and Anopheles gambiae in relation to malaria occurring in infants. Annals of Tropical Medicine and Parasitology 1947; 41:380-407.
15. Macdonald G. The analysis of infection rates in diseases in which superinfection occurs. Trop Dis Bull 1950; 47(10):907-915.
16. Aron JL, May RM. The population dynamics of malaria. In: Anderson RM, ed. Population Dynamics and Infectious Disease. London, UK: Chapman and Hall 1982; 139-179.
17. Fine PEM. Superinfection—a problem in formulating a problem. Bureau of Hygiene and Tropical Diseases 1975; 72:475-488.

18. Dietz K, Molineaux L, Thomas A. A malaria model tested in the african savannah. Bull World Health Organ 1974; 50(3-4):347-357.
19. Bailey NTJ. The biomathematics of malaria. Oxford: Oxford University Press, 1982.
20. Dietz K. Density-dependence in parasite transmission dynamics. Parasitol Today 1988; 4(4):91-97.
21. Dietz K. Mathematical models for transmission and control of malaria. In: Wernsdorfer W, McGregor I, eds. Principles and Practice of Malaria. Edinburgh, UK: Churchill Livingstone 1988; 1091-1133.
22. Muirhead-Thomson RC. The distribution of anopheline mosquito bites among different age groups; a new factor in malaria epidemiology. Br Med J 1951; 1(4715):1114-1117.
23. Port GR, Boreham PFL, Bryan JH. The relationship of host size to feeding by mosquitoes of the Anopheles gambiae giles complex (Diptera: Culicidae). Bull Entomological Res 1980; 70:133-144.
24. Takken W, Knols BGJ. Odor-mediated behavior of Afrotropical malaria mosquitoes. Annual Reviews of Entomology 1999; 44:131-157.
25. Smith T, Charlwood JD, Takken W et al. Mapping the densities of malaria vectors within a single village. Acta Trop 1995; 59(1):1-18.
26. Lindsay SW, Emerson PM, Charlwood JD. Reducing malaria by mosquito-proofing houses. Trends Parasitol 2002; 18(11):510-514.
27. Lindsay SW, Snow RW. The trouble with eaves; house entry by vectors of malaria. Trans R Soc Trop Med Hyg 1988; 82(4):645-646.
28. Lacroix R, Mukabana WR, Gouagna LC et al. Malaria infection increases attractiveness of humans to mosquitoes. PLoS Biol 2005; 3(9):e298.
29. Smith DL, McKenzie FE, Snow RW et al. Revisiting the basic reproductive number for malaria and its implications for malaria control. PLoS Biol 2007; 5(3):e42.
30. Dietz K. Models for vector-borne parasitic diseases. Lecture Notes in Biomathematics 1980; 39:264-277.
31. Dye C, Hasibeder G. Population dynamics of mosquito-borne disease: effects of flies which bite some people more frequently than others. Trans R Soc Trop Med Hyg 1986; 80(1):69-77.
32. Kingsolver JG. Mosquito host choice and the epidemiology of malaria. The American Naturalist 1987; 130:811-827.
33. Smith DL, Dushoff J, McKenzie FE. The risk of a mosquito-borne infection in a heterogeneous environment. PLoS Biol 2004; 2(11):e368.
34. Le Menach A, McKenzie FE, Flahault A et al. The unexpected importance of mosquito oviposition behaviour for malaria: nonproductive larval habitats can be sources for malaria transmission. Malar J 2005; 4(1):23.
35. Ross R. An application to the theory of probabilities to the study of a priori pathometry—part I. Proc Roy Soc Lond Part 1 A 1916; 92:204-230.
36. Box GEP, Draper NR. Empirical Model-Building and Response Surfaces: Wiley, 1987.
37. Molineaux L, Dietz K, Thomas A. Further epidemiological evaluation of a malaria model. Bull World Health Organ 1978; 56(4):565-571.
38. McKenzie FE. Why model malaria? Parasitol Today 2000; 16(12):511-516.
39. Levins R. The strategy of model building in population biology. American Scientist 1966; 54:421-431.
40. Hay SI, Smith DL, Snow RW. Measuring malaria endemicity from intense to interrupted transmission. Lancet Infect Dis 2008; 8(6):369-378.
41. Smith DL, Guerra CA, Snow RW et al. Standardizing estimates of the Plasmodium falciparum parasite rate. Malar J 2007; 6:131.
42. Corran P, Coleman P, Riley E et al. Serology: a robust indicator of malaria transmission intensity? Trends Parasitol 2007; 23(12):575-582.
43. Macdonald G. Epidemiological basis of malaria control. Bull World Health Organ. 1956; 15(3-5):613-626.
44. Snow RW, Marsh K. The consequences of reducing transmission of plasmodium falciparum in Africa. Adv Parasitol 2002; 52:235-264.
45. Smith DL, Dushoff J, Snow RW et al. The entomological inoculation rate and Plasmodium falciparum infection in african children. Nature 2005; 438(7067):492-495.
46. Woolhouse ME, Dye C, Etard JF et al. Heterogeneities in the transmission of infectious agents: implications for the design of control programs. Proc Natl Acad Sci USA 1997; 94(1):338-342.
47. Kirby MJ, Green C, Milligan PM et al. Risk factors for house-entry by malaria vectors in a rural town and satellite villages in the gambia. Malar J 2008; 7:2.
48. Lindsay SW, Jawara M, Paine K et al. Changes in house design reduce exposure to malaria mosquitoes. Trop Med Int Health 2003; 8(6):512-517.
49. Le Menach A, Takala S, McKenzie FE et al. An elaborated feeding cycle model for reductions in vectorial capacity of night-biting mosquitoes by insecticide-treated nets. Malar J 2007; 6:10.

50. Hasibeder G, Dye C. Population dynamics of mosquito-borne disease: persistence in a completely heterogeneous environment. Theor Popul Biol 1988; 33(1):31-53.
51. Struik SS, Riley EM. Does malaria suffer from lack of memory? Immunol Rev 2004; 201:268-290.
52. Snow RW, Marsh K. The consequences of reducing transmission of Plasmodium falciparum in africa. Adv Parasitol 2002; 52:235-264.
53. Carter R, Chen DH. Malaria transmission blocked by immunisation with gametes of the malaria parasite. Nature 1976; 263(5572):57-60.
54. Githeko AK, Brandling-Bennett AD, Beier M et al. The reservoir of it Plasmodium falciparum malaria in a holoendemic area of western Kenya. Trans R Soc Trop Med Hyg 1992; 86:355-358.
55. Aron JL. Mathematical modeling of immunity to malaria. Mathematical Biosciences 1988; 90:385-396.
56. Aron JL. Dynamics of acquired immunity boosted by exposure to infection. Mathematical Biosciences 1983; 64:249-259.
57. Aron JL. Acquired immunity dependent upon exposure in an SIRS epidemic model. Mathematical Biosciences 1988; 88:37-47.
58. Smith T, Hii JL, Genton B et al. Associations of peak shifts in age—prevalence for human malarias with bednet coverage. Trans R Soc Trop Med Hyg 2001; 95(1):1-6.
59. Gupta S, Snow RW, Donnelly CA et al. Immunity to noncerebral severe malaria is acquired after one or two infections. Nat Med 1999; 5(3):340-343.
60. Koella JC, Antia R. Epidemiological models for the spread of antimalarial resistance. Malar J 2003; 2:3.
61. Klein EY, Smith DL, Boni MF et al. Clinically immune hosts as a refuge for drug-sensitive malaria parasites. Malar J 2008; 7:67.
62. Smith TA. Estimation of heterogeneity in malaria transmission by stochastic modelling of apparent deviations from mass action kinetics. Malar J 2008; 7:12.
63. Tatem AJ, Rogers DJ, Hay SI. Estimating the malaria risk of African mosquito movement by air travel. Malar J 2006; 5:57.
64. Tatem AJ, Hay SI, Rogers DJ. Global traffic and disease vector dispersal. Proc Natl Acad Sci USA 2006; 103(16):6242-6247.
65. Killeen GF. Following in Soper's footsteps: northeast Brazil 63 years after eradication of Anopheles Gambiae. Lancet Infect Dis 2003; 3(10):663-666.
66. Packard RM, Gadehla P. A land filled with mosquitoes: Fred L. Soper, the Rockefeller Foundation and the Anopheles gambiae invasion of Brazil. Med Anthropol 1997; 17(3):215-238.
67. Packard RM, Gadelha P. A land filled with mosquitoes: Fred L. Soper, the Rockefeller Foundation and the Anopheles gambiae invasion of Brazil. Parassitologia 1994; 36(1-2):197-213.

Chapter 2

Vector Transmission Heterogeneity and the Population Dynamics and Control of Lymphatic Filariasis

Edwin Michael* and Manoj Gambhir

Abstract

A long-standing gap in lymphatic filariasis epidemiology is quantifying the potential effect that heterogeneous infection processes occurring in the major mosquito vector genera may have on parasite transmission and control. Although previous studies have focussed on examining the forms of the density dependent mechanisms regulating larval infection in various mosquito genera, there has been little work done thus far in investigating how such differential processes might interact with density-dependent processes occurring in other stages of the parasite life cycle to influence overall transmission dynamics between areas exposed to different transmitting vector populations. Here, we explore the impact that differences in vector genus-related larval infection dynamics may have on filariasis transmission and control using newly derived parasite transmission models incorporating the forms of the density-dependent processes regulating larval infection in the two major vectors transmitting filariasis, viz. culicine and anopheline mosquitoes. The key finding in this work is that filarial infection thresholds, system resilience, transmission dynamics and parasite response to control efforts, can all be influenced by the prevailing transmitting mosquito genus. In particular, we show that infection thresholds may be raised, system resilience to perturbations lowered and effects of repeated mass treatments in eliminating infection enhanced in anopheline filariasis compared to culicine filariasis, as a direct result of the occurrence and action of multiple positive density-dependent mechanisms influencing infection in this vector-parasite system, such as the "facilitation" function regulating larval infection dynamics in the vector and the inverse probability function governing adult worm mating in the host. These findings indicate that anopheline filariasis may be easier to eradicate than culicine filariasis for a given precontrol infection level, although the actual intensity of interventions required to achieve eradication may in fact be similar to that for culicine filariasis because of the higher infection levels generated as a result of the "facilitation" process in *Anopheles* transmission areas.

Introduction

It has long been suggested that variable within-vector host nonlinearities, including density-dependent processes affecting parasite larval establishment and development and vector mortality, could act as major factors underlying the observed heterogeneous transmission dynamics of vector-borne diseases.[1,2] Addressing the impacts of this heterogeneity has critical resonance for

*Corresponding Author: Edwin Michael—Department of Infectious Disease Epidemiology, Faculty of Medicine (St. Mary's campus), Imperial College London, Norfolk Place, London W2 1PG, U.K. Email: e.michael@imperial.ac.uk

Modelling Parasite Transmission and Control, edited by Edwin Michael and Robert C. Spear.

studies of the population dynamics and control of the vector-borne helminthic disease, lymphatic filariasis, in particular, given the long-held belief that the forms of the density-dependent processes acting on larval infection dynamics may differ significantly between the taxonomic group of mosquitoes involved in parasite transmission.[3-8] The key empirical finding from this body of work is that the density-dependent mechanism regulating larval infection in one major group of filarial mosquito vectors, viz. culicine mosquitoes and to some extent *Aedes* mosquitoes,[8] may be of the negative density dependent or "limitation" form, whereas larval infection dynamics in the other main parasite-transmitting mosquito vector, the *Anopheles* group of mosquitoes, is thought to be governed by a positive feedback or "facilitation" process.[3,4,8] The latter mechanism is particularly interesting because it could lead to an unstable equilibrium population size, below which the parasite would be driven to extinction.[3] As discussed previously, this outcome would clearly make anopheline-transmitted filariasis easier to eradicate through intervention efforts,[5-7] a conclusion which has reinforced the impression that differences in the prevailing transmitting vector genus could largely underlie the inconsistencies observed in the success of filariasis control programmes in different endemic areas.[3,9,10]

In reality, however, as pointed out recently[8,11] to fully understand the impact of these vector genus-specific infection processes on overall filariasis transmission and control, it is also necessary to consider how these processes interact dynamically with other density dependent processes occurring in other parts of the parasite life cycle.[1,2] This is because without considering the action of such multiple density-dependent factors, it will be unclear as to what extent vector-specific infection processes alone will contribute to overall parasite transmission regulation, particularly given the likely stronger negative feedback processes, e.g., acquired immunity, which may regulate infection in human hosts.[9,12] Furthermore, recent work has also underscored the crucial impact that the simultaneous presence and interactions between multiple density-dependent processes acting at different points in the life cycle of a population can have on the dynamics of parasite system persistence, including the values of parasite extinction thresholds.[13,14] These findings suggest that a composite analysis investigating the simultaneous effects of infection processes in both the vector and human host will be crucial to efforts for determining how vector-parasite combinations occurring in different endemic regions will interact with the efforts presently being proposed to achieve the global control of this debilitating parasitic disease.

Here, we review recent work in modelling lymphatic filariasis transmission[15] in order to systematically investigate the impact that vector genus-specific density-dependent processes may have on overall parasite transmission, persistence and control. Specifically, we focus on the incorporation and simultaneous analyses of the "limitation" and "facilitation" functions found to govern larval infection, in addition to density-dependent mechanisms thought to regulate adult worm infection in the human host, such as worm mating probability and acquired immunity, in the two major parasite-vector species combinations implicated in the transmission of filariasis, viz. one in which the vector intermediate host are culicine mosquitoes and the other in which they are anopheline. We begin the chapter by first providing details of the life history of the parasite, characteristics and locality of the major transmitting vectors and particulars underlying the specification and quantification of the functional forms for the "limitation" and "facilitation" mechanisms regulating larval development in culicine versus anopheline vectors respectively. This is followed by details of methods that allow successful incorporation of these functions into overall transmission models for describing culicine- and anopheline-transmitted filariasis population dynamics, highlighting in particular the key need for and derivation of expressions that describe the actions of these functions at the host population level. Model analyses are then described for investigating the influence of vector genus-specific density-dependent processes on system stability, infection\extinction thresholds and parasite population dynamics. The import of these vector genus-specific transmission dynamics for the control of lymphatic filariasis from the application of the currently-proposed mass chemotherapy intervention regimens are then examined using model simulations. We end by underscoring the usefulness of these models and future work to improve their utility in resolving remaining questions regarding filariasis transmission dynamics and control.

Lymphatic Filariasis Disease and Parasite Life Cycle

Lymphatic filariasis is a major cause of acute and chronic morbidity of humans in tropical and subtropical areas of Asia, Africa, the Western Pacific and some parts of the Americas, with over some 1.2 billion people thought to live in areas where they are at risk of infection.[16] Of the estimated 120 million cases of people living with active filarial infection currently, 91% are caused by the filarial worm, *Wuchereria bancrofti,* while *Brugia malayi* and *B. timori* infections account for the other 9%.[17,18] *Brugia timori* is only known to be endemic in Timor and the Flores islands of the Indonesian archipelago.[19] These lymphatic-dwelling parasites cause damage to the lymphatic system which leads to lymphoedema, genital pathology (especially hydrocoeles) and elephantiasis in some 41 million men, women and children.[17,20] A further 76 million, most often with the transmission stage, microfilariae (Mf), in their blood, are thought to have hidden internal damage to their lymphatic and renal systems. Indeed, recent estimates made by the WHO suggest that lymphatic filariasis may be the second leading known cause of permanent and long-term disability worldwide.[21]

The filarial parasites have biphasic life cycles involving the definitive mammalian host and various species of mosquito vectors. *W. bancrofti* seems to be exclusively a human parasite, whereas *Brugia* spp. are zoonotic in limited situations. Given the involvement of vectors, parasite transmission is indirect, with human infection initiated by the deposition of the third-stage (L3) infective larvae on the skin of the host following the bite of an infective mosquito. The deposited L3 larva penetrates the skin at the site of the bite and migrates to the lymphatic system of the host where they mature over 12 months or so into adult male and female worms.[22] Most estimates of the life span of the adult parasite suggest a period of 5-10 years.[23,24] The lymphatic-dwelling filariae are dioecious and undergo ovoviviparous reproduction resulting in the release of Mf from adult females, which circulate in the host bloodstream. The life-cycle is completed when these Mf are ingested with the bloodmeal taken by female mosquitoes of a susceptible vector species and develop over 10 to 14 days into the infective-stage L3 larvae capable of infecting a host.

Mosquito Vectors of Lymphatic Filariasis

Both the *W. bancrofti* and *B. malayi* filarial parasites are unique among the various mosquito-transmitted parasites in that larval development can take place in several genera of mosquitoes. A great diversity of mosquito spp. in the genera *Culex*, *Anopheles*, *Aedes* and *Mansonia* are capable of acquiring and supporting the development of Mf to infective L3 larvae. Indeed, based on the geographic distribution of these vectors, three main zones of filariasis transmission are recognized: the South Pacific islands and some limited areas of South East Asia, where *Aedes* vectors predominate; West Africa, Papua New Guinea, Vanuatu and Solomon islands where *Anopheles* mosquitoes are principal vectors; and China, South East Asia, Egypt, East Africa, the Caribbean and Latin America where the infection is transmitted mainly by *Culex* species, although *A. gambiae* and *A. funestus* form the predominant vectors that transmit filariasis in rural East Africa (Table 1). The variations in the biological characteristics of each vector genus listed in Table 1, including the differential vector-specific forms of the density-dependent processes regulating larval infection noted above, provide the bases for the long-held belief that the taxonomic group of mosquitoes involved in filariasis transmission may determine not only parasite transmission dynamics but also the effectiveness of any control strategy aimed at interrupting transmission in an area.

Vector-Parasite Infection Relationships

The impact that vector genera differences can have on filariasis transmission and control was first pointed out by Pichon et al,[3] who suggested that this heterogeneous effect may arise from variations in the form of density dependent processes acting on parasite uptake and development in the different filaria-transmitting vector genera. In particular, these workers and others (see study references given in Snow et al[8]) used experimental data on Mf uptake to L3 development obtained from mosquitoes fed on human volunteers to show that if x is the number of Mf ingested by a mosquito and y is the number which survive to become infective, then limitation is the

Table 1. Important biological characteristics of the major filariasis vectors

Characteristics	*Culex Quinquesfasciatus*	*Anopheles* spp.	*Aedes* spp.
Locality	China, South East Asia, Egypt, East Africa, Urban West Africa, Caribbean region, Latin America	West Africa, rural East Africa, Papua New Guinea, Vanuatu, Solomon islands	South Pacific islands, South East Asia
Breeding sites	Pools, domestic water supplies, septic tanks	Brackish water, swamps, streams, puddles	Tyres, drums, wells, cisterns, tins, crab holes, tree holes
Biting time	Night	Night	Day
Biting location	Indoors	Indoors and Outdoors	Outdoors
Infection process	Limitation	Facilitation	Limitation or proportional

negative feedback process occurring when the parasite yield y/x decreases linearly with x, whereas facilitation is the positive feedback process occurring when y/x increases with x. The fits of these linearized models to data were first used to show as far back as 1974 that limitation may be most commonly associated with filarial infections of culicine mosquitoes, whilst facilitation may be associated with filarial infections of anopheline mosquitoes.[3] Despite this, (1) variability in the results of fits to individual studies, (2) statistical problems connected with the fitting of linearized versions of essentially nonlinear functions to data[25-27] and (3) the lack of appropriately parameterized functional forms for the Mf-L3 relationship (but see Subramanian et al[28]), has meant that the mathematical development of filariasis transmission models incorporating density-dependent regulatory functions reflecting variations in vector genus-related infection dynamics has thus far been hampered, with the result that current models are based on only one form of the functional relationship describing the Mf to L3 development process, viz. that related to the limitation function in culicine mosquitoes.[29,30]

Quantifying the Mf-L3 Functional Response in Vector Populations

Given the above uncertainties, a major focus of recent work in modelling lymphatic filariasis transmission has been on quantifying the suggested relationships occurring between Mf uptake and L3 development in culicine and anopheline mosquitoes in the form of functional response equations that allow ready incorporation into overall filarial parasite transmission dynamics models.[8,15,28] Here, we describe particulars of our recent work in this area based primarily on specifying and fitting the appropriate functional response models to published mosquito feeding experiment data, as follows.[15]

First, for *Culex* mosquitoes, given that a limitation response occurs which results in a saturation in the production of L3 larvae as Mf loads in bloodmeals increase,[8] the Mf-L3 uptake and development data can be fitted to the function:[30]

$$L = \kappa_{S1}\left(1 - e^{-r_1 m / \kappa_{S1}}\right) \quad (1)$$

where m is the Mf density in the human host (20 μL blood), κ_{S1} is the maximum limiting value of L3 numbers developing in the mosquito and r_1 controls the rate at which the L3 development rises with Mf ingested. The fit of this function to data collected from all available published studies[8,31] is shown in Figure 1A. Note that while this is the appropriate functional form for the uptake of Mf and the development of L3 mosquitoes in a single mosquito from one bloodmeal, a fundamental need, as remarked above, is to account for the effect of the distribution of Mf among hosts in order to quantify the average L3 output in a community if we are to model population-level effects. In order to do this in a manner consistent with field observations,[28,31] the average L3 level may be calculated in the mosquito populations by assuming that they take bloodmeals from a human host community in whom the mean Mf level is M and is negative binomially distributed with shape parameter k. This population-averaged uptake function (L_{pop}) can be derived as (see Gambhir and Michael[15]):

$$L_{pop} = \kappa_{S1}(1 - f(M)) \tag{2}$$

where:

$$f(M) = \left(1 + \frac{M}{k}\left(1 - e^{-r_1/\kappa_{S1}}\right)\right)^{-k} \tag{3}$$

This function is very similar to the individual uptake curve, but there is now a dependence on the shape parameter k, which determines the level of parasite aggregation in the human hosts.

For anopheline mosquitoes, a closely-related function to that used to model the *Culex* uptake response, but describing behaviour consistent with facilitation in the part of their Mf-L3 uptake/development curves corresponding to low Mf densities, can be given by the equation:

$$L = \kappa_{S2}\left(1 - e^{-r_2(m-T)/\kappa_{S2}}\right)^2 \tag{4}$$

As for the *Culex* function, κ_{S2} is the maximum limiting value of L3 numbers developing in the mosquito and r_2 controls the rate at which L3 development rises with Mf ingested. This function has two features that distinguish it from the *Culex* function above. The first is the fact that it is raised to the power of two, an operation that introduces a concavity into the shape of the function prior to its limitation-associated saturation at very high Mf loads. The second is the offset parameter (T) appearing in the exponent of the exponential function, which ensures that the uptake value rises smoothly from its zero point only when the Mf density in a bloodmeal is

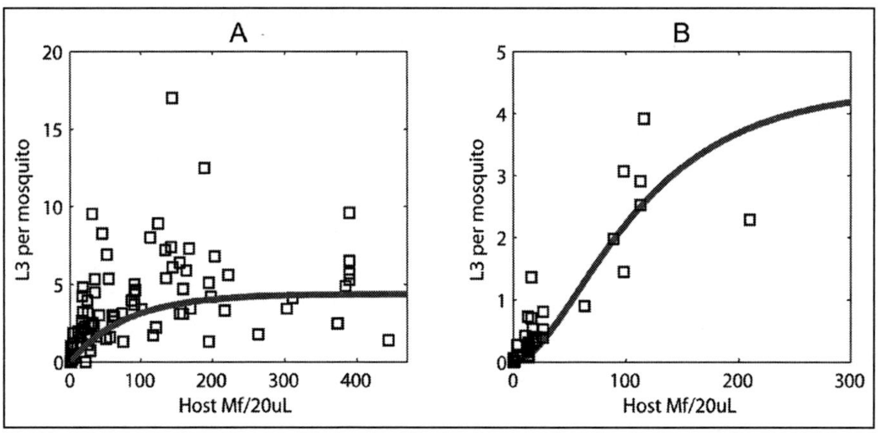

Figure 1. Functional forms relating microfilaria (Mf) uptake (from 20 μL host blood) and L3 development per mosquito in the two vector genus studied in this work. The squares denote observed data for each vector respectively (sources in Gambhir and Michael[15]). The curve fitted to the (A) *Culex* data is a limiting function of Mf (equation 1 in the text) whereas the *Anopheles* curve in (B) describes a development response that begins with a facilitation phase which then approaches an upper limit at higher Mf uptakes (equation 4 in the text).

greater than a threshold density given by the value described by T. Biologically, this functional behaviour may be associated with the action of the anopheline cibarial armature or teeth, which prevents Mf from passing undamaged into the mosquito gut at intakes lower than the threshold T[32] but which allows the undamaged passage of the majority of ingested Mf at higher ingestion loads owing to the protection afforded due to the entanglement of the first passing Mf about the cibarial teeth. The best-fit of the model to published data is shown in Figure 1B. Again, this function needs to be averaged over the host population leading to the derivation of the following population-averaged function:[15]

$$f(M) = \sum_{m=0}^{T} P_{nb}(m;k,M) + \left[\frac{2e^{r_2 T/\kappa_{S2}}}{\left[1+\frac{M}{k}\left(1-e^{-r_2/\kappa_{S2}}\right)\right]^k} - \frac{e^{2r_2 T/\kappa_{S2}}}{\left[1+\frac{M}{k}\left(1-e^{-2r_2/\kappa_{S2}}\right)\right]^k} \right] \times \sum_{m=T}^{\infty} P_{nb}(m;k,M) \tag{5}$$

where $P_{nb}(m; k, M)$ is the negative binomial probability mass function, with mean M and aggregation parameter k. Table 2A provides the best-fit parameter values for both the *Culex* and *Anopheles* functions given in Equations 1 and 4.

Derivation of Vector-Specific Models of Lymphatic Filariasis Transmission

The basic deterministic dynamical model for lymphatic filariasis transmission primarily consists of a series of coupled differential and partial differential equations for three state variables describing the changes in numbers over time of three key parasite life stages—worm burden (W), Mf count (M) and stage L3 larvae (L)—and one state variable describing the acquisition and loss of immunity to parasites in human hosts (I).[30,33,34] Structurally, these equations are divided into those which describe changes in parasitic infection in the human definitive hosts (W, M, I) and the L3 larval equation (L), which is used to calculate the change in larval density in the mosquito intermediate hosts. The equations pertaining to changes in parasite stages in the human host population describe change as a function of host age as well as time, since model inputs—such as exposure to mosquito biting—vary over age.[33,35] By contrast, because the processes governing the rate of change of the average L3 larval density within the mosquito population operate on much shorter timescales than those in the human host population, the model assumes that the mosquito larval density comes to equilibrium as soon as the corresponding human host parasite and immunity levels are computed. The coupled partial differential and differential equations comprising the basic dynamical model are thus:

$$\frac{\partial W}{\partial t} + \frac{\partial W}{\partial a} = \lambda \frac{V}{H} \psi_1 \psi_2 s_2 h(a) L^* e^{-\beta I} - \mu W \tag{6}$$

$$\frac{\partial M}{\partial t} + \frac{\partial M}{\partial a} = \alpha \phi(W, k) W - \gamma M \tag{7}$$

$$\frac{\partial I}{\partial t} + \frac{\partial I}{\partial a} = W \tag{8}$$

$$\frac{\partial L}{\partial t} = \lambda \kappa g \int \pi(a)(1 - f(M))da - \sigma L - \lambda \psi_1 L \tag{9}$$

Table 2A. Parameter values used in the model for the vector uptake functions. These values were obtained through nonlinear least squares fits the data collated by Snow and Michael[31]

Parameter	Value	Standard Error
κ_{S1}	4.406	0.362
r_1	0.019	0.058
κ_{S2}	4.395	0.332
r_2	0.055	0.004
T	0	N/A

Table 2B. Descriptions and values of the parameters of the vector-specific filariasis models. Typical values of some of the parameters cannot be given due to their multi-valued or dynamic nature[15]

Parameter Symbol	Definition	Value
λ	Number of bites per mosquito	10 per month
V/H	Ratio of number of vectors to hosts	Various
ψ_1	Proportion of L3 leaving mosquito per bite	0.414
ψ_2	Proportion of L3 leaving mosquito that enter host	0.32
s_2	Proportion of L3 entering host that develop into adult worms	0.2
β	Strength of acquired immunity	0.112
μ	Death rate of adult worms	0.0104 per month
α	Production rate of Mf per worm	0.33 per month
γ	Death rate of Mf	0.1 per month
g	Proportion of mosquitoes which pick up infection when biting an infected host	0.37
σ	Death rate of mosquitoes	5 per month
$k(M)$	Aggregation parameter from negative binomial distribution	$k_0 + k_1M$ (0.0029 + 0.0236M)
$h(a)$	Parameter adjusting rate at which individuals of age a are bitten	Linear rise from 0 at age zero to 1 at 10 years
L^*	Equilibrium value of the larval density (see Equation 5)	—
$\phi(W, k)$	Mating probability as a function of worm burden W and aggregation parameter k	—
$\pi(a)$	Probability that an individual is of age a	—
$f(M)$	Variable component of the population-averaged Mf uptake and L3-larval development function. This is a function of the average Mf level M (see Equations 8 and 10 below for specific functional forms)	—

The above equations are similar to those previously published for the lymphatic filarial transmission model described by Norman et al,[30] but with a major difference in that here a mating probability ($\phi(W, k) = 1 - (1 + W/2k)^{-(1+k)}$,[36]) has been incorporated into the mean Mf count (M) equation to account for the production of Mf as a function of density-dependent worm mating within the human hosts. This probability function is dependent upon the worm burden W and the shape parameter k of the negative binomial distribution describing parasite dispersion among hosts, with the form selected to account for the dioecious and polygamous nature of *W. bancrofti* worms.[9,36] The dependence of the shape parameter of the negative binomial distribution on the average worm burden is considered to be identical to that used to describe the relationship between Mf prevalence and intensity, i.e., it is a linear function of worm burden with gradient k_1 and with zero intercept k_0. Since the rates associated with the vectors are considerably faster than the time taken in the development of worm and Mf burdens, the L3 density is assumed to instantaneously equilibrate. By setting the right hand side of Equation 9 to zero and rearranging, the following expression is obtained for the instantaneous value of the L3 larval density, L^*:

$$L^* = \frac{\lambda \kappa g \int \pi(a)(1 - f(M))da}{\sigma + \lambda \psi_1} \qquad (10)$$

here, the parameter κ denotes the saturation value of the vector Mf uptake function as detailed above (equations 2 and 4, i.e., K_{s1} denoting the saturation value for the *Culex* uptake/development function and K_{s2} the corresponding value for the *Anopheles* function). This expression thus readily allows the incorporation of the culicine and anopheline larval infection dynamics into the basic filariasis model in order to mimic vector-specific transmission dynamics via both substitutions for κ as describe above and by substituting the $f(M)$ term in equation 10 with equations 3 and 5. Table 2B provides a description of the equation parameters along with the values used in these models (details in Gambhir and Michael[15]).

Impact of Vector-Specific Infection Processes on Parasite System Stability, Persistence and Extinction

System Equilibria, Transitions and Stable States

The impacts of the combined vector and host density-dependent processes investigated in this study on the persistence and extinction dynamics of the filarial parasite system can be examined by conducting a numerical analysis of the stability of the solutions of the model equations based on varying initial values of L^* (details in Gambhir and Michael[15]). The results from these analyses for both the culicine and anopheline models are shown in Figure 2. The graphs depict the steady states of host Mf infection prevalence (converted from modeled Mf loads using the expression: $P(M, k) = 1 - (1 + M/k)^{-k}$, where P is the prevalence, M is the Mf density and k is the aggregation parameter of the negative binomial distribution[15,30]), as well as the nature of their transitions that appear along the vector biting rate gradient with and without the inclusion of worm mating probabilities in the respective models. As noted previously,[15] these results, i.e., the existence of multiple parasite system regimes or states demonstrate the likely occurrence of complex dynamics in the transmission of filariasis. Here, however, we focus on the impact of vector genus transmission heterogeneity in giving rise to this complexity. Thus, in the culicine case, when worm mating probabilities were not included and where the L3 uptake and development function shows only limiting behaviour, the results show that the system gives rise to just one threshold—a vector threshold biting rate (TBR) below which the only stable equilibrium is at the zero parasite level (Fig. 2A). Above this transmission threshold, a transcritical system bifurcation appears to occur leading to the existence of endemic stable infection levels that increase in magnitude smoothly and reversibly as biting rates increase. Thus, no worm breakpoints and hence existence of alternate infection states are possible, i.e., only positive endemic states occur above the TBR, the system losing its stable endemic infection state smoothly to converge into a zero-infection state as the vector biting rate is reduced below the TBR (Fig. 2A). The effects

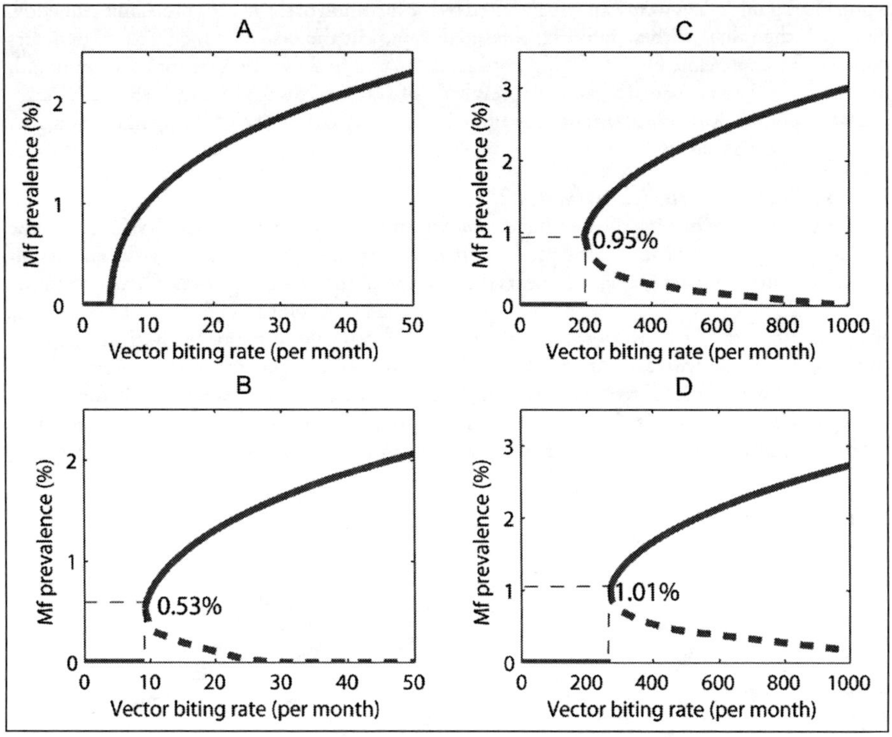

Figure 2. The effect of varying the vector biting rate on the equilibrium Mf prevalence among human hosts when the vector intermediate host was culicine and when the worm mating probability was (A) and was not (B) included in the model and when the intermediate host was anopheline with (C) no mating and with (D) mating probabilities included. Inclusion of the mating probability introduced a set of breakpoints (dotted line) in the case of *Culex* and it raised and increased the range of the breakpoints in the case of *Anopheles*. The labelled Mf values on the y-axes correspond to the maximum breakpoints, whereas additionally in each graph the solid curve (A) and vertical dashed drop lines (B-D) crossing the x-axes denote the threshold biting rates (TBRs) estimated in each scenario corresponding to (A) 4, (B) 9, (C) 197, (D) 271 vector bites per month. From Gambhir and Michael.[15]

of introducing the worm mating function into the culicine model is shown in Figure 2B. In contrast to the situation when the function was excluded, it is clear that this can bring about a discontinuous jump in the Mf prevalence at the TBR giving rise to the appearance of three equilibria, two stable ones separated by an unstable one, mimicking thus the occurrence of a subcritical bifurcation in the system dynamics. The unstable equilibrium points of the model comprise a set of positive-valued worm breakpoints—the largest of which was found at the TBR—producing an unstable dynamic boundary or separatrix with increasing biting rates across which the system converge to either a zero or endemic stable equilibrium state. Thus, at vector biting rates above the TBR, bistable infection states, one endemic and the other zero-infection, can exist depending upon whether initial L^* load values can give rise to infection levels above or below the unstable breakpoint values (Fig. 2B). By contrast, in the anopheline-mediated system in which the L3 development function was of a facilitation form, a subcritical bifurcation was observed in which a discontinuous jump in the Mf prevalence at the TBR occurs regardless of whether the worm mating function is included or not (Figs. 2C,D). The values of the vector biting rate at which the nonzero worm breakpoints exist were also far greater in this case than for the culicine system (Figs. 2B vs C,D). In addition, both the TBR and the maximum worm breakpoint values were found to be raised by the inclusion of the mating probability function. Note that the values of these infection thresholds represent the first such model-based estimates obtained for anopheline filariasis.[9,10] Their values compared to those obtained for these thresholds in culicine filariasis (maximal breakpoint prevalence of 1.01% versus 0.53% and TBR of 271 versus just 9 per month) also confirm that parasite elimination thresholds are likely to be significantly higher in anopheline filariasis.[9,13]

System Resilience to Perturbations

The results of perturbing the *Culex* and *Anopheles* models by varying initial L3 values from the vicinity of the respective unstable equilibria are illustrated in Figure 3. Each age-dependent Mf prevalence curve in these graphs represents the solution of the model equations for a given initial L3 seed value set to a value either above or beneath the breakpoint level for a vector biting rate above the estimated TBR. The results depict that an increasing perturbation below the breakpoint curve leads to an eventual age-Mf prevalence of zero across all ages, whereas a similar perturbation above the breakpoint leads to stabilisation at the equilibrium endemic level (Fig. 3). The sizes of the regions between the unstable and either stable equilibrium state are also depicted in the figure for each model and clearly provide a qualitative measure of the extents of the respective basins of

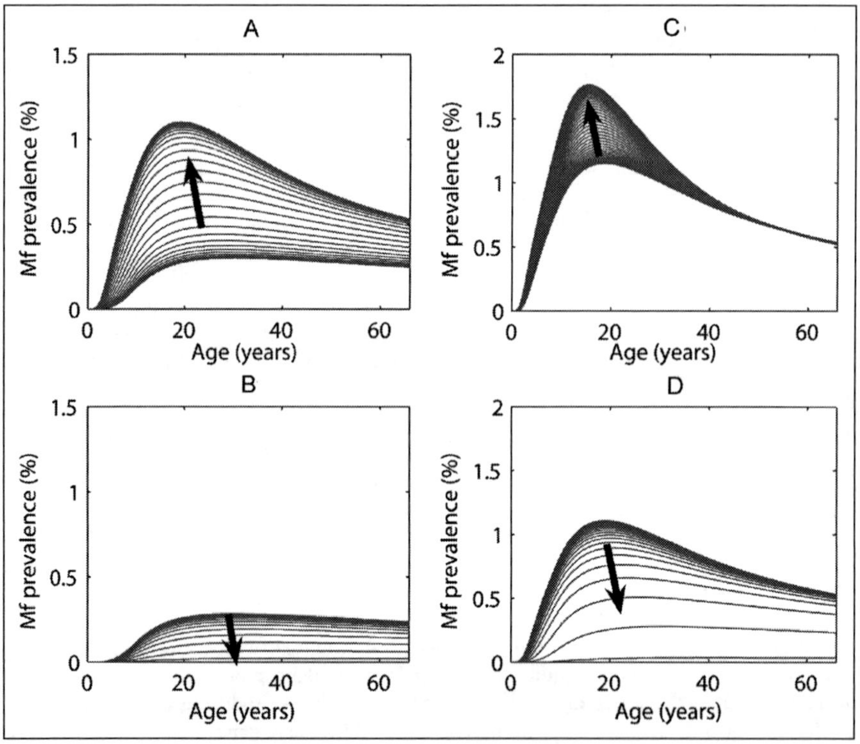

Figure 3. The direction of change of the Mf prevalence age-profile among human hosts when the intermediate host was culicine and the initial Mf prevalence was (A) above and (B) below the breakpoint value (0.21%) with a vector biting rate of 11 per month; and when the intermediate host was anopheline with initial Mf prevalence (C) above and (D) below the breakpoint value (0.82%) with a vector biting rate of 280 per month. Each age-dependent curve represents a perturbation around the initial unstable equilibrium curve with the black arrows indicating the direction in which these curves are likely to travel on the way to the stable equilibrium. From Gambhir and Michael.[15]

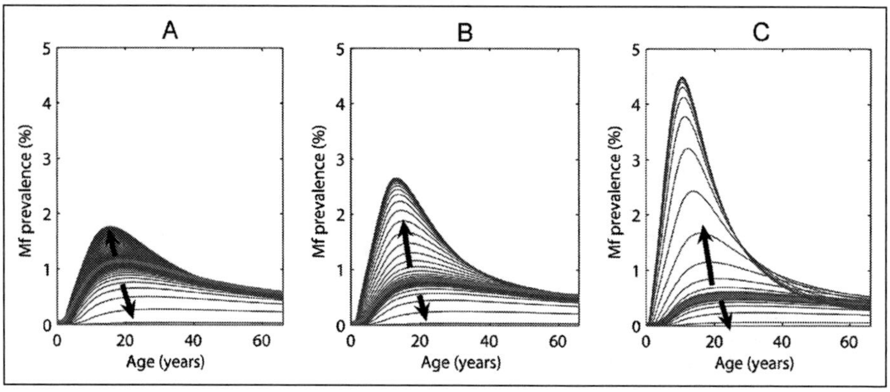

Figure 4. The direction of change of the Mf prevalence age-profile among human hosts at different anopheline vector biting rates. Results are shown for vector biting rates of (A) 280 (just above the TBR of 271), (B) 350 and (C) 600. In each case, the blue curves represent initial perturbations around the Mf age-profile corresponding to the breakpoint Mf prevalence (red curve) and the black arrows indicate the direction in which the curves are likely to travel on the way to attaining the zero or endemic stable equilibrium. From Gambhir and Michael.[15] A color version of this image is available at www.landesbioscience.com/curie

attraction to either stable state. Given that they in turn provide a measure of the maximum amount each system can be changed before losing its ability to recover to either stable state, these results also graphically offer an insight into the likely resilience of the two parasite systems to each of the stable state. For example, the results show that in culicine filariasis, the basin of attraction to the endemic equilibrium state is much larger in size compared to the case in anopheline filariasis, while the opposite is true in the case of the basins of attraction to the zero infection state (Fig. 3A,B versus C,D). The larger basin of attraction to the endemic state but the smaller basin of attraction to the zero state thus suggests that culicine filariasis may be more resilient to perturbations of the system away from the endemic attractor compared to anopheline filariasis, i.e., the former may be more resilient to extinction compared to the latter.

Figure 4 portrays the effects of perturbing the anopheline system near the vicinity of the unstable worm breakpoint values (by varying initial L3 values) on the corresponding Mf age-profiles for three increasing values of the vector biting rate. For each value of the biting rate, the system was initialised as before, above and below the breakpoint Mf prevalence level (given by the red curves in Fig. 4) and the direction in which the profiles move was then tracked (indicated by the arrows in the figure). The results show that as the biting rate increases, the relative sizes of the basins of attraction for the zero-parasite and endemic equilibrium can change considerably. In particular, the range of Mf prevalence values that led to the zero equilibrium diminished while those that led to the endemic equilibrium (upper curve) grew, indicating that system resilience to perturbations even for anopheline filariasis will be greater at higher vector biting rates.

System Hysteresis

The existence of multistable states separated by an unstable boundary when the TBR has been exceeded introduces the possibility of the occurrence of bistability or hysteresis in the transmission dynamics of lymphatic filariasis.[37] Figure 5A illustrates this phenomenon in terms of the equilibrium Mf prevalence versus vector biting rate relationship for the model in which the vector is culicine. The bottom black arrow shows the path taken by the equilibrium Mf prevalence when the biting rate is increased to the point on the far right (the 0.1% Mf breakpoint). Until this point is reached, the Mf steady state remains at zero. There is a jump (a subcritical bifurcation) at this point to the endemic equilibrium stage (upward pointing red arrow) and the steady state remains on this stable branch as the biting rate is increased further. Decreasing the biting rate to a point at

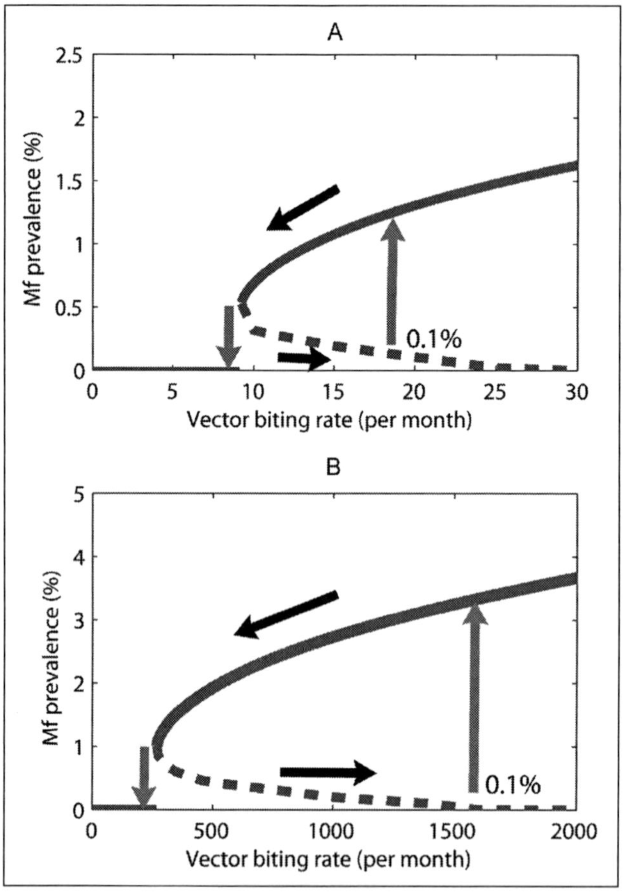

Figure 5. Hysteresis loops in the Mf prevalence/vector biting rate plane for (a) culicine and (b) anopheline intermediate hosts, showing the two asymmetrical ways by which a shift between alternative Mf stable states can occur with varying vector biting rates. If the parasite system is on the lower zero state but at high vector biting rates and thus close to the worm breakpoint bifurcation boundary, a slight incremental change in Mf levels may bring it beyond the birfurcation (say at 0.1% Mf prevalence) and induce a drastic shift of the system to its endemic equilibrium (rightmost red arrow). If one attempts to restore the parasite-free equilibrium state by reducing the vector biting rate (black leftward arrow), the system shows hysteresis. A backward shift to the parasite-free equilibrium (leftmost red arrow) will occur only if the vector biting rate is reduced far enough to reach the TBR bifurcation point. The hysteresis loop is wider in extent for the anopheline model compared to culicine-transmitted filariasis. From Gambhir and Michael.[15] A color version of this image is available at www.landesbioscience.com/curie.

which the endemic state appeared, however, will not return the system to the zero Mf prevalence state. For this to happen, the biting rate needs to be reduced further backward to the TBR occurring at the maximal breakpoint (see top black arrow), at which point the endemic state loses its stability and jumps back to the origin or zero state (downward pointing red arrow) and remains there as the biting rate is decreased further. These different paths followed by the parasite system for increasing versus decreasing the vector biting rate constitute the hysteresis loop effect. Figure 5B shows the equivalent loop for the system in which the vector is anopheline. Here, the loop has a considerably greater range in the space, indicating both that the emergence or re-invasion of the filarial endemic state for a similar input of infected hosts (0.1% Mf prevalence) will occur at a significantly larger vector biting rate and the subsequent state transition to the zero state will require a greater reduction of vector numbers compared to the culicine case (Fig. 5A).

Impact of Vector-Specific Infection Processes on Age Patterns of Infection

The results depicted in Figure 6 show that apart from effects on parasite system characteristics, including infection thresholds and dynamics of system resilience to perturbations, vector genus-specific infection processes may also play an important role in contributing to the differential community age-patterns of filarial infection observed between the culicine versus anopheline vectored filariasis endemic areas.[9,10] The curves in the figure represent the equilibrium age-Mf prevalence patterns generated by the culicine and anopheline-specific filariasis transmission models for proportionately increasing rates of the monthly vector biting intensity and show that largely as a result of the greater levels of L^* produced in the case of anopheline filariasis, for proportional increases in vector biting higher age-prevalences of infection will be produced in *Anopheles*-mediated filariasis transmission areas compared to culicine areas. They also indicate the likely greater occurrence of convexity in the age-patterns of infection in anopheline compared to culicine filariasis for proportionately similar vector biting rates. This is because of the greater cumulative experience of L3 infection levels in the anopheline transmission communities, which would give rise to the earlier generation of significant amounts of host acquired immunity in these populations compared to populations exposed to culicine mosquitoes. Field support for these differential infection patterns have come partially from work carried out by Michael and Bundy[12] and Michael et al,[9] who clearly showed that infection prevalences could be significantly higher for a given vector biting rate in communities exposed to *Anopheles* compared to *Culex* vectors.

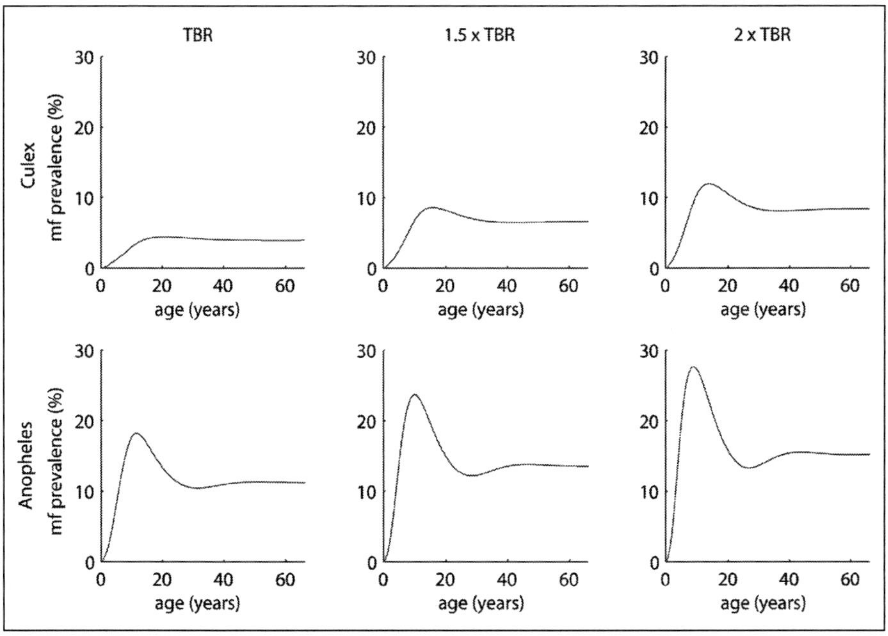

Figure 6. The equilibrium age-Mf prevalence curves predicted by the culicine (upper panel) and anopheline (lower panel) filariasis models for proportionally increasing monthly vector biting rates. Results are shown at just above the TBR and at 1.5 and 2 times the TBR rates estimated for each vector (see text).

The Impact of Vector Genus on the Dynamics of Filariasis Control

The impact of the interaction between the culicine- and anopheline-vector infection processes with a key presently proposed mass drug treatment regimen on the dynamics of filariasis infection elimination is modeled and portrayed in Figure 7. The curves in the figures show the simulated yearly decline in mean Mf prevalence in the cases of anopheline and culicine filariasis respectively as a result of annual combined diethylcarbamazine (DEC)/albendazole (ALB) mass drug treatments (efficacy values as listed in the label to the figure) applied at 80% drug coverage to communities with different precontrol infection prevalences. Two features of these results are immediately apparent with respect to the impact of vector-specific population dynamics on filariasis control by mass chemotherapy. First, a notable finding is that for a similar precontrol infection prevalence, mean community Mf prevalences will drop much more steeply to meet parasite elimination thresholds in the case of anopheline filariasis compared to the more gradual decline in infection trends generated for the case of culicine filariasis. This will clearly reduce the number of treatment cycles required to achieve infection elimination in the case of anopheline filariasis for a given precontrol infection level (compare curves in Fig. 7A versus 7B) and is a direct outcome of the two different forms of density-dependent regulation of L3 output occurring in culicine versus anopheline vectors. Essentially, this result reflects the fact that as Mf load drops progressively over time in the host community as a result of repeated drug treatment, the severe negative density-dependent regulation of L3 output will ease in the case of culicine filariasis leading to a proportionately higher output of L3 at lower community Mf loads whereas the opposite result, viz. a lower L3 output as Mf load drops, would occur in the case of anopheline filariasis. The second notable feature of the results in Figure 7 is that this effect on the duration of annual mass treatments required to achieve parasite elimination will be further enhanced given the likely higher worm breakpoint value that may occur in anopheline (the 1.01% Mf prevalence breakpoint threshold depicted by the dot-dashed horizontal line in Fig. 7B) compared to culicine filariasis (the 0.53% Mf prevalence threshold depicted by the dot-dashed line in Fig. 7A). The dramatic effect that this difference in parasite elimination thresholds and the rates of decline induced in mean community Mf prevalences as a result of drug treatment in anopheline versus culicine filariasis is highlighted by the simulation results shown in the graphs for the high precontrol community prevalence of 20%. These show that while at least 8 annual combined DEC/ALB mass treatments at 80% population coverage will be required to reduce Mf prevalence from an initial level of 20% to under the breakpoint or elimination threshold of 0.5 Mf % in the case of culicine filariasis, the corresponding duration of annual mass treatments in the case of anopheline filariasis to reach the elimination target of 1.01 Mf % will only be in the region of 6 years. The absolute number of annual mass treatments required will be even lower than this for eliminating anopheline filariasis compared to filariasis transmitted by culicine vectors when the precontrol community infection levels are lower than a prevalence of 20% (Fig. 7).

Conclusion

A recurring theme in this volume is the potential complexity occurring in the transmission dynamics of parasitic infections, the key processes of which require careful elaboration from data and the impact of which needs the development of suitable models that are able to incorporate and address interactions between such processes. In this chapter, we have illustrated the importance of both these requirements for investigating a long-standing topic in the transmission of lymphatic filariasis, viz. the impact that vector-specific larval infection dynamics can have on the overall transmission dynamics of this macroparasitic disease. This has not only allowed us for the first time to successfully derive and analyze dynamic deterministic models for describing filariasis transmission by the two major mosquito vectors implicated in the transmission of this disease, viz. culicine and anoheline mosquitoes respectively; it has also enabled the first reliable study of the differential impact that having one or the other of these mosquitoes as the dominant transmitting vector can have on overall filariasis transmission dynamics and parasite control in host populations.

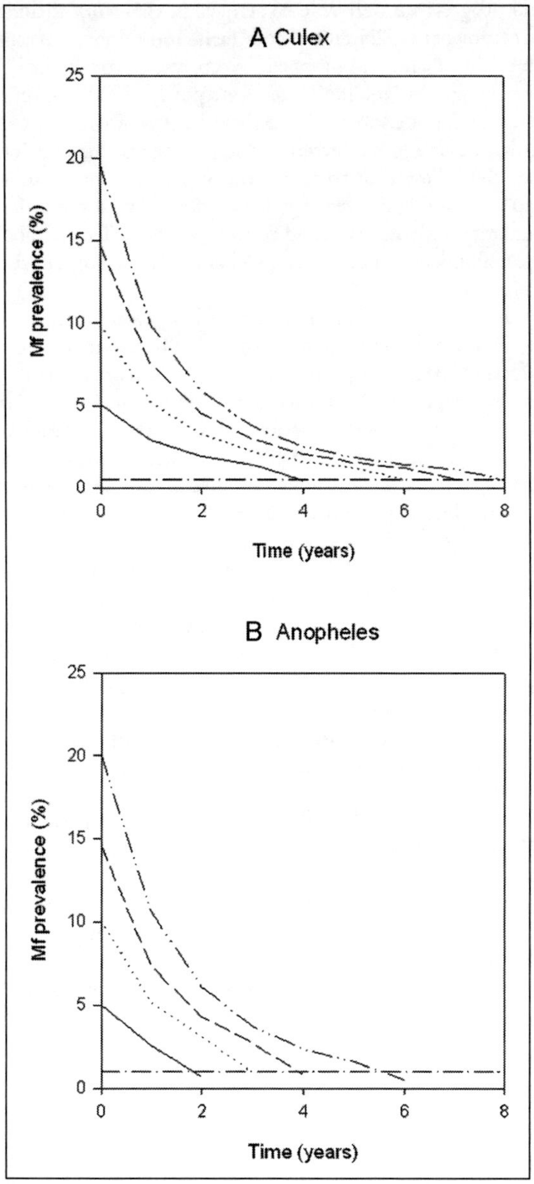

Figure 7. Annual changes in overall community Mf prevalence (scaled to 1 ml blood sampling volume) as predicted by the deterministic models for (A) culicine and (B) anopheline filariasis transmission following an annual intervention program with the DEC/ALB drug regimen. Predictions in terms of the number of annual cycles of treatment required to reach endpoints of 0.5% Mf prevalence for culicine (horizontal dashed line in (A)) and 1.0% Mf prevalence for anopheline filariasis (horizontal dashed line in (B)) are shown for a treatment coverage 80% for various baseline community Mf infection prevalences. All figures are given at the scale of 1 ml blood sampling volume. Drug efficacy values are as follows: percentage worm kill (55%), percentage Mf cured (95%) and months Mf reduced (6 months).[34]

The key technical progress we were able to achieve in this work include (1) the successful specification and estimation of the "limitation" and "facilitation" functions acting to regulate larval uptake and development in culicine and anopheline vectors respectively and (2) the efficient incorporation of these contrasting negative ("limitation") and positive ("facilitation") density-dependent functions into existing basic models that contain the other major within-host density-dependent factors acting to regulate adult worm infection for describing the transmission of filariasis in host populations. This has thus allowed us to satisfactorily resolve a long-standing methodological gap that has impeded quantifying the heterogeneous effect of vector-specific density-dependent processes on overall filariasis transmission and control, viz. the effective modelling of the simultaneous effects of the major density-dependent mechanisms acting on infection at various points or stages of the parasite life-cycle.

The results from the analyses of these new models have clearly demonstrated the considerable effect that vector-related diversity in larval infection dynamics can have on filarial system dynamics, worm infection patterns and the dynamics of parasite control in host populations due to repeated mass chemotherapy. The fundamental result that emerges is that this differential vector-related impact is primarily due to variations occurring between each vector-parasite-host combination in the presence and number of positive density-dependent mechanisms regulating parasite numbers at different points in the life cycle. Our analysis of the parasite system dynamics indicate that while in culicine filariasis, in which only one positive density-dependent mechanism—the adult worm mating probability—is likely to operate significantly, relatively low infection thresholds may exist, such thresholds may be significantly higher in the case of anopheline filariasis chiefly as a result of the occurrence of at least two positive density-dependences, one facilitating larval output in the vector and the other governing the mating probability of adult worms in the host. Indeed, we show that not only may this difference in the number of positive density-dependences between the two vector-parasite systems have a differential impact on parasite extinction thresholds, but also that possibly as a result of the weaker effect of mating probability (as opposed to the combined effect of the multiple positive density-dependent factors acting in anopheline filariasis) compared to a comparatively stronger negative effect of acquired immunity, culicine filariasis may be more resilient to control perturbations than anopheline filariasis (see Fig. 3). We show that this is essentially an outcome of both the likely occurrence of a larger extent of endemic infection relative to the worm elimination threshold and the action of strong negative density-dependent feedback mechanisms—which by causing state variables to return towards their original values can produce greater stability in parasite transmission—occurring in the culicine system. Similarly, another important finding of relevance to enhancing parasite control or eradication and sustaining such states is that while hysteresis dynamics, whereby infection emergence and elimination may occur at different points along the vector biting rate (Fig. 5), could occur for both culicine and anopheline filariasis, the vector biting range over which these dynamics may occur would be longer for the latter filarial system making including vector control into intervention strategies to maintain the controlled state more important for this system compared to the culicine case.[15]

Our modelling of the impact that culicine and anopheline vector-specific infection dynamics could have on filariasis control by annual mass chemotherapy with the combined DEC/ALB drug regimen, a key antifilariasis intervention, has supported previous more empirically-dependent conclusions that anopheline filariasis may be easier to eradicate compared to culicine filariasis. However, our analysis based on the explicit vector-specific filariasis transmission models described here, has revealed that the greater effect of this control method in achieving the elimination of anopheline filariasis is due to both the likely higher worm breakpoint or elimination threshold that may exist for this system as well as the steeper fall in infection expected over time with repeated mass treatments as infection levels in the community fall. We have shown here that this steeper fall in particular is an outcome of the lower output of L3 larvae expected with low Mf burdens as a result of the operation of the facilitation function governing larval uptake and development in these mosquitoes. However, it is important to recognize that while these results apply for a given precontrol infection level, in reality the actual overall duration of required mass treatments in anopheline areas compared to *Culex* transmission areas will depend crucially on the prevailing levels of infection prevalence observed between the two areas. In particular, if on average higher levels of infection are likely to be generated and observed in *Anopheles* areas (as a result of higher L3 ouput facilitated in anopheline mosquitoes (Fig. 6)), then it is possible that despite the faster rate of decline induced on infection by mass chemotherapy, the higher precontrol levels observed for anopheline filariasis would make the overall duration of treatments required for achieving parasite elimination for this system to be similar to that required in a *Culex* region. This result indicates that a careful consideration of both transmission and control dynamics due to vector infection heterogeneity needs to be taken into account when calculating the desired magnitude of control required to eliminate either filariasis.

Although the present work has demonstrated how addressing vector-specific transmission dynamics via the derivation and analysis of appropriate mathematical models incorporating infection processes in both specific vectors and in the human host can aide the dissection of the complex nature of the transmission and control of filarial disease, the results have also indicated several lines of future research. First, it is clear that a more complete investigation of parasite transmission dynamics will require an examination of how stochastic effects influencing transmission processes would affect the findings presented here.[38-40] We believe that the dynamical aspects of the current results would still hold broadly in a stochastic context but a quantification of the probabilities associated with extinction events—calculated by examining the outcomes of a large ensemble of model runs—would, however, perhaps be more realistic in guiding control programme design and management. Second, our research has also highlighted the crucial need for improved detection and determination of parameter values for all relevant components and processes occurring in parasite transmission ecology, especially those associated critically with density-dependent mechanisms governing infection transmission in different host-parasite-vector systems, if we are to more fully understand the stability, resilience and extinction dynamics of parasitic systems. In particular, we indicate here a critical need to identify and quantify more reliable mating probability functions for macroparasites, perhaps via the application of novel molecular ecological tools to parasite samples in order to reveal patterns in worm mating behaviours.[41,42] We also indicate a pressing need to identify and quantify the effects of other possible positive density-dependent mechanisms that may mediate filariasis transmission, such as the role that infection tolerance processes could play in regulating infection at high vector biting rates.[13,43] Although our focus here was on clarifying the dynamics and control of lymphatic filarial infection, we further also suggest that similar studies will play important roles in improving fundamental understanding of the invasion, growth, persistence and extinction dynamics of other parasitic diseases. Given the current global interest to achieve the eradication of helminth infections, it is clear that undertaking realistic parasite transmission model development and analyses, particularly from the perspective of quantifying the impact of complex population dynamics, in close conjunction with empirical studies has now become a pressing research priority in quantitative parasitology.

Acknowledgements

J.W. Kazura of Case Western Reserve University and M. Rao of the National Institutes of Health, USA, provided insightful comments on the work described in this chapter. This work was financially supported by NIH grant no. RO1 AI069387-01A1.

References

1. Dietz K. Density-dependence in parasite transmission dynamics. Parasitol Today 1988; 4:91-97.
2. Dye C, Williams BG. Nonlinearities in the dynamics of indirectly-transmitted infections (or, does having a vector make a difference?). In: Grenfell BT, Dobson AP, eds. Ecology of Infectious Diseases in Natural Populations. Cambridge, UK: Cambridge University Press, 1995:260-279.
3. Pichon G. Relations mathematiques entre le nombre des microfilaires ingerees et le nombre de parasites chez differents vecteurs naturels ou experimentaux de filarioses. Cah ORSTOM ser Entomol Med Parasitol 1974; 12:199-216.

4. Southgate BA, Bryan JH. Factors affecting transmission of Wuchereria bancrofti by anopheline mosquitoes. 4. Facilitation, limitation, proportionality and their epidemiological significance. Trans R Soc Trop Med Hyg 1992; 86:523-530.
5. Dye C. Does facilitation imply a threshold for the eradication of lymphatic filariasis? Parasitol Today 1992; 8:109-110.
6. Wada Y, Kimura E, Takagi M et al. Facilitation in Anopheles and spontaneous disappearance of filariasis: has the concept been verified with sufficient evidence? Trop Med Parasitol 1995; 46:27-30.
7. Pichon G. Limitation and facilitation in the vectors and other aspects of the dynamics of filarial transmission: the need for vector control against Anopheles-transmitted filariasis. Ann Trop Med Parasitol 2002; 96(Suppl 2):S143-152.
8. Snow LC, Bockarie MJ, Michael E. Transmission dynamics of lymphatic filariasis: vector-specific density dependence in the development of Wuchereria bancrofti infective larvae in mosquitoes. Med Vet Entomol 2006; 20:261-272.
9. Michael E, Malecela-Lazaro MN, Kabali C et al. Mathematical models and lymphatic filariasis control: endpoints and optimal interventions. Trends Parasitol 2006; 22:226-233.
10. Michael E, Malecela-Lazaro MN, Kazura JW. Epidemiological modelling for monitoring and evaluation of lymphatic filariasis control. Adv Parasitol 2007; 65:191-237.
11. Basanez MG, Remme JH, Alley ES et al. Density-dependent processes in the transmission of human onchocerciasis: relationship between the numbers of microfilariae ingested and successful larval development in the simuliid vector. Parasitology 1995; 110:409-427.
12. Michael E, Bundy DA. Herd immunity to filarial infection is a function of vector biting rate. Proc R Soc Lond B Biol Sci 1998; 265:855-860.
13. Duerr HP, Dietz K, Eichner M. Determinants of the eradicability of filarial infections: a conceptual approach. Trends Parasitol 2005; 21:88-96.
14. Berec L, Angulo E, Courchamp F. Multiple Allee effects and population management. Trends Ecol Evol 2006; 22:185-191.
15. Gambhir M, Michael E. Complex ecological dynamics and eradicability of the vector borne macroparasitic disease, lymphatic filariasis. PLoS One 2008; in press.
16. WHO. Global programme to eliminate lymphatic filariasis. Wkly Epidemiol Rec 2007; 82:361-380.
17. Michael E, Bundy DA, Grenfell BT. Re-assessing the global prevalence and distribution of lymphatic filariasis. Parasitology 1996; 112:409-428.
18. Michael E, Bundy DA. Global mapping of lymphatic filariasis. Parasitol Today 1997; 13:472-476.
19. Sasa M. Human Filariasis. Tokyo: University of Tokyo Press, 1976.
20. WHO. Lymphatic filariasis: progrogress of disability prevention activities. Wkly Epidemiol Rec 2004; 47:417-424.
21. WHO. World Health Report. Geneva, Switzerland: World Health Organization, 1995.
22. Ottesen EA. The filariases and tropical eosinophilia. In: Warren KS, Mahmood AAF, eds. Tropical and Geographical Medicine. New York: McGraw-Hill Inc., 1990:407-429.
23. Vanamail P, Subramanian S, Das PK et al. Estimation of age-specific rates of acquisition and loss of Wuchereria bancrofti infection. Trans R Soc Trop Med Hyg 1989; 83:689-693.
24. Michael E. The population dynamics and epidemiology of lymphatic filariasis. In: Nutman TB, ed. Lymphatic Filariasis. London: Imperial College Press, 2000:41-81.
25. Bates DM, Watts DG. Nonlinear Regression Analysis and its Applications. New York: John Wiley and Sons, 1988.
26. Brown D, Rotheray P. Models in Biology: Mathematics, Statistics and Computing. Chichester, England: John Wiley and Sons, 1993.
27. Juliano SA. Nonlinear curve fitting. Predation and functional response curves. In: Scheiner SM, Gurevitch J, eds. Design and Analysis of Ecological Experiments. 2nd Edition ed. Oxford, UK: Oxford University Press, 2001:178-196.
28. Subramanian S, Krishnamoorthy K, Ramaiah KD et al. The relationship between microfilarial load in the human host and uptake and development of Wuchereria bancrofti microfilariae by Culex quinquefasciatus: a study under natural conditions. Parasitology 1998; 116:243-255.
29. Plaisier AP, Subramanian S, Das PK et al. The LYMFASIM simulation program for modeling lymphatic filariasis and its control. Methods Inf Med 1998; 37:97-108.
30. Norman RA, Chan MS, Srividya A et al. EPIFIL: the development of an age-structured model for describing the transmission dynamics and control of lymphatic filariasis. Epidemiol Infect 2000; 124:529-541.
31. Snow LC, Michael E. Transmission dynamics of lymphatic filariasis: density-dependence in the uptake of Wuchereria bancrofti microfilariae by vector mosquitoes. Med Vet Entomol 2002; 16:409-423.

32. Bryan JH, Southgate BA. Factors affecting transmission of Wuchereria bancrofti by anopheline mosquitoes. 2. Damage to ingested microfilariae by mosquito foregut armatures and development of filarial larvae in mosquitoes. Trans R Soc Trop Med Hyg 1988; 82:138-145.
33. Chan MS, Srividya A, Norman RA et al. Epifil: a dynamic model of infection and disease in lymphatic filariasis. Am J Trop Med Hyg 1998; 59:606-614.
34. Michael E, Malecela-Lazaro MN, Simonsen PE et al. Mathematical modelling and the control of lymphatic filariasis. Lancet Infect Dis 2004; 4:223-234.
35. Michael E, Ramaiah KD, Hoti SL et al. Quantifying mosquito biting patterns on humans by DNA fingerprinting of bloodmeals. Am J Trop Med Hyg 2001; 65:722-728.
36. May RM. Togetherness among Schistosomes: its effects on the dynamics of the infection. Math Biosciences 1977; 35:301-343.
37. May RM. Thresholds and breakpoints in ecosystems with a multiplicity of stable states. Nature 1977; 269:471-477.
38. May R. Stability and Complexicity in Model Ecosystems. Princeton: Princeton University Press, 1973.
39. Wang YH, Gutierrez AP. An assessment of the use of stability analyses in population ecology. Journal of Animal Ecology 1980; 49:435-452.
40. Lande R, Engen S, Saether B-E. Stochastic Population Dynamics in Ecology and Conservation. Oxford: Oxford University Press, 2003.
41. Avise JC. Molecular Markers, Natural History and Evolution. Sunderland, Massachusetts: Sinauer Associates, Inc. Publishers, 2004.
42. Freeland JR. Molecular Ecology. Chichester, England: John Wiley and Sons Ltd., 2005.
43. King CL. Transmission intensity and human immune responses to lymphatic filariasis. Parasite Immunol 2001; 23:363-371.

Chapter 3

Modelling Multi-Species Parasite Transmission

Andrea Pugliese*

Abstract

Some models are presented for the dynamics of a host population with two parasite species. The models differ in two main aspects: whether they include direct competition among parasites and whether the analysis is based on some approximation and which one. If the analysis is not constrained by a priori assumptions about parasite distributions, it is found that species coexistence is very unlikely without some kind of direct competition among parasites; on the other hand, coexistence generally occurs when inter-specific competition is lower than intraspecific, similarly to standard theory for free-living species. If hosts differ in their predisposition to infection, but not in an identical way towards the two parasite species, then species coexistence becomes feasible even if inter-specific competition is as strong as intraspecific; in this case, coexistence becomes easier as the variance in predisposition increases. These models do not yield universal predictions for patterns of parasite distributions; an analysis of the mechanisms of interaction in each specific system is necessary for that.

Introduction

Models for host-macroparasite interaction have a relatively long history, starting from the pioneering work of Kostizin[1] and with the two seminal papers by Anderson and May[2] making a strong impact also on empirical research.[3,4] On the other hand, very few authors have studied models with several species of parasites, despite the fact that parasite communities are routinely found and examined in empirical research. This is presumably due to the much higher complexity of the resulting mathematical models (see below) and the difficulties in extending to multi-species the approach (approximation via the negative binomial assumption) that has been so fruitful in the analysis of single species models.

None the less, several interesting models have been developed over the years. In this chapter, I will give a personal review of the subject, mainly focused on the subject of coexistence: what are the factors that lead to species coexistence? In so doing, I will quickly review some examples of dynamic models for two parasite species competition. In the final section, I briefly discuss whether these models give any general insights to understanding for parasite community ecology.

The models studied assume that parasites have only one host (i.e., they are monoxenic) and that infections occur through free-living larvae. I believe that most results would apply to more complex systems as well.

Simple Models for Multispecies Parasite Dynamics

The main difficulty in modelling parasite dynamics is that one cannot simply divide the host population into infected and not, but one has to describe and predict the distribution of para-

*Andrea Pugliese— Dipartimento di Matematica, Università di Trento Via Sommarive 14 38050 Povo, Italy. Email: pugliese@science.unitn.it

Modelling Parasite Transmission and Control, edited by Edwin Michael and Robert C. Spear.

sites among hosts, since the effect of parasites on hosts and on other parasite species depend on the number present in that host and perhaps also in features of their establishment. The typical approach used in one-species macroparasite models has been to choose a priori that the parasite distribution is negative binomial, generally (following Anderson and May[2]) with fixed aggregation parameter k, but also with a varying aggregation parameter[5,6] and to then obtain an equation for the temporal dynamics of the mean parasite burden and, depending on the model, of other variables, such as host and/or free-living larvae density.

To my knowledge, the first model on two macroparasite species interacting with one host species has been proposed by Dobson.[7] His model followed this approach, assuming that each parasite species follows the negative binomial distribution with fixed aggregation parameter (k_1 and k_2) and that the two distributions are independent; he then derived a model for the dynamics of the two parasite densities, assuming that the two species (and indeed all individual parasites) do not interact except in that each parasite contributes to the death of a host harboring both species. His main result is that there is an ample parameter region in which both species will coexist: the smaller the parameters k_i are (meaning the more aggregated their distributions), the larger the coexistence region will be.

That model has been extended to communities of parasites,[8] to allow for interference or facilitation between parasites[9,10] and to allow for logistic host growth.[11] The result about parasite coexistence has proved to be robust with respect to all these changes.

Even two species identical in all parameters and differing in just one (for instance, the rate of egg production) can coexist; thus a completely inferior competitor (the one with a lower egg production) can survive. This result is actually rather puzzling and seems to be in contrast with all the theory of competitive exclusion.[12] Competition theory does allow for several species to coexist on a single resource, for instance because of competitive balance shifts along a temporal cycle, or because of a colonization-competition trade-off in a metapopulation setting. In all cases, coexistence of competitors seems always to require the existence of some trade-off between traits.

The problem seems to lie in the a priori assumptions made. It will be addressed here by examining conditions for coexistence in a model based only on explicit assumptions about the interaction mechanisms.

Structure and Parameters of Models

The models analyzed here concern the interaction of a host population with two species of monoxenic parasites with infections occuring through free-living larvae. In contrast to the models outlined in the previous section, the model considered here does not contain a priori assumptions about parasite distributions and allows for many types of competition among parasites in the same host. The model is deterministic, with main variables the density of hosts carrying i 1-parasites and j 2-parasites, denoted as $p_{ij}(t)$. Other authors[13,14] start from a stochastic model, but then, to obtain analytic results, use some approximations leading to so-called hybrid models[15] similar to those discussed here.

The system of differential equations satisfied by $p_{ij}(t)$ is rather cumbersome and can easily be obtained with some book-keeping: it is written explicitly in the Appendix. It can be derived by noting that $p_{ij}(t)$ may increase because some hosts that were carrying a different number of parasites switch to having exactly i 1-parasites and j 2-parasites ($p_{00}(t)$ increases also because of new births) and vice versa may decrease because some hosts carrying i 1-parasites and j 2-parasites switch to a different number of parasites (or die). Listing the possible transitions and their rates, as in the following Table 1, is then enough to specify the system.

In words, it is assumed here that adult parasites affect (additively) hosts' mortality and (multiplicatively) hosts' fertility. Moreover, parasites within a host interact directly by increasing (additively) the mortalities (according to the matrix τ that differentiates intraspecies and inter-species effects), decreasing (multiplicatively) the fertilities (according to the matrix r) and decreasing the probability of establishment of an infecting larva (through the matrix γ). Finally, new infections occur through encounters (at rate β) with free-living larvae.

Table 1. List of the variables of the model and of the events being modeled together with their rates

Variable or Event	Host Transition	Rate or Symbol
Host density		$N = \Sigma_{i,j} p_{ij}$
Density of hosts carrying i 1-parasites and j 2-parasites		p_{ij}
Density of free-living stages of species 1 [2]		L_1 $[L_2]$
Birth from a host carrying i 1-parasites and j 2-parasites	$\varnothing \to (0,0)$	$b(N)(1-\xi_1)^i(1-\xi_2)^j$
Death of a host carrying i 1-parasites and j 2-parasites	$(i, j) \to \varnothing$	$d + \alpha_1 i + \alpha_2 j$
Density-dependence in host fertility (K carrying capacity of hosts)		$b(N) = b\left(\frac{d}{b}\right)^{N/K}$
Death of adult 1- [2-] parasites	$(i, j) \to (i-1, j)$ $[(i, j) \to (i, j-1)]$	$\sigma_1 + i\tau_{11} + j\tau_{12}$ $[\sigma_2 + i\tau_{21} + j\tau_{22}]$
Rate of parasite establishment	$(i, j) \to (i+1, j)$ $[(i, j) \to (i, j+1)]$	$\beta_1 L_1 \psi_1 (1-\gamma_{11})^i (1-\gamma_{12})^j$ $[\beta_2 L_2 \psi_2 (1-\gamma_{21})^i (1-\gamma_{22})^j]$
Birth from an adult 1- [2-] parasites	$\frac{d}{dt}L_1 = + \left[\frac{d}{dt}L_2 = +\right]$	$h_1(1-r_{11})^{i-1}(1-r_{12})^j$ $[h_2(1-r_{21})^i(1-r_{22})^{j-1}]$

The dynamics of free-living larvae has to be specified; new ones are produced from adult parasites (see Table 1), while they are removed through deaths (at rate δ_1 [or δ_2]) or encounters with hosts. One has then the differential equation

$$\frac{d}{dt}L_1 = h_1 \sum_{i,j} p_{ij} i(1-r_{11})^{i-1}(1-r_{12})^j - \delta_1 L_1 - \beta_1 L_1 N \tag{1}$$

and analogously for L_2. A usual simplification[2] is the assumption of fast dynamics of larvae, so that they are at quasi-equilibrium with adult parasites. From (1), one obtains

$$L_1 = \frac{h_1 \sum_{i,j} p_{ij} i(1-r_{11})^{i-1}(1-r_{12})^j}{\delta_1 + \beta_1 N} \tag{2}$$

Finally, substituting (2) in the expression (Table 1) for the rate of adult parasite establishment, one obtains that, for a host carrying *i* 1-parasites and *j* 2-parasites, this is equal to

$$\varphi_1(1-\gamma_{11})^i(1-\gamma_{12})^j \quad \text{with } \varphi_1 = \frac{h_1 \psi_1 \sum_{k,l} k(1-r_{11})^{k-1}(1-r_{12})^l p_{kl}}{c_1 + N} \quad \text{and } c_1 = \delta_1 / \beta_1. \tag{3}$$

In the transitions listed, several relevant phenomena of host-parasite interactions have been neglected, such as nonlinear effects of parasite abundance on host mortality, facilitation of parasite establishment through parasite-caused impairment of immune response,[16] context-dependent (through the sequence of infection by different species) parasite competition.[17] The system is already complicated enough as it is and indeed it will be simplified to allow analysis; moreover, the main interest of the chapter lies in parasite competition rather than in facilitation.

The resulting system of differential equations is clearly very difficult to study, not least for its size: since the number of adult parasites in a host, *i* and *j*, can in principle be any number, it is a doubly infinite system; even if we restrict these numbers to a maximum, say 100, we would still have a system of 10,000 differential equations. The idea of simplifying the system by some kind of moment closure, either through a negative binomial assumption,[7] or through a normal approximation,[14] is clearly very appealing, although it is then necessary to understand whether the results are an artifact of the approximation.

A different approach is to limit the study to the computation of the invasion criteria for each species; these are sometimes possible to compute without any approximation.[18] In this way it is not possible to infer the overall dynamics of the system, but at least one can compute exactly the parameter region that allow for species coexistence.

Invasion Criteria

While analyzing the dynamics of a complex nonlinear model, such as that including hosts and two parasite species is very difficult, it is often possible to study the (linearized) dynamics close to an equilibrium. In particular, the computation of the invasion coefficient (i.e., the growth rate of one parasite species in a population close to the equilibrium where hosts coexist with a first population species) is sometimes feasible. Basically, this is an extension of the basic reproduction ratio (R_0) of a parasite in a parasite-free population, a quantity fundamental in models for microparasites,[19] but that can be adapted for macroparasites as well.[11,20]

In some cases it is then possible to characterize a quantity R_1^2 representing the basic reproductive ratio of parasite 2 invading a population at the equilibrium E_1 where hosts coexist with parasite 1. If R_2^1, the population of parasite 2 is able to initially increase and will establish itself; on the other hand, if R_1^2, that population will decrease and will eventually get extinct.

Coexistence will be deemed to occur when both R_1^2 and R_2^1 are greater than 1, i.e. when each parasite species is able to invade an equilibrium with only the other species present, together with the host. This principle is well established in theoretical population biology and can be justified, under some technical assumptions, through persistence theory.[21] One could extend the same principle to more than 2 species, through the computation of the invasion coefficient of parasite species *n* into an equilibrium with the first *n - 1* present; however, this is generally very difficult because it requires finding explicitly an equilibrium with more than 1 parasite species.

Generally the basic reproduction ratio R_1^2 can be defined of the average number of established adult 2 parasites produced by a newly established parasite 2 during its expected life time.[18,20] We can split this number in at least three components:

$$R_1^2 = (\textit{average number of larvae produced over an adult parasite life time}) \\ \bullet (\textit{probability that larva is ingested by a host}) \bullet (\textit{probability of a successful establishment}) \quad (4)$$

All these quantities will depend, according to the model used, on the features of the equilibrium E_1 since parasite survival and fertility may depend on how many 1 parasites are present in the same host (and perhaps also on host density); the probability that a larva is ingested by a host will depend on host density; the probability of a successful establishment may depend on host immune response, hence on the burden (as a surrogate of the host's previous exposure) of 1 parasites.

The Model without Direct Interactions

This has been the case mainly analyzed in the literature, mainly because it is the simplest, so that it can also serve as a reference for studying the effect of direct interactions. All the terms in Table 1 referring to parasite interactions, the matrices γ, r and τ are set equal to 0 after tau.

Pugliese[18] computes the invasion coefficient for that case. As discussed above, this requires to obtain some features of the equilibrium with only 1 parasite species present and then to compute the component of formula (4) for the basic reproductive ratio R_1^2.

Without repeating the technical steps presented there, we sketch the main ideas. First of all, at an equilibrium, if parasite interactions are neglected, the probability distribution of parasite burden follows some simple law, depending on hosts' age: precisely, if parasite infections occur one at the time and hosts do not differ in their resistance to infection, parasite distribution for each age is Poisson[22] with mean

$$X(a) = \frac{\varphi}{\sigma + \alpha}(1 - e^{-(\sigma+\alpha)a}), \quad (5)$$

where σ and α are parasite mortality and parasite-induced host mortality (see Table 1) and ϕ is the equilibrium level of the parasite establishment rate (see Eqn. 3). Alternatively (generally data do not support the assumption of a Poisson distribution of parasite load, even when accounting for host's age[23,24]), one can allow for multiple infections or heterogeneity in hosts' susceptibility to infection, obtaining mixtures of Poisson with levels of aggregation comparable to observed ones.[22]

One can then compute the components of (4). Since $\gamma_{ij} = 0$, the probability of a successful establishment is equal to ψ_1. Since larvae may either die (at rate δ) or be ingested by hosts (at rate βN), the probability of being ingested is $\frac{\beta N}{\delta+\beta N} = \frac{N}{c+N}$ with $c = \delta/\beta$ (see Eqn. 3) and N is host population size at equilibrium; when the equilibrium includes parasite species 1, this will be denoted as $\bar{N}_1$.

The only part of (4) that requires lengthy computations is the average number of larvae produced over an adult parasite life time. First of all, one notes that in this model ($r_{ij} = 0$), fertility is constant (h_2); then one has

$$R_1^2 = \frac{h_2 \psi_2 \bar{N}_1}{c_2 + \bar{N}_1} T_1^2 \quad (6)$$

where T_1^2 represents the expected lifetime of a 2-parasite that has just infected an average host in a population at the equilibrium E_1 where hosts coexist with 1-parasites. This can be computed as the average of the expected lifetime of a 2-parasite establishing itself in a host of a given age and parasite burden; the average will be weighed using the probability $\bar{N}_1(s)/\bar{N}_1$ of surviving to age *s*, (according to the stationary age density) and the Poisson assumption with (5) as its mean, or the somewhat more complicated distributions following from multiple infections, for heterogeneous hosts.

From the Poisson assumption, one obtains, after some algebra,

$$R_1^2 = \frac{h_2 \psi_2}{c_2 + \bar{N}_1} \int_0^\infty \frac{1 - e^{-(\alpha_2+\sigma_2)s}}{\alpha_2 + \sigma_2} \bar{N}_1(s) ds. \tag{7}$$

Formula (7) shows that the reproduction ratio of parasite 2 increases with its fertility (h_2), with its probability of establishment (ψ_2), decreases with larval death rate (δ_2 through the parameter c_2), with its adult death rate (σ_2) and its induced host death rate (α_2). All this is rather intuitive and would not require modelling.

More interesting is the transformation of (7) into an expression containing the corresponding parameters for parasite 1; the expression is particularly simple if $c_1 = c_2$, which means that the parameters relative to the larval stages are the same for the two parasite species. One then obtains

$$R_1^2 = \frac{h_2 \psi_2}{h_1 \psi_1} \frac{\alpha_1 + \sigma_1}{\alpha_2 + \sigma_2} \frac{\int_0^\infty (1 - e^{-(\alpha_2+\sigma_2)s}) \bar{\pi}_1(s) ds}{\int_0^\infty (1 - e^{-(\alpha_1+\sigma_1)s}) \bar{\pi}_1(s) ds}, \tag{8}$$

where $\bar{\pi}_1(s)$ represents the probability of surviving to age *s*, at the equilibrium E_1 of coexistence with parasite 1.

Formula (8) shows clearly a principle: a parasite that is superior to the resident one (higher fertility and lower death rate and induced host death rate) will always be able to invade and will never be invaded by the other. Mutual invasibility can occur only if there exists a trade-off between traits, namely one parasite has a higher fertility (say $h_2 > h_1$), but suffers also from a higher mortality or higher damages caused to its host ($\alpha_2 + \sigma_2 > \alpha_1 + \sigma_1$). When a trade-off occurs, Pugliese[18] shows that coexistence is possible but only when the parameters are very precisely balanced: for given values of $\alpha_2 + \sigma_2 > \alpha_1 + \sigma_1$, there exists a very narrow interval of values h_2 (with $h_2 > h_1$) that gives rise to coexistence.

This conclusion is in strong contrast with that obtained by Dobson[7] and others,[9-11] that parasite coexistence is easy and is especially facilitated by aggregation in parasite distribution. Given that the model with single infections and homogeneous hosts gives rise to very little aggregation (which is caused by the mixture of Poisson with different mean, because of hosts' age), it may be not surprising that coexistence is very unlikely in this model.

In Pugliese[18] the issue was tackled by analyzing models including mechanisms that built more aggregation into parasite distributions. Precisely, two models were analyzed: one with multiple infections of larvae (with a Poisson distribution of mean λ), the other with heterogeneity in hosts' susceptibility to infection, i.e., ψ is not constant among hosts, but has a distribution with means $\bar{\Psi}_1$ and $\bar{\Psi}_2$ for the two parasite species.

In both models, the invasion coefficient can be computed from (4) along lines similar to those leading to (7) and (8). The computations are more involved and are not reported here. Eventually, one arrives at an expression of the type

$$R_1^2 = \frac{h_2 \bar{\Psi}_2}{h_1 \bar{\Psi}_1} F(\alpha_2, \sigma_2, \alpha_1, \sigma_1) \tag{9}$$

where the function *F* depends on all details of the model, but has two fundamental properties:

a. *F* equals 1 if all parameters are the same ($F(\alpha, \sigma, \alpha, \sigma) = 1$), which means that if two parasite species differ only in their fertility, the one with higher fertility outcompetes the other;

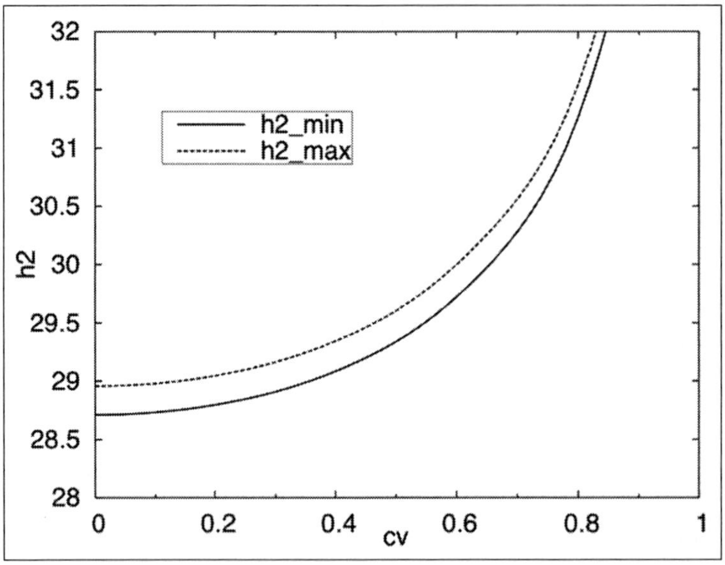

Figure 1. The values of h_2 as a function of the coefficient of variation in host susceptibility to parasites for which a second parasite species could coexist with a first parasite species. Other parameter values are $\alpha_1 = 5$, $\alpha_2 = 0.5$, $\sigma_1 = \sigma_2 = 2$, $h_1 = 65$, $c_1 = c_2 = 1$, $d = 0.5$, $b = 1$, $K = 1000$. Figure adapted from Theoretical Population Biology, vol. 57, pp. 145-165, Figure 6, © 2000 Elsevier, with permission.

b. F is decreasing with α_2 and σ_2 (the reproductive ratio decreases, if a parasite suffers higher mortality, or induces a higher death rate on the host) and is increasing with α_1 and σ_1 (it is easier for a parasite to invade, if the resident parasite suffers higher mortality, or induces a higher death rate on the host).

These two properties imply that a trade-off between fertility, on the one hand and parasite- or host-induced mortality, on the other hand, is necessary for two parasites to coexist.

The coexistence region could be examined only by numerically computing the function F in (9). It was found (see Fig. 1) that the width of the coexistence region was basically independent of the degree of aggregation induced by the model: one could assume a very large heterogeneity in host susceptibility and thus a very aggregated parasite distribution, but still the potential for parasite coexistence was very limited, basically the same as with homogeneous hosts and little aggregation.

The rationale for the difference in this result from what obtained by Dobson is discussed at length in Pugliese.[18] Basically, Dobson[7] assumes that parasite distributions are aggregated and independent. On the other hand, if aggregation arises from the fact that some hosts are more susceptible to infection (from whichever parasite species), it is clear that both species of parasites will be found in the most susceptible hosts, so that parasite distribution will be positively correlated; the stronger is host heterogeneity (and thus parasite aggregation), the more positive the correlation will be. Indeed, Dobson and Roberts,[9] studying a negative binomial approximation with fixed correlation coefficient, show that positive correlation coefficients hinder coexistence.

Clearly, things would be different if some hosts were more susceptible to parasites of species 1 and others to parasites of species 2; this is explored in references 13-14 (see below), but it is a different explanation: coexistence arises because of differential host susceptibility, not because of aggregation per se.

Competition among Parasites

The results obtained in the previous section are very elegant mathematically, providing a subtle reason for coexistence among parasites without direct interactions: the shift in the age-dependence of host mortality when at equilibrium with either parasite species. On the other hand, they are somewhat disconcerting biologically, because coexistence is very unlikely, while routinely parasites of several species are found in the same host populations and individuals.[25] Moreover, it refers to any kind of parasites: even parasites colonizing different organs could coexist only under very restrictive conditions, which seems plain nonsense.

The point is that in the model the only density-dependence mechanism, that keeps parasite load from growing to infinity, is the induced host mortality. Hence, parasite density at equilibrium will be at a sufficient high level to overall induce a significant host mortality. In turn, the high level of host mortality will make it very difficult for a second parasite species to invade, unless it has a higher reproduction number than the resident one and then displaces it.

In short, to make a realistic multiparasite model, it is essential to introduce competition between parasites, whether they are of the same species or of different ones. In Table 1, three levels of competition are considered, translated as rates depending on the number of parasites in a host: parasite mortality will increase, while parasite fertility and probability of successful establishment will decrease.

Empirical evidence exist for all these facts;[26,27] the number of parasites in a host may be relevant because of its impact on host resources available for other parasites, or because of the immune response induced in the host; in the latter case, it might be better modelling immune response, as depending on the history of infection of individual hosts,[28] rather than on the current parasite load, but this would be rather more complex and current parasite load may be a reasonable proxy for history of infection.

Explicit mechanisms of parasite competition have been introduced in models for competition between two parasite species by Bottomley et al,[13,14] using the methods developed by Isham and coworkers.[22,29] They consider separately the effect of parasite load on probability of establishment and on parasite fertility; here I consider the same cases, neglecting the effect on parasite mortality, mainly because of the mathematical complications of this. First, I use the same approach as in previous Section, that leads to an exact computation of the invasion coefficient, hence to finding the conditions for coexistence as mutual invasibility, only for the case of competition acting on parasite fertility. Then, I show the approach by Bottomley et al[13,14] that involves the computation of approximate equations for the first two moments of parasite distribution, thus yielding conditions for coexistence but also the dynamic pattern of parasite densities.

Parasite Fertility Depending on Available Resources

Let r_{ij} measure the effect of the load of parasite species j on the fertility of parasite species i; precisely, I assume (see Table 1) that the fertility of one parasite of species 1, living in a host that harbors i parasites of species 1 and j of species 2, is $h_1(1-r_{11})^{i-1}(1-r_{12})^j$. Note that I use a multiplicative effect (and not an additive one) of parasites, to avoid the fertility becoming negative at high parasites load; as long as $i\,r_{11}$ and $j\,r_{12}$ are not too large, that expression can be approximated as $h_1(1-(i-1)\,r_{11}-j\,r_{12})$, yielding a more usual expression.

Most of the analysis outlined for the model without direct interactions still applies. One can use (4) to find the invasion coefficient for species 2 in a population where the host coexists with species 1. In the equilibrium with only species 1, equation (5) still holds; in this case, however, the average number of larvae produced by an adult parasite is not simply the product of the fertility rate times its expected lifetime, because the fertility depends on the number of other parasites present. The computations (that will be presented elsewhere) can be performed easily using functional-analytic methods.[30] The final result can be written as

$$R_1^2 = \frac{h_2\psi_2}{c_2+\bar{N}_1}\frac{1}{\alpha_2+\sigma_2}\int_0^\infty (1-e^{-(\alpha_2+\sigma_2)s})\exp\{-\tfrac{\varphi_1}{\alpha_1+\sigma_1}(1-e^{-(\alpha_1+\sigma_1)s})r_{21}\}\bar{N}_1(s)ds, \tag{10}$$

or, if $c_1 = c_2$, as

$$R_1^2 = \frac{h_2\psi_2}{h_1\psi_1}\frac{\alpha_1+\sigma_1}{\alpha_2+\sigma_2}\frac{\int_0^\infty (1-e^{-(\alpha_2+\sigma_2)s})\exp\{-\frac{\varphi_1}{\alpha_1+\sigma_1}(1-e^{-(\alpha_1+\sigma_1)s})r_{21}\}\bar{\pi}_1(s)ds}{\int_0^\infty (1-e^{-(\alpha_1+\sigma_1)s})\exp\{-\frac{\varphi_1}{\alpha_1+\sigma_1}(1-e^{-(\alpha_1+\sigma_1)s})r_{11}\}\bar{\pi}_1(s)ds}, \tag{11}$$

where φ_1 is the rate at which (at equilibrium) new 1-parasites establish themselves (see Eqn. 3), while $\bar{\pi}_1(s)$ is as in (8).

Expressions (10) and (11), that extend (7) and (8), are rather cumbersome, but one can easily use it on a computer to find the parameter values that allow for coexistence, i.e., those for which both $R_1^2 > 1$ and $R_2^1 > 1$ (see Figs. 2 and 3).

Moreover, one can easily understand some particular cases:

if all parameters of the two parameter species are identical, then $R_1^2 = 1$; hence, if $r_{21} < r_{11}$ (i.e., interspecific is lower than intraspecific competition), $R_1^2 > 1$. One then obtains in this context the classical result that two competing species that have the same demographic parameters but use somewhat different resources can always coexist.

In the extreme case where $r_{21} = 0$ (i.e., the presence of species 1 has no effect on parasites of species 2) and $\alpha_1 = 0$ (hence population density is at its carrying capacity, K, independently of the presence of parasites), expression (10) simplifies to

$$R_1^2 = \frac{h_2\psi_2}{c_2+K}\frac{K/L}{\alpha_2+\sigma_2}\int_0^\infty (1-e^{-(\alpha_2+\sigma_2)s})\pi(s)ds$$

where L is the average length of host life. In this case, $R_1^2 > 1$ is simply the condition for species 2 to be able to persist with the host at its carrying capacity. Then two species that do not interact and do not increase host mortality can always coexist, provided each can persist with the host.

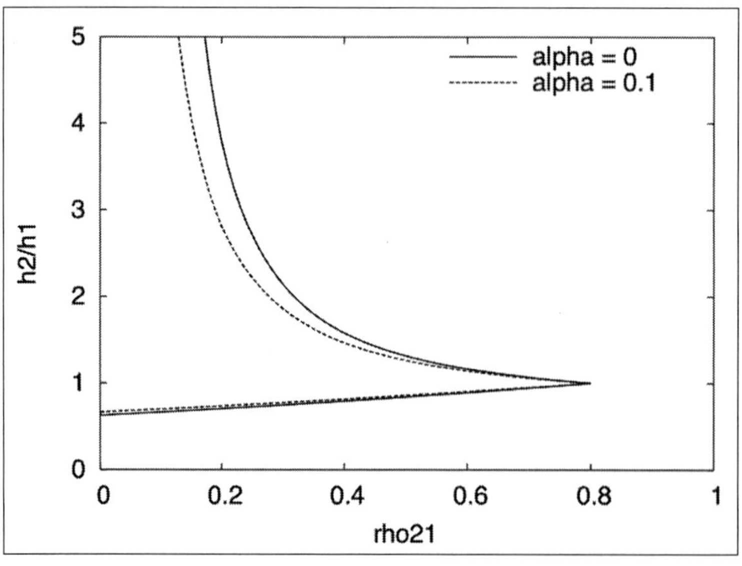

Figure 2. The region in the parameter space ($r_{21} = r_{12}$, h_2/h_1) that allows species coexistence (to the left of the chevron-shaped curves) for $\alpha_1 = \alpha_2 = 0$ (solid curve) or $\alpha_1 = \alpha_2 = 0.1$ (dashed curve). Other parameter values are $r_{11} = r_{22} = 0.8$, $\sigma_1 = \sigma_2 = 2$, $h_1 = 4$, $c_1 = c_2 = 1$, $d = 0.5$, $b = 1$, $K = 1000$.

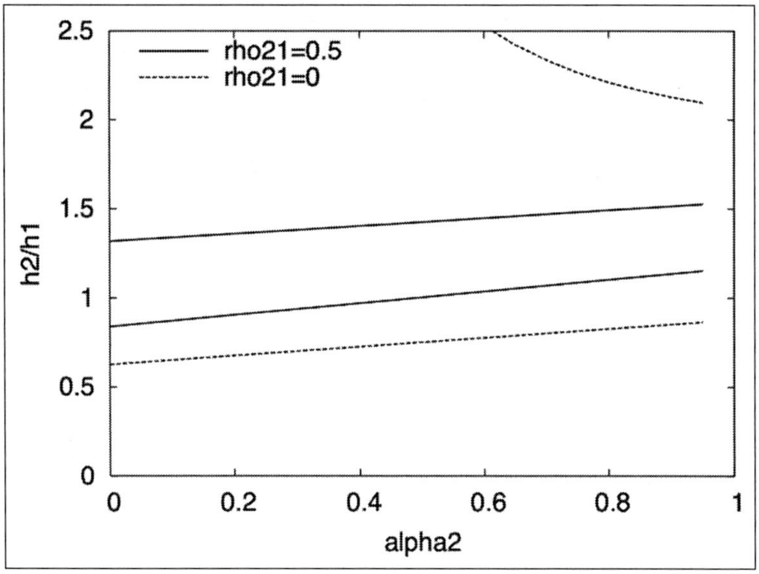

Figure 3. The region in the parameter space (α_2, h_2/h_1) that allows species coexistence (between the curves) for $r_{12} = r_{21} = 0.5$ (solid curve) or $r_{12} = r_{21} = 0$ (dotted curve). Other parameter values are $r_{11} = r_{22} = 0.8$, $a_1 = 0$, $\sigma_1 = \sigma_2 = 2$, $h_1 = 4$, $c_1 = c_2 = 1$, $d = 0.5$, $b = 1$, $K = 1000$.

Generally, if $\alpha_1 = 0$, computations are easier, since the presence of parasites does not affect host demography. Then (10) can be written as

$$R_1^2 = \frac{h_2\psi_2}{c_2 + K}\frac{K/L}{\alpha_2 + \sigma_2}\int_0^\infty (1 - e^{-(\alpha_2+\sigma_2)s})\exp\{-\tfrac{\varphi_1}{\sigma_1}(1 - e^{-\sigma_1 s})r_{21}\}\pi(s)ds.$$

Figure 2 shows how the coexistence region depends on the ratios of parasite fertilities and the strength of their interspecific competition. In the figure, the two parasite species have the same value of all parameters except the fertility h_i and have a common inter-specific competition coefficient $r_{21} = r_{12}$. It can be seen that, when inter-specific competition is very low, the two species will coexist for almost all feasible fertility values; on the other hand, the ratio of fertilities must become very close to 1 for the two species to coexist, as the inter-specific competition coefficient approaches 0.8, the value of the intraspecific competition coefficient. When $r_{21} = r_{12} = r_{11} = r_{22}$, coexistence is impossible, as seen in Figure 2 and shown by (11). It can also be seen from Figure 2 that the coexistence regions are similar for $\alpha_1 = 0$ and $\alpha_1 = 0.1$ (a rather strong parasite-induced mortality, comparing with the other parameter values).

Figure 2 compares parasite species differing in only one demographic parameter; in Figure 3, the case of a trade-off between fertility and parasite-induced mortality is analyzed, as in ref. 18. One can again see the strong influence of the interspecific competition coefficient $r_{12} = r_{21}$ on the width of the coexistence region, although coexistence becomes anyway more difficult when the competitor is extremely lethal.

The trade-off between fertility and parasite-induced mortality was shown[18] to be sufficient for species coexistence, in absence of intra- or inter-specific competition, although the rates had to be balanced very carefully. One may wonder whether this can happen also with inter-specific competition equal to the intraspecific one. However, repeating the computations shown in Figure 3 with $r_{12} = r_{21} = r_{11} = r_{22} = 0.8$, one sees that coexistence is impossible for any values of α_2 or h_2/h_1. Vice versa, one finds a region where both $R_1^2 < 1$ and $R_2^1 < 1$, i.e., both monospecific equilibria are uninvadable. Namely, a (narrow) parameter region may exist where, even if inter-specific

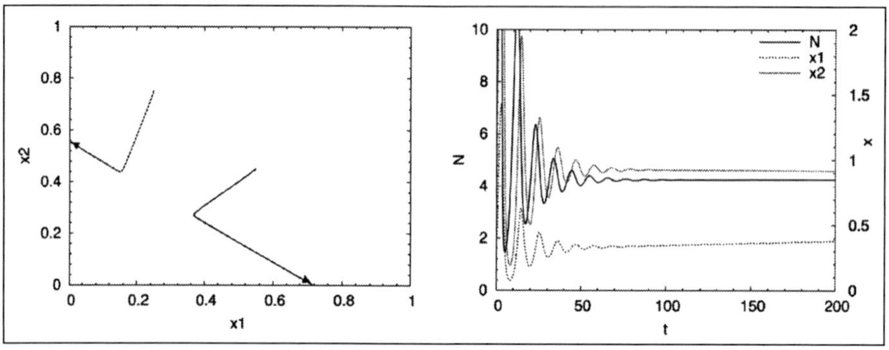

Figure 4. Simulations of the system in the variables p_{ij}, with equal inter-specific and intraspecific competition coefficients. In the left panel, parameter values are $r_{11} = r_{22} = r_{12} = r_{21} = 0.5$, $\alpha_1 = 0.1$, $\alpha_2 = 0.5$, $\sigma_1 = \sigma_2 = 2$, $h_1 = 4$, $h_2 = 4.55$, $c_1 = c_2 = 1$, $d = 0.5$, $b = 1$, $K = 1000$. Axes represent average parasite loads, $X_1 = \Sigma_{i,j}\, ip_{ij}/N$ and $X_2 = \Sigma_{i,j}\, jp_{ij}/N$. The solid line shows a simulation starting from (0.55, 0.45), the dotted line one starting from (0.25, 0.75). In both cases $N(0) = 800$. In the right panel, parameter values are the same except $r_{11} = r_{22} = r_{12} = r_{21} = 0.01$, $h_2 = 4.492$. On the x-axis, time; on the y-axis, host population density (scale on the left) $N = \Sigma_{i,j}\, p_{ij}$ and average parasite loads (scale on the right) X_1 and X_2.

competition is exactly equal to intraspecific one (and could even be slightly lower), the system exhibits competitive exclusion dependent on initial densities, as it happens for Lotka-Volterra systems when inter-specific competition is higher than the intraspecific one.

In Figure 4, two contrasting examples are shown of simulations with inter-specific competition exactly equal to intraspecific one; in the left panel, competition is rather strong ($r_{11} = r_{22} = r_{12} = r_{21} = 0.5$) and values of α_i and h_i have been found, for which both exclusion equilibria are attractive, as discussed above; in the right panel, competition is very low ($r_{11} = r_{22} = r_{12} = r_{21} = 0.01$) and values of α_i and h_i have been found that allow for species coexistence, like in the case without competition.

The simulations in the left panel of Figure 4 show a very simple behavior, typical of two-dimensional Lotka-Volterra competition systems: fast convergence to a one-dimensional manifold connecting the equilibria and then slow convergence to an equilibrium, along the manifold. In this case then parasite competition follows the standard patterns of competition theory and it becomes reasonable the search for some simple approximating system.

In the simulation shown in the right panel of Figure 4, after some initial oscillations host population and parasite loads approach a stable coexistence equilibrium. It may be noted that the population density reached (around 4) is extremely lower than the carrying capacity (1,000); this is not meant to be realistic and is due to the choice of very high induced mortalities and the very low level of parasite competition.

The previous analysis can be applied, in a relatively simple way, when parasite resource competition affects only parasite birth rate. Judging from preliminary simulations, it seems likely, that similar results will hold also when parasite competition increases death rates (through parameters r of Table 1) or decreases establishment probability (through parameters γ of Table 1). The computations outlined in this Section do not easily extend however, since parasite distribution will not be Poisson for fixed age (as in (5)), so that no explicit formulae can be found, although the ideas of invasion criteria still apply.

Instead, normal approximations can be obtained,[14] as shown in the following Section.

Normal Approximations

Bottomley et al[13,14] have studied parasite competition through normal approximations. The idea is very simple: from the equations in the variables p_{ij}, or directly from computations of the possible instantaneous changes in parasite load, one can find equations satisfied by the first moments of the parasite distributions (means, variances, covariances...). Unfortunately, as usual in most complex models in ecology, the equations for lower moments have terms including higher moments; to obtain a closed low-dimensional system, one then needs some form of "moment closure".[31] A simple approach is the normal approximation,[22] i.e., to assume that higher moments can be expressed in terms of the first and second moments, according to the same relations that hold for a normal distribution. The assumption may not seem very adequate for parasites, since normal distributions are continuous and include a negative part; the negative binomial distribution used since Anderson and May[2] does not have these problems, but cannot be easily generalized to two or more variables;[10] moreover, Bottomley et al[14] show, through simulations, that the normal approximations works reasonably well, as long as parameter values are not extreme.

The use of the normal approximation requires a correction in the laws used for density-dependence. Precisely, it becomes more convenient assuming that parasite fertility decreases with the number of parasites according to an additive law $h_1(1-(i-1)r_{11}-jr_{12})$ and similarly the probability of establishment is $\psi_1(1-i\gamma_{11}-j\gamma_{12})$. For the rule to be reasonable, it should be assumed that fertilities (or probability of establishment) are 0 when the quantities are negative. The following analysis is feasible only without that restriction; however, as long as the parameters r and γ are not too large, ignoring the restriction does not make a big difference.[14]

The equation (1) for the larvae still hold, together with the quasi equilibrium approximation (2) that, with the change to the additive law, changes (3) into

$$\varphi_1 = \frac{h_1\psi_1 N(<x_1> - r_{11}(<x_1^2> - <x_1>) - r_{12}<x_1x_2>)}{c_1 + N} \tag{12}$$

where x_i is the i-parasite load and $<\,>$ represents the average.

It is also easy writing an equation for the total host density N

$$N' = N(b(N) - d - \alpha_1<x_1> - \alpha_2<x_2>) \tag{13}$$

One can then write equations for $<x_i>$ obtaining

$$\begin{aligned} <x_1>' &= \varphi_1(1-\gamma_{11}<x_1> - \gamma_{12}<x_2>) - \sigma_1<x_1> - \rho_{11}<x_1^2> - \rho_{12}<x_1x_2> - \alpha_1<x_1^2> \\ &\quad - (b(N) - \alpha_1<x_1> - \alpha_2<x_2>)<x_1> \\ <x_2>' &= \varphi_2(1-\gamma_{21}<x_1> - \gamma_{22}<x_2>) - \sigma_2<x_2> - \rho_{22}<x_2^2> - \rho_{21}<x_1x_2> - \alpha_2<x_2^2> \\ &\quad - (b(N) - \alpha_1<x_1> - \alpha_2<x_2>)<x_2>. \end{aligned} \tag{14}$$

Competition at Establishment. No Induced Mortality

In the case where the parasites do not induce mortality ($\alpha_i = 0$) and parasite competition acts only by reducing the probability of establishment ($r_{ij} = \tau_{ij} = 0$), equations (14) become a closed system. In fact, N is then fixed at the carrying capacity K with $b(K) = d$ and (14) reduces (dropping for ease of notation the brackets) to

$$\begin{aligned} X_1' &= \frac{h_1\psi_1 K X_1}{c_1 + K}(1-\gamma_{11}X_1 - \gamma_{12}X_2) - (\sigma_1 + \mathrm{d})X_1 \\ X_2' &= \frac{h_2\psi_2 K X_2}{c_2 + K}(1-\gamma_{21}X_1 - \gamma_{22}X_2) - (\sigma_2 + \mathrm{d})X_2 \end{aligned} \tag{15}$$

Equations (15) have exactly the form of a Lotka-Volterra competition system. Hence, according to the values of the coefficients, one can have the four possible outcomes: equilibrium coexistence of the two species, competitive exclusion of species 1, competitive exclusion of species 2, contingent competitive exclusion (either species may be excluded depending on initial densities).

First of all, it is necessary to assume that each parasite species alone is able to persist with the hosts at the carrying capacity K. This condition can be written in terms of basic reproductive numbers as

$$R_0^1 = \frac{h_1\psi_1 K}{(c_1+K)(\sigma_1+d)} > 1 \quad \text{and} \quad R_0^2 = \frac{h_2\psi_2 K}{(c_2+K)(\sigma_2+d)} > 1. \tag{16}$$

Then, the conditions for coexistence are

$$\frac{\gamma_{21}}{\gamma_{11}} < \frac{1-(R_0^2)^{-1}}{1-(R_0^1)^{-1}} \quad \text{and} \quad \frac{\gamma_{12}}{\gamma_{22}} < \frac{1-(R_0^1)^{-1}}{1-(R_0^2)^{-1}} \tag{17}$$

In case R_0^1 and R_0^2 are much larger than 1, conditions (17) simply mean that inter-specific competition is lower than intraspecific one.

If both conditions are reversed (intraspecific competition lower than inter-specific one), one obtains contingent competitive exclusion.

Finally, if one of (17) holds and the other not, strict competitive exclusion of one species occurs.

Note that, while (15) are obtained from (14) without approximations, still they are not exact for the complete system because they have been derived neglecting the constraints (see above) that only the positive values of $h_1(1-(i-1)r_{11}-jr_{12})$ and $\psi_1(1-i\gamma_{11}-j\gamma_{12})$ had to be considered.

Competition Acting on Parasite Fertility

One has to add equations for the second moments, since they appear in right hand sides of (14) unless $r_{ij}=0$.

Using as variables the variances and covariance ($V_1 = \langle x_1^2\rangle - \langle x_1\rangle^2$ $V_2 = \langle x_2^2\rangle - \langle x_2\rangle^2$, $C_{12} = \langle x_1 x_2\rangle - \langle x_1\rangle\langle x_2\rangle$), one obtains, after some lengthy calculations, equations satisfied by them, involving the third moments. The normal approximation allows to write the third moments in terms of the first two, using the relation

$$\langle UVW\rangle = \langle UV\rangle\langle W\rangle + \langle UW\rangle\langle V\rangle + \langle VW\rangle\langle U\rangle - 2\langle U\rangle\langle V\rangle\langle W\rangle \tag{18}$$

exact if (U, V, W) follows a multivariate normal distribution.

The expressions in the general model are very complex and add little insight, though they could be used for numerical computation. I restrict to the case of competition acting only on parasite fertility, the case analyzed above through invasion coefficient and also studied[13,14] through normal approximations.

First of all, φ_1 and φ_2 are given by (12) that is rewritten now, using V_1, V_2 and C_{12} as

$$\varphi_1 = \frac{h_1\psi_1 N}{c_1+N}\left(x_1(1+r_{11}) - r_{11}(V_1+x_1^2) - r_{12}(C_{12}+x_1x_2)\right) \tag{19}$$

and analogously for φ_2. Then equations (14) can be rewritten for this case as

$$X_1' = \varphi_1 - \sigma_1 X_1 - \alpha_1 V_1 - b(N)X_1 + \alpha_2 X_1 X_2 \tag{20}$$

Finally, writing the equations for the second moments and applying also (18), one obtains:

$$V_1' = \varphi_1 - \sigma_1(2V_1 - X_1) + \alpha_2 X_1^2 X_2 - (V_1 - X_1^2)b(N) \tag{21}$$

$$C_{12}' = -(\sigma_1+\sigma_2)C_{12} - \alpha_1 X_1(C_{12}+X_1X_2) - \alpha_2 X_2(C_{12}+X_1X_2) - (C_{12}-X_1X_2)b(N) \tag{22}$$

Equations (13) – (20) – (21) – (22) (with the analogous ones for x_2 and V_2) are a closed system describing parasite competition. One can use them to find the equilibrium with only one species present and then find the invasion conditions for the second species; mutual invasibility could then be considered as denoting coexistence. Unfortunately, analytical computations are still rather difficult and it is generally necessary to resort to numerical computations.

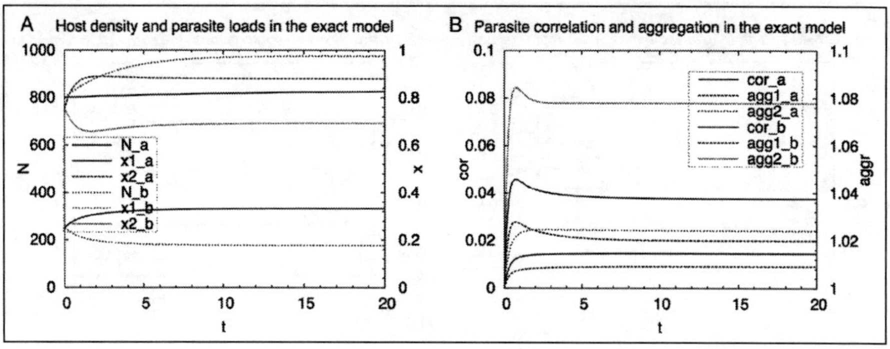

Figure 5. Host density and parasite loads vs time (left panel); correlation between the distribution of the two species and aggregation (= variance/mean) of each vs time (right panel) in two simulations of the infinite system. Simulation "A" has $b = 0.2$, $d = 0.1$; simulation "B" has $b = 1$, $d = 0.5$. All other parameter values are the same: $r_{11} = r_{22} = 0.8$, $r_{12} = r_{21} = 0.4$, $\alpha_1 = \alpha_2 = 0.01$, $\sigma_1 = \sigma_2 = 2$, $h_1 = 4$, $h_2 = 5$, $c_1 = c_2 = 1$, $K = 1000$. Initial conditions are $N(0) = 800$ and Poisson independent distribution for each parasite of means 0.25 and 0.75, respectively.

The normal approximation works effectively only when host death rates are small.[14] In the limiting case where $\alpha_1 = \alpha_2 = 0$ and also $b(N) = d$, at equilibrium $x_1 = V_1$ and $C_{12} = 0$; then, using these relations in (20), the equations for x_1 and x_2 take the Lotka-Volterra form (15) and one can easily analyze them.

One can also compare numerically system (13) – (20) – (21) – (22) with the exact system presented in the Appendix; two examples of this comparison are shown in Figures 5 and 6.

From the figures it can be seen, first of all, that, in both cases, simulations converge quickly to a coexistence equilibrium (inter-specific competition is half the intraspecific one). Moreover, the same qualitative trends exist for both exact and approximate models; increasing host death and birth rate (while maintaining all other parameters) results in decreased parasite loads, hence in higher host density. Similarly, parasite aggregation and correlation increase with host death and birth rate, since in these models parasite aggregation (and correlation) results from the hidden variable age: for each given age, parasite distributions are Poisson (no aggregation) and independent, but differ in their means; the mixture of these distributions results into a (little) aggregated distribution.

On the other hand, the absolute values predicted for parasite loads are rather different, especially for the more abundant species (2) and so is host density, since it will suffer from parasite-induced mortality. It must be remembered, however, that a multiplicative law for parasite fertility is used in the exact model and an additive law in the normal approximation; the latter results in a lower fertility when there are at least 3 parasites in a host, so it is no wonder that they yield quantitative different results.

Competition and Host Heterogeneity

As can be seen from Figures 5 and 6 (right panels), the previous models yield a very low correlation between parasite distribution and a very low aggregation in each. Then, on the one hand, one could feel justified in assuming a priori that parasite distributions are independent; on the other hand, it seems necessary to allow for some aggregation in the distributions, to have a model closer to reality. The approach by Dobson[7] and others[11] is to use independent negative binomial distributions for each species, or with a fixed correlation coefficient.[9,10]

Bottomley et al[13,14] have instead modeled a mechanism that produces parasite aggregation, studying its effects on species coexistence, according to the detailed assumptions used. Precisely, they assume that the parameter ψ (establishment probability) is not a constant, but varies among

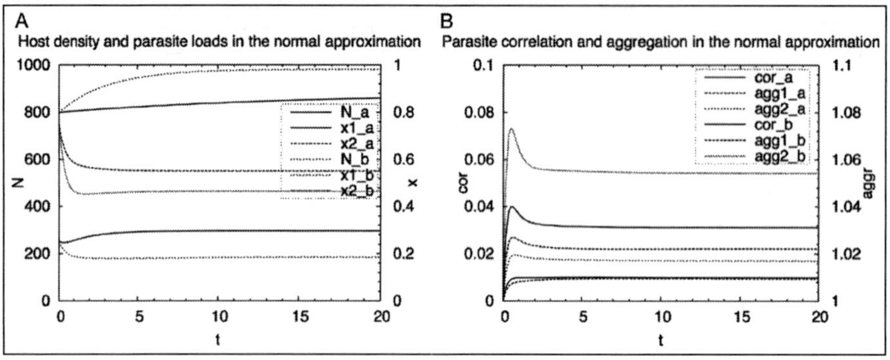

Figure 6. Host density and parasite loads vs time (left panel); correlation between the distribution of the two species and aggregation (= variance/mean) of each vs time (right panel) in two simulations of the approximating system (13)-(20)-(21)-(22). Simulations "a" and "b", parameters and initial conditions as in Figure 5.

hosts according to some given distribution (like in Fig. 1); the equations for the second moments will then involve mean and variance of this distribution. Note that, since ψ is a probability, it must lie between 0 and 1, so that its variance cannot be large; however, it is also possible to assume that the encounter rates of hosts with larvae (parameter β in Table 1) is variable among hosts; by a redefinition of parameters one can include this variation in ψ and then its variance can take any value, as will be assumed here.

If some hosts have a higher predisposition to infection (measured by their value of ψ), they are more likely to get infected, so that a correlation will build up between ψ and parasite load x. The covariance between ψ and x, $C_{\psi x}$ will then be a variable of the system.

To keep things simple, I will restrict the analysis to the case where parasites do not induce mortality ($\alpha_i = 0$) and parasite competition acts only by reducing the probability of establishment ($r_{ij} = \tau_{ij} = 0$). Then, host population density is fixed at K and equations for parasite loads x_i do not depend on variances (see (15)).

The resulting competition system will assume given (different distributions) for ψ_1 and ψ_2, summarized by their means $\bar{\psi}_1$ and $\bar{\psi}_2$, variances V_1^{ψ} and V_2^{ψ} and covariance ($C^{\psi\psi}$). Variables of the system will be parasite loads (x_1 and x_2) and the covariances between ψ_i and x_j ($C_{ij}^{\psi x}$). Through some steps,[14] one arrives at the following system of equations

$$\frac{d}{dt}X_1 = \frac{h_1 K}{c_1 + K} X_1\left(\bar{\psi}_1 - \gamma_{11}(C_{11}^{\psi x} + \bar{\psi}_1 X_1) - \gamma_{12}(C_{12}^{\psi x} + \bar{\psi}_1 X_2)\right) - (\sigma_1 + d)X_1$$

$$\frac{d}{dt}C_{11}^{\psi x} = \frac{h_1 K}{c_1 + K} X_1\left(V_1^{\psi} - \gamma_{11}(\bar{\psi}_1 C_{11}^{\psi x} + V_1^{\psi} X_1) - \gamma_{12}(\bar{\psi}_1 C_{12}^{\psi x} + V_1^{\psi} X_2)\right) - (\sigma_1 + d)C_{11}^{\psi x} \quad (23)$$

$$\frac{d}{dt}C_{12}^{\psi x} = \frac{h_2 K}{c_2 + K} X_2\left(C^{\psi\psi} - \gamma_{21}(\bar{\psi}_2 C_{11}^{\psi x} + C^{\psi\psi} X_1) - \gamma_{22}(\bar{\psi}_2 C_{12}^{\psi x} + C^{\psi\psi} X_2)\right) - (\sigma_2 + d)C_{11}^{\psi x}$$

with analogous equations for x_2, $C_{21}^{\psi x}$ and $C_{22}^{\psi x}$.

In order to study coexistence in model (23), one can, as discussed above, compute the equilibrium E_i with only one species present and find the conditions for invasion from the other species. As shown in Bottomley et al,[14] this can be written as reproduction numbers

$$R_1^2 = \frac{h_2 K}{c_2 + K} \frac{\bar{\psi}_2 - \gamma_{21}(C_{21}^{\psi x} + \bar{\psi}_2 x_1)}{\sigma_2 + d}\bigg|_{E_1} = R_0^2[1 - \gamma_{21}(\rho \frac{cv_2^{\psi}}{cv_1^{\psi}}(\frac{1-(R_0^1)^{-1}}{\gamma_{11}} - x_1) + x_1)]\bigg|_{E_1} \quad (24)$$

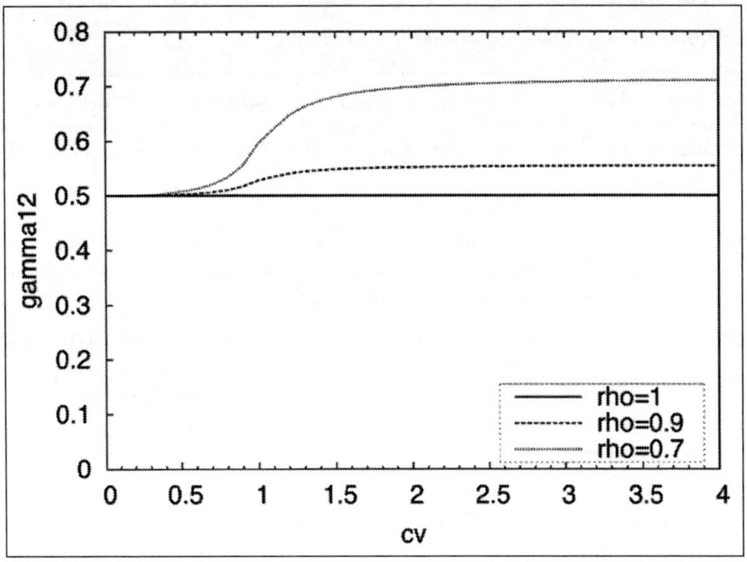

Figure 7. The maximum value of inter-specific competition $\gamma_{12} = \gamma_{21}$ that allows for coexistence, vs $cv_1^\psi = cv_2^\psi$. for different values of ρ. Other parameter values are $\gamma_{11} = \gamma_{22} = 0.5$, $\alpha_1 = \alpha_2 = 0$, $\sigma_1 = \sigma_2 = 2$, $h_1 = h_2 = 15$, $c_1 = c_2 = 1$, $K = 1000$, $d = 0.5$.

where all quantities are computed at the equilibrium E_1. Moreover, R_0^2 and R_0^1 are the basic reproduction number given in (16), using the means of ψ_1 and ψ_2, ρ is the correlation coefficient between ψ_1 and ψ_2 and cv_i^ψ are the coefficients of variation (= standard deviation/mean) of ψ_i.

The interest of the analysis of this model lies mainly in understanding the role of ρ, cv_1^ψ and cv_2^ψ. In fact, high values of cv_1^ψ and cv_2^ψ correspond to highly aggregated parasite distributions at equilibrium; indeed, Dobson[7] had found that aggregation promotes parasite coexistence.

One can start the analysis by a simple case: when $\rho = 1$, $cv_1^\psi = cv_2^\psi$ the conditions for coexistence $R_1^2 > 1$ and $R_2^1 > 1$ become identical to the conditions (17) found without heterogeneity. Hence, when there is perfect correlation among hosts between predispositions to each parasite species, a high aggregation (as long as it is the same in both species) has no effect on coexistence. This conclusion is similar to what found by Pugliese[18] in the exact invasion analysis of a model without direct interactions (see Fig. 1). When cv_1^ψ and cv_2^ψ are different, the conclusion is not as straightforward; one can see that the invasion of a parasite is hampered, if its variation in predisposition is much larger than that of the resident parasite, while is facilitated if its much lower; on the whole, still aggregation does not promote parasite coexistence.

Decreasing the correlation ρ among host predisposition to parasites makes coexistence easier; this can be seen from (24), since the coefficient of ρ is negative. Intuitively, it is clear that, if some hosts are more predisposed to parasite 1 and others to parasite 2, parasite coexistence becomes easier.

Less intuitive is the fact that there is an interaction between the effects of the two parameters. When $\rho < 1$ and $cv_1^\psi = cv_2^\psi$, increasing the coefficients of variations (still keeping $cv_1^\psi = cv_2^\psi$) makes it easier satisfying the invasion conditions. This follows indirectly from (24), since higher cv_1^ψ results in lower x_i at equilibrium. Hence, in these circumstances the conclusion that aggregation promotes coexistence may be justified. A quantitative example is shown in Figure 7; it can be seen that an imperfect correlation in predisposition ($\rho < 1$) and high coefficients of variation allow for the coexistence of parasite species that are identical in all demographic parameters, but with inter-specific competition higher than intraspecific one; the effect is not very large, though, unless correlation is rather low.

Table 2. Summary of models examined in this chapter

Biological Assumption	Mathematical Method	Mechanisms of Coexistence
No direct interactions. Induced mortality[7]	Assumption of independent negative binomial distributions	Aggregation of parasite distributions
As above[9]	Negative binomial distributions with fixed correlation coefficient	As above and favored by negative correlation coefficient
As above[18]	Exact computation of invasion coefficients	Trade-off fertility-survival, with subtle effects on host mortality schedules that can allow coexistence
Competition at establishment. No induced mortality[13]	Normal approximations	Inter-specific competition lower than intraspecific
As above with variance in predisposition to infection[13]	Normal approximations	Imperfect correlation between predisposition to infection from the different parasite species
Density-dependence in parasite fertility.[13] Induced mortality allowed (this chapter)	Normal approximations.[13] Exact computation of invasion coefficients (this chapter)	Inter-specific competition lower than intraspecific

Conclusion

Several models for parasite competition have been presented. All are rather complex and do not allow for an easy analysis. For this reason, only some special cases have been considered in this chapter and more extensive analyses would be necessary before drawing clear conclusions. In Table 2 I present a summary of the models considered and the mechanisms yielding coexistence in each.

Still, it is possible to state some general, though preliminary, results. Parasite coexistence generally requires intraspecific competition; if parasite populations are controlled only indirectly through their effects on hosts (through mortality or fertility) and do not compete directly for host resources, then at equilibrium host density will be severely reduced and parasite coexistence would require precise balances of demographic parameters that appear rather unlikely. Such a trade-off between parasite fertility and survival (including the host's) would result in a coexistence of metapopulation type, with one species quicker at reaching new hosts and another surviving longer on the colonized ones.

On the other hand, if parasite populations are controlled by their competition for host resources, or other types of interactions among them, then coexistence occurs as long as inter-specific competition is lower than intraspecific one, following a dynamics reminiscent of Lotka-Volterra equations.

A specific feature of host-parasite interactions is host predisposition for infection, that is generally considered the most relevant mechanism generating aggregated parasite distributions, a feature universally observed in empirical surveys.[24] If there exists only a generic predisposition for all parasite species, then this has no effects on the feasibility of parasite coexistence. On the other hand, if the predisposition to different parasite species differs among hosts, then coexistence becomes more likely, as quite obvious intuitively; in this case, a higher variance in predisposition generates both a higher aggregation in parasite distribution and a larger parameter region for coexistence; in this limited sense, it may stated that aggregation promotes coexistence.

Are there general predictions about patterns that can be observed in natural population? At this level of generality, it seems very difficult; for instance, predictions about correlations between different parasite species depend on the mechanisms allowing for coexistence. Positive (but low) correlation coefficients are produced simply by the mixture of hosts of different age (and other groups); strong intraspecific competition, coupled with a low inter-specific one, will make them more positive, as well as a strong variance in generic predisposition to infection. On the other hand, predisposition to different parasite species differing among hosts will force towards negative correlation.

Finally, the dynamics exhibited by the models displayed in this chapter is very simple: quick convergence towards an equilibrium. This indeed is a typical behavior of competition systems and one may wonder whether interactions with hosts can modify this. The examples shown in this chapter have considered a rather limited parameter range, that allowed for a low-dimensional truncation of the infinite system and/or for a reasonable normal approximation. It is known, on the other hand, that host-parasite interactions may induce cycles, especially when parasites affect host fertility,[2] although variance in host predisposition to infection and parasite competition decrease the likeliness of cyclic behavior.[32] Simple dynamics with quick convergence to equilibrium has been the rule in the examples examined for this chapter; however, it seems likely that more extensive numerical explorations will uncover examples of complex dynamics in host-2 parasite systems.

Acknowledgements

I thank Roberto Rosà for reading a first draft of this chapter and suggesting several improvements; Michael Edwin for inviting me to write this chapter and being always very supportive.

References

1. Kostizin VA. Symbiose, parasitisme et évolution (ètude mathèmatique). Paris: Hermann, 1934. English translation. In: Scudo F, Ziegler J, eds. The Golden Age Of Theoretical Ecology. Berlin: Springer-Verlag, 1978:369-408.
2. Anderson RM, May RM. Regulation and stability of host-parasite populations interactions I-II. J Anim Ecol 1978; 47:219-247, 249-267.
3. Hudson PJ, Rizzoli A, Grenfell BT et al. Ecology Of Wildlife Diseases. Oxford: Oxford Univ Press, 2002.
4. Poulin R. Evolutionary ecology of parasites. Princeton: Princeton Univ Press, 2007.
5. Adler FR, Kretzschmar M. Aggregation and stability in parasite-host models. Parasitol 1992; 104:199-205.
6. Pugliese A, Rosà R. Analysis of a model for macroparasitic infection with variable aggregation and clumped infections. J Math Biol 1998; 36:419-447.
7. Dobson AP. The population dynamics of competition between parasites. Parasitol 1985; 91:317-347.
8. Dobson AP. Models For Multi-Species Parasite-Host Communities. In: Esch GW, Bush AO, Aho JM, eds. Parasite communities: patterns and processes. London: Chapman and Hall, 1990:261-288.
9. Dobson AP, Roberts MG. The population-dynamics of parasitic helminth communities. Parasitol 1994; 109:S97-S108.
10. Roberts MG, Dobson AP. The population dynamics of communities of parasitic helminths. Math Biosci 1995; 126:191-214.
11. Gatto M, De Leo G. Interspecific competition among macroparasites in a density-dependent host population. J Math Biol 1998; 37:467-490.
12. Armstrong RA, McGehee R. Competitive exclusion. Amer Nat 1980; 115:151-170.
13. Bottomley C, Isham V, Basanez MG. Population biology of multispecies helminth infection: interspecific interactions and parasite distribution. Parasitol 2005; 131:417-433.
14. Bottomley C, Isham V, Basanez MG. Population biology of multispecies helminth infection: competition and coexistence. J Theor Biol 2007; 244:81-95.
15. Nåsell I. Hybrid Models Of Tropical Infections. Berlin: Springer, 1985.
16. Behnke JM, Gilbert FS, Abu-Madi MA et al. Do the helminth parasites of wood mice interact? J Animal Ecol 2005; 74:982-993.
17. Jackson JA, Pleass RJ, Cable J et al. Heterogenous interspecific interactions in a host—parasite system. Int J Parasit 2006; 36:1341-1349.
18. Pugliese A, Coexistence of macroparasites without direct interactions. Theor Pop Biol 2000; 57:145-165.

19. Diekmann O, Heesterbeek JAP, Metz JAJ. On the definition and the computation of the basic reproduction ratio R0 in models for infectious diseases in heterogeneous populations. J Math Biol 1990; 28:365-382.
20. Pugliese A, Tonetto L. Thresholds for macroparasite infections. J Math Biol 2004; 49:83-110.
21. Hutson V, Schmitt K. Permanence and the dynamics of biological systems. Math Biosci 1992; 111:1-71.
22. Isham V. Stochastic models of host-macroparasite interaction. Ann Appl Prob 1995; 5:720-740.
23. Pacala SW, Dobson AP. The relation between the number of parasite/host and host age: population dynamic causes and maximum likelihood estimation. Parasitol 1988; 96:197-210.
24. Shaw DJ, Dobson AP. Patterns of macroparasite abundance and aggregation in wildlife populations: a quantitative review. Parasitol 1995; 111S:111-133.
25. Poulin R. Species richness of parasite assemblages: evolution and patterns. Ann Rev Ecol Syst 1997; 28:341-358.
26. Smith G, Grenfell BT. The population biology of ostertagia ostertagi. Parasitol Today 1985; 1:76-81.
27. Behnke JM, Bajer A, Sinski E et al. Interactions involving rodent nematodes: experimental and field studies. Parasitol 2000; 122:S39-S49.
28. Woolhouse MEJ. A theoretical framework for the immunoepidemiology of helminth infection. Parasite Immunology 1992; 14:563-578.
29. Chan MS, Isham V. A stochastic model of schistosomiasis immuno-epidemiology. Math Biosci 1998; 151:179-198.
30. Moschen MP, Pugliese A. The threshold for persistence of parasites with multiple infections. Comm Pure Applied Analysis, in press.
31. Keeling MJ, Rand DA, Morris AJ. Correlation models for childhood epidemics. Proc R Soc Lond B 1997; 264:1149-1156.
32. Rosà R, Pugliese A. Aggregation, stability and oscillations in different models for host-macroparasite interactions. Theor Popul Biol 2002; 61:319-334.

Appendix. The Model Analyzed

The system in the variables $p_{ij}(t)$ arising from the assumptions shown in Table 1 is:

$$\begin{aligned}
p'_{ij}(t) = {} & \varphi_1(1-\gamma_{11})^{i-1}(1-\gamma_{12})^j p_{i-1,j}(t) + \varphi_2(1-\gamma_{21})^i(1-\gamma_{22})^{j-1} p_{i,j-1}(t) + (i+1)(\sigma_1+\tau_{11}(i+1)+\tau_{12}j)p_{i+1,j}(t) \\
& + (j+1)(\sigma_2+\tau_{21}i+\tau_{22}(j+1))p_{i,j+1}(t) - [\varphi_1(1-\gamma_{11})^i(1-\gamma_{12})^j + (1-\gamma_{21})^i(1-\gamma_{22})^j \\
& + i(\sigma_1+\alpha_1+\tau_{11}i+\tau_{12}j) + j(\sigma_2+\alpha_2+\tau_{21}i+\tau_{22}j) + d]p_{i,j}(t) \\
p'_{00}(t) = {} & (\sigma_1+\tau_{11})p_{1,0}(t) + (\sigma_2+\tau_{22})p_{0,1}(t) - (\varphi_1+\varphi_2++d))p_{00}(t) + b(N)\sum_{i,j} p_{i,j}(t)(1-\xi_1)^i(1-\xi_2)^j
\end{aligned}$$

where $\varphi_1 = \beta_1 L_1 \psi_1$ and, by convention, $p_{-1,j}(t) = p_{i,-1}(t) = 0$.

The system is completed by equations (1) for L_1 and L_2.

In the quasi-equilibrium approximation, L_1 and L_2 are given by (2) and φ_1 [φ_2] by (3).

CHAPTER 4

Metapopulation Models in Tick-Borne Disease Transmission Modelling

Holly Gaff* and Elsa Schaefer

Abstract

Human monocytic ehrlichiosis (*Ehrlichia chaffeensis*), or HME, is a tick-transmitted, ricksettisal disease with growing impact in the United States. Risk of a tick-borne disease such as HME to humans can be estimated using the prevalence of that disease in the tick population. A deterministic model for HME is explored to investigate the underlying dynamics of prevalence in tick populations, particularly when spatial considerations are allowed. The dynamics of HME in a single spatial patch are considered first to determine which model components are most important to predicting disease dynamics in a local ecology. The model is then expanded to spatially-explicit patches on which patch connectivity, the surrounding environment and boundary effects are studied. The results of this investigation show that predicting risk of this disease to humans is determined by many complicated interactions. Areas that would be endemic in isolation may or may not sustain the disease depending on the surrounding habitat. Similarly, control efforts are shown to be far more effective when applied in wooded habitats than in neighboring grassy habitats. Boundary assumptions which describe the reality of increasing habitat fragmentation additionally play a large role in predicting the endemicity of an HME outbreak. Overall, HME and all tick-borne diseases are complex, nonlinear systems that have just begun to be explored.

Introduction

Tick-borne diseases are increasingly affecting human health throughout the world. These diseases include Lyme disease, Rocky Mountain spotted fever, human babesiosis, ehrlichiosis, anaplasmosis, tick-borne relapsing fever, Colorado tick fever and tick paralysis. Each of these tick-borne diseases has a unique life history that includes a combination of three elements. First, one or more tick species serves as a competent vector of the disease, i.e., the tick can transmit the disease. Second, the given tick species has a preferred host or hosts that are reservoir hosts for the pathogen, i.e., the host can transmit the disease back to ticks but is not affected by the infection. Finally, the disease is caused by a pathogen with its own dynamics and these pathogens consist of viruses, protozoa and bacteria including many ricksettia. While tick-borne diseases have great variety in their structure, one feature all of these diseases share is the erratic nature of human outbreaks, which makes the prediction and prevention of future outbreaks quite difficult. Recent literature displays increasing research efforts in this area[1] and there is a growing understanding of the underlying dynamics of tick-borne diseases. This increased comprehension has in turn supported the development of mathematical models to further explore tick-borne disease.

*Corresponding Author: Holly Gaff—Community and Environmental Health, College of Health Sciences, Old Dominion University, 3133A Health Sciences Building, Norfolk, Virginia 23529, USA. Email: hgaff@odu.edu

Modelling Parasite Transmission and Control, edited by Edwin Michael and Robert C. Spear.

Human monocytic ehrlichiosis (causative agent: *Ehrlichia chaffeensis*) is a tick-transmitted, rickettsial disease that has recently increased substantially in the USA from 142 reported cases in 2001 to 506 reported cases in 2005.[2,3] HME, as first reported in reference 4, produces a variety of clinical symptoms ranging from a mild illness to a severe, life-threatening disease with a case fatality rate of approximately 3%. In the United States, HME cases are most frequently reported in the Southeastern and south-central States,[5] but the range is rapidly expanding northward.[6]

The lone star tick (*Amblyomma americanum*) is suspected to be the major species that transmits HME[7,8] and the white-tailed deer (*Odocoileus virginianus*) has been identified as a major reservoir host for *E. chaffeensis*.[9,10] The lone star tick is a hard-bodied Ixodid tick with four life stages: egg, larva, nymph and adult. With the exception of the transition from egg to larva, the tick is obligated to acquire a bloodmeal to molt from each life stage to the next. Unlike many Ixodid tick species, the lone star tick uses the same hosts for each life stage and in many areas, the majority of their blood meals are from the white-tailed deer.[11] This simplified life history allows the exploration of dynamics between a single tick species and a single host species. The system may allow us to gain insights into tick-borne diseases that might otherwise be missed because of the additional complications of multiple host species.

This chapter focuses on the analysis of a mathematical model for HME transmission, which is based on a previously published model by Gaff and Gross.[12] We do not repeat their work, but rather, after a brief introduction to the model, we focus our attention on exploring the dynamics resulting from various spatial configurations of that model. These investigations reveal the importance of metapopulation structures and local environmental parameters in predicting the dynamics of HME and in evaluating the efficacy of control efforts, such as the reduction of the tick population through acaricide use. In particular, we will illustrate the importance of correctly modelling local environmental parameters, patch connectivity, the surrounding environment and boundary assumptions in HME analysis efforts.

Methods

Metapopulation Modelling

Population biologists have been using metapopulation models in some form for nearly one hundred years,[13] with the word "metapopulation" first appearing in 1969.[13] In such models, a patch of land is marked as either inhabitable or uninhabitable. Generally, inhabitable patches have limited connectivity to other inhabitable patches and the focus is on the probability of extinction or survival across a given set of patches. Landscape ecology has emerged more recently[14] and supports the approach of viewing a landscape in continuous patches with differing ecologies. A model using landscape ecology will have four major components: variations in patches, variations in the surrounding environment, boundary effects and patch connectivity.[14] Because tick population dynamics are very sensitive to the local environment, this latter modelling approach is quite useful. Within landscape ecology, there is the so-called patch-matrix theory in which a spatial region is divided into patches and the dynamics in the model occur within each patch while interaction such as migration or colonization is allowed between patches.[14] In the patch-matrix theory, the patch structure is static throughout the time that the model is considered; however, other landscape ecology models may actually consider variable patch structures in which the spatial patterns and patch relationship change over time.[15] In either setting, patches may be considered to have different sizes and elaborately-shaped boundaries. Recent studies have considered the effect of forest fragmentation on tick-host or other species interactions and determined that patch sizes, patch boundaries, patch interactions and direct and indirect consequences of patch placement at the edge of a grid have complex dynamics that must be considered carefully in future studies.[16,17]

In this chapter, we consider a temporally static rectangular grid of uniformly-shaped patches. Each patch will have unique parameter values that reflect its local ecology. Migration will be allowed to each patch's nearest neighbors based on the gradient in population levels. Because our patch and habitat assumptions will be spatially explicit but not necessarily spatially realistic, we may want to consider our model to be an example of metapopulation ecology rather than landscape ecology.[18]

Let's return for a moment to the four components of metapopulation ecology and discuss their role in our investigations.

Variations within Patches

We will investigate the importance of the choice of parameter values to the dynamics of HME outbreaks within a single patch. In general, tick populations fare much better in wooded areas than in grassy areas as a result of exposure to heat and desiccation in open areas such as grasslands. Many of our investigations will explore the importance of the effect of division between grassy and wooded areas in our spatial models on the endemicity of disease.

Patch Connectivity

The amount of patch connectivity is critical in our explorations. It isn't surprising that if migration between neighboring patches is high, then the individual spatial characteristics are quickly lost to a population with converging characteristics. It is difficult and important to determine an appropriate and realistic scheme for patch migration to capture the essential dynamics of a study.

Variations in the Surrounding Environment

With migration between patches in our model, we explore the importance of the surrounding environment to the sustainability of disease. For example, if an endemic area is surrounded by nonendemic areas, under what conditions can the disease be sustained when migration to surrounding nonendemic patches is allowed?

Boundary Effects

It has been demonstrated that patch edges are critical in spatial dynamics.[17,22] We illustrate some boundary effects and contrast several choices of boundary conditions.

Description of the Mathematical Model

As we explained in the introduction, our model is a simplified representation of the tick-host-HME system using a single host population, a single life stage and a single pathogen. The dynamics are modeled using differential equations on a discrete spatial grid consisting of environmental patches. White-tailed deer (*O. virginianus*) serve as our generic host population and lone-star ticks (*A. americanum*) as the tick population. The population densities of the host and tick population in each patch are denoted by N_i and V_i respectively and the densities of the HME-infected and infectious populations are denoted by Y_i (hosts) and X_i (ticks). Both hosts and ticks are considered susceptible when not infected. Once infected with HME ticks are assumed to be infected for life.

$$\frac{dN_i}{dt} = \beta_i\left(\frac{K_i - N_i}{K_i}\right)N_i - b_iN_i + \sum_j m_{ij}(N_j - N_i)$$

$$\frac{dV_i}{dt} = \hat{\beta}_iV_i\left(\frac{M_iN_i - V_i}{M_iN_i}\right) - (\hat{b}_i + \delta_i)V_i + \sum_j m_{ij}(V_j - V_i)$$

$$\frac{dY_i}{dt} = A_i\left(\frac{N_i - Y_i}{N_i}\right)X_i - \beta_i\frac{N_iY_i}{K_i} - (b_i + v_i)Y_i + \sum_{N_j<N_i} m_{ij}\frac{Y_i}{N_i}(N_j - N_i) + \sum_{N_j>N_i} m_{ij}\frac{Y_j}{N_j}(N_j - N_i)$$

$$\frac{dX_i}{dt} = \hat{A}_i\left(\frac{Y_i}{N_i}\right)(V_i - X_i) - \hat{\beta}_i\frac{V_iX_i}{M_iN_i} - (\hat{b}_i + \delta_i)X_i + \sum_{V_j<V_i} m_{ij}\frac{X_i}{V_i}(V_j - V_i) + \sum_{V_j>V_i} m_{ij}\frac{X_j}{V_j}(V_j - V_i)$$

Table 1. Variable and parameter names used in the model for each 10,000 m² patch i

Name	Description	Baseline Values
N_i	Host population density	Initially 10 hosts/patch
V_i	Tick population density	Initially 3000 ticks/patch
Y_i	Density of hosts infected with disease	Initially 8 hosts/patch
X_i	Density of ticks infected with disease	Initially 80 ticks/patch
β_i	Growth rate for hosts	0.20/month
$\hat{\beta}$	Growth rate for ticks	0.75/month
K_i	Carrying capacity for hosts	20 ticks/patch (woods) 15 ticks/patch (transition) 10 ticks/patch (grass)
M_i	Maximum number of ticks per host	200/host
b_i	External death rate of hosts	0.01/month
$\hat{b}_i$	External death rates of ticks	0.01/month (woods) 0.05/month (transition) 0.1/month (grass)
δ_i	Acaracide-induced death rate for ticks	0/month
A_i	Transmission rate from ticks to hosts	0.02/month
$\hat{A}_i$	Transmission rate from hosts to ticks	0.07/month
v_i	Recovery rate of hosts	0/month
m_{ij}	Migration rate between neighboring patches	0.01

Parameter values are taken from reference 12.

Parameter and function names and values are summarized in Table 1. The first equation describes the population dynamics of the host population in patch i, for which there is logistic growth with carrying capacity, K_i, as well as an external death rate, b_i, caused by hunting or removal. The last term in the equation models the migration from the given patch to all other patches at a rate dependent on the population differential between patches and the connectivity between patches. The second equation describes the tick population dynamics, which are similar to the host dynamics with the carrying capacity for the ticks as the product of the maximum number of ticks per host, M_i and the number of hosts, N_i. In addition, the external death rate for ticks includes both weather, $\hat{b}_i$ and acaricide, δ_i, components. The first terms in the third and fourth equations describe the new infections for the host and tick populations respectively, with transmission rates given by A and $\hat{A}$. The next terms reflect the logistic growth and external deaths of the infected host and tick populations and the last terms are the migration terms, which show the proportion of infected hosts and ticks that migrate as given with the rates in the first two equations. The third equation also includes a host recovery rate, v_i.

Variations within Patches

Prior to performing our metapopulation experiments, we assess the relative importance of each input parameter's influence on the results of the experiment. To accomplish this task, we use the Latin hypercube sampling (LHS) technique that is described in detail by references 19 and 20.

We begin by choosing the parameters in Table 1 that we believe are most likely to vary within individual ecologies and we also choose biologically feasible ranges for the parameters. Note that in our LHS runs we combine the natural and acaracide-related tick death rate into a single term, $\hat{b}$. We then randomly sample each of the nine chosen parameters uniformly across their feasible ranges to create 600 parameter sets. We select numerical outputs that can quantify the behavior of the model in each 30-year simulation: (1) the maximum percent of infected ticks during all time steps, a measure of the intensity of an outbreak and (2) R_0, the basic reproduction number, which is a measure of the endemicity of an outbreak.[12] If R_0 is greater than one, a disease is endemic and if R_0 is smaller than one, the disease will die out.

After verifying a monotone relationship between each input parameter and each output measure, we compute the partial rank correlation coefficients (PRCC) to determine how much each input parameter contributes to changes in our output measures. In Figure 1 we show the graphs of the residues from multiple linear regressions performed on the input parameter ranks compared to the reproductive number ranks. By ranks we mean that actual values have been replaced with the integer values that correspond to their relative ascending orders. The corresponding PRCC values are shown above each graph with respective statistical significances. The illustration shows the correlation between a strong PRCC value and a linear relationship between input and output ranks. In this analysis, 600 runs with random parameter samplings on a uniform distribution are performed and the PRCC values for runs with several hundred more or fewer samplings return similar values. The strength of the measured PRCC correlation depends strongly on the chosen parameter distributions; in particular, a distribution chosen with a large range will have a stronger PRCC result. The ranges chosen for this analysis are considered to be the most biologically feasible.

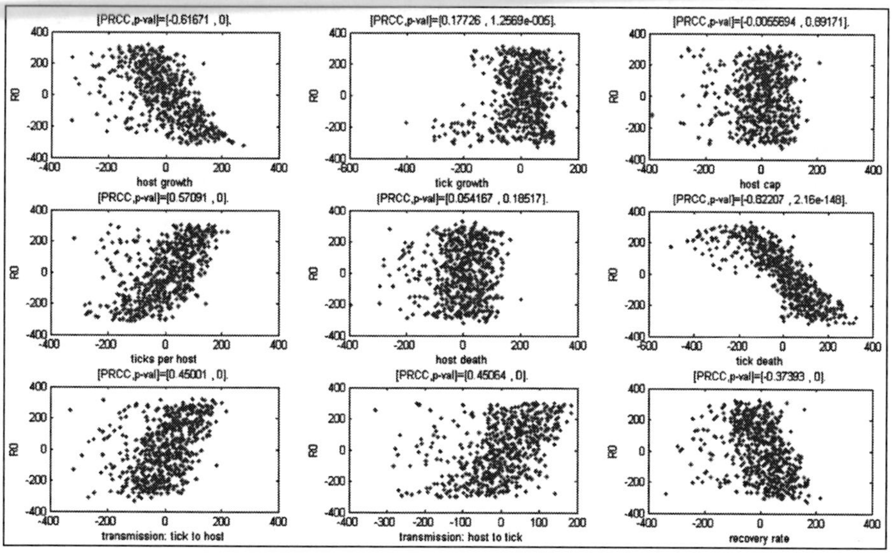

Figure 1. LHS for basic reproductive number on a single patch. Each input parameter is sampled 600 times and subsequently 600 randomly-mixed parameter sets are created. The basic reproductive number R_0 is computed for each of the 600 parameter sets. Then each parameter set and the resulting R_0 values are replaced with their integer rank and multiple linear regression is performed on the ranks of each input parameter versus R_0. The residues are pictured along with PRCC and p-values which indicates the significance of the nonzero PRCC. We see that a stronger linear trend between input and output ranks corresponds to a PRCC-value near 1 (direct correlation) or –1 (inverse correlation). These results are summarized along with other PRCC data in Table 2.

Table 2. PRCC values for selected parameters and outcome measures

Parameters Varied for a Single Patch Model									
Parameter	β	$\hat{\beta}$	K	M	b	$\hat{b}$	A	$\hat{A}$	ν
min, max	0.05, 0.3	0.25, 0.9	5, 50	150, 500	0.01, 0.1	0.001, 0.5	0.013, 0.03	0.045, 0.1	0.001, 0.12
Maximum Percent of Infected Ticks in any Month Over 30 Years									
PRCC	−0.73*	−0.77*	−0.10	0.64*	0.08	−0.80*	0.59*	0.80*	−0.45*
Basic Reproductive Number R_0									
PRCC	−0.62*	0.18*	−0.0056	0.57*	0.054	−0.82*	0.45*	0.45*	0.37*
Parameters Varied for a Two Patch Model									
Parameter	β_i	$\hat{\beta}_i$	M_i	$\hat{b}_i$	$\hat{b}_2$	A_i	$\hat{A}_i$	ν_i	m_{ij}
min, max	0.05, 0.3	0.25, 0.9	150, 500	0.001, 0.5	0.001, 0.5	0.013, 0.03	0.045, 0.1	0.001, 0.12	0, 0.25
Maximum Percent of Infected Ticks in Patch 1 in Any Month Over 30 Years									
PRCC	−0.74*	−0.82*	−0.67*	−0.70*	−0.35*	0.55*	0.83*	−0.18*	0.0061
Maximum Percent of Infected Ticks in Combined Patches in Any Month Over 30 Years									
PRCC	−0.65*	−0.81*	0.56*	−0.36*	−0.35*	0.47*	0.74*	−0.18*	0.032
Basic Reproductive Number R_0 in Patch 1									
PRCC	−0.70*	0.13	0.61*	−0.75*	−0.40*	0.53*	0.50*	−0.19*	0.099
Basic Reproductive Number R_0 in Combined Patches									
PRCC	−0.67*	−0.03	0.58*	−0.57*	−0.58*	0.53*	0.49*	−0.15*	−0.0069

The PRCC measures close to 1 indicate a strong direct correlation and those near −1 indicate strong inverse correlation. The significances of the nonzero PRCC values are checked with the student T-distribution. Superscript *notations denote PRCC values that are significant at the $\alpha = 0.001$ level.

Referring to the PRCC measurements for the one-patch disease model shown in Table 2, we see that all parameters with the exception of host carrying capacity and host external death contribute significantly to both the intensity and endemicity of HME outbreaks. We note that $\hat{A}$, the transmission rate from hosts to ticks, is much more important to the intensity than to the endemicity. Of particular interest is the tick growth rate, $\hat{\beta}$, which contributes weakly, yet directly, to the basic reproductive number and contributes inversely to the outbreak intensity. Biologically this result suggests that a large number of ticks would cause a larger epidemic immediately, but would dilute the infection in the long term. Likewise, host recovery, v, decreases the immediate intensity of an outbreak but encourages endemicity; however, the impact of v is fairly weak in comparison to the other parameters and we assume no host recovery as our baseline value.

We note additionally that the one parameter with extremely strong PRCC values for both the intensity and endemicity of an HME outbreak is $\hat{b}$, the external death rate of ticks. The intensity, but not the endemicity, is also strongly influenced by host and tick birth rates, β and $\hat{\beta}$, as well as by the disease transmission rate, $\hat{A}$, from hosts to ticks. Looking forward to our simulations with multiple patches, we note that because the value of $\hat{b}$ will vary in the wooded and grassy patches, we would expect patch dynamics to be important in predicting long-term disease behavior while the short-term intensity of an HME outbreak may not show as much sensitivity to patch dynamics if three of the four the most influential parameters are constant along our patches.

We presume the variables that are biologically the most sensitive to changing in various geographic locations are $\hat{b}$, the external death rate for ticks and A, the transmission rate from ticks to hosts. Both of these parameters are found to be significant both to the intensity and endemicity of HME outbreaks in the LHS sensitivity analysis. Differing weather patterns would influence the tick death rate, $\hat{b}$ and to a lesser extent the tick growth rate, $\hat{\beta}$. Variability in species diversity of hosts, with some species more susceptible to the disease than others, would influence the value of A and, to a lesser extent, $\hat{A}$. In Figure 2 we illustrate the effects of varying these two parameters, which are most likely to change within a single patch.

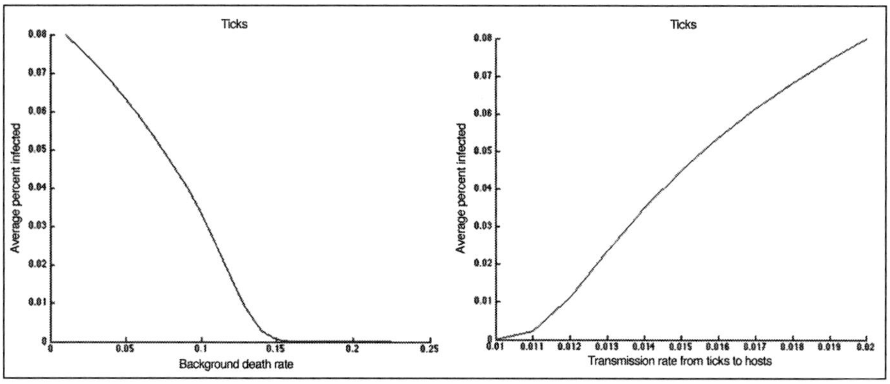

Figure 2. Importance of the tick death rate and host susceptibility. The graph on the left shows the long-term annual average of the percent of infected ticks. The vertical axis is the average percent of infected ticks over the last eight years of a 30-year HME simulation. We see that as the environmentally induced death rate of ticks, $\hat{b}$, increases the percentage of hosts and ticks with HME diminishes. The graph on the right shows the long-term annual average of the percent of infected ticks. We see that as the host susceptibility, A, increases the percentage of hosts and ticks with HME grows as well. This is an important parameter to consider when measuring the influence of host diversity.

Patch Connectivity

Following the set-up of reference 12, we began with two patch types: grassy patches and wooded patches. In our baseline setup, both grassy and wooded areas are endemic for HME. However, the parameters for the wooded areas are more favorable to growth of both the tick population and the infection. In particular, two parameters differ from grassy to wooded areas, the host carrying capacity, K_i and the tick death rate, $\hat{b}_i$. In our one-patch sensitivity analyses, we found that the value for the carrying capacity was not particularly important in predicting disease dynamics but that the tick death rate did have strong influence. We begin this section with a simple consideration of two connected patches that share all parameter values except for the tick death rate, $\hat{b}_i$ and we explore how their connectedness influences disease levels.

We apply the LHS sensitivity analysis to the two-patch model and we allow our significant parameters from the one-patch model to vary in the same ranges. We have two significant new assumptions in this experiment. First, the tick death rate, $\hat{b}_i$, is allowed to vary separately in the individual patches, while the other parameters vary uniformly in both patches. Secondly, we introduce migration as given in the mathematical model and we vary the value of m_{ij} between 0 and 0.25. The results of the analysis are presented in Table 2. We first observe that the parameters with strong PRCC values in the one-patch simulations have similarly strong values in the two-patch simulations and that the variations in migration levels were insignificant compared to the variations in these parameters. Secondly, we observe that by allowing the tick death rate, $\hat{b}_i$, to vary individually in each patch, we effectively cut its PRCC value in half as a contributor to the overall dynamics of the infection. Of course, this makes intuitive sense because the parameter is only contributing to half of the area. The parameter's contribution to the dynamics in its patch of origin is cut only slightly by its interaction with neighboring patches. We can reconfirm in the sensitivity analysis for the two-patch model that the effect of the tick birth rate, $\hat{\beta}_i$, on the disease endemicity is not significant and the value of host recovery v is statistically significant but small. We also confirm that the dynamics of changing the tick death rate, $\hat{b}_i$, between patches is more influential on the value of the basic reproductive number than to the maximum percent of infected ticks overall.

Though we note above that the PRCC values for the migration rates are insignificant in the LHS analysis, we caution against the conclusion that migration is not important. We showed, for example, in reference 21 that when two patches interact with migration the patches very quickly lose the individual characteristics that were caused by their local ecology. Thus, the results of the sensitivity analysis merely indicate that global changes in the landscape ecology are more important than the connectivity of the individual patches; however, in a given ecology, the patch connectivity is indeed extremely important.

To continue our exploration of the importance of patch connectivity, we now move to our consideration of the interaction of a wooded patch and a grassy patch with variable birth and death rates oscillating near the baseline values given in Table 1. The oscillating values are explained in reference 12 and are intended to mimic tick life history as well as seasonal climatic effects. We vary the patch connectivity between the grassy patch and the wooded patch. Figure 3 graphs the migration rate versus the long-term annual average proportion of infected of hosts and ticks in the combined patches. The graph indicates that as migration between patches increases, the individual characteristics at each patch become diminished. Note that Table 1 records our baseline migration level of $m_{ij} = 0.01$ and at this level metapopulation effects created by environmentally unique patches can be observed.

One unexpected result from this study is the importance of location for control efforts. One method frequently used to control HME in a given area is through the use of acaracides, chemical agents applied topically or systemically to hosts that kill ticks. In our model, the application of acaracide is equivalent to raising the value of the parameter δ, or equivalently increasing $\hat{b}$. At the bottom of Figure 3, we consider two scenarios for a two-patch system. With a solid lines, we indicate the longtime average annual percent of infected ticks in the grassy (blue) and the wooded (red) patch if the acaracide control is used only in the grassy patch and with a dotted line we show the same outputs but with the assumption that the acaracide control is used only in the wooded patch. The figure shows that without regard to migration rates, control applied in the wooded patch effectively reduces the prevalence in both patches to near zero. On the other hand, if the control is applied only in the grassy patch the prevalence remains high in both patches. It is perhaps counterintuitive that increases in migration rates contribute significantly to the reduction of HME prevalence in both patches; i.e., the patch in which the acaracide control is applied can benefit more from the acaracide if migration is allowed. Patch connectivity can play a significant role in the disease dynamics.

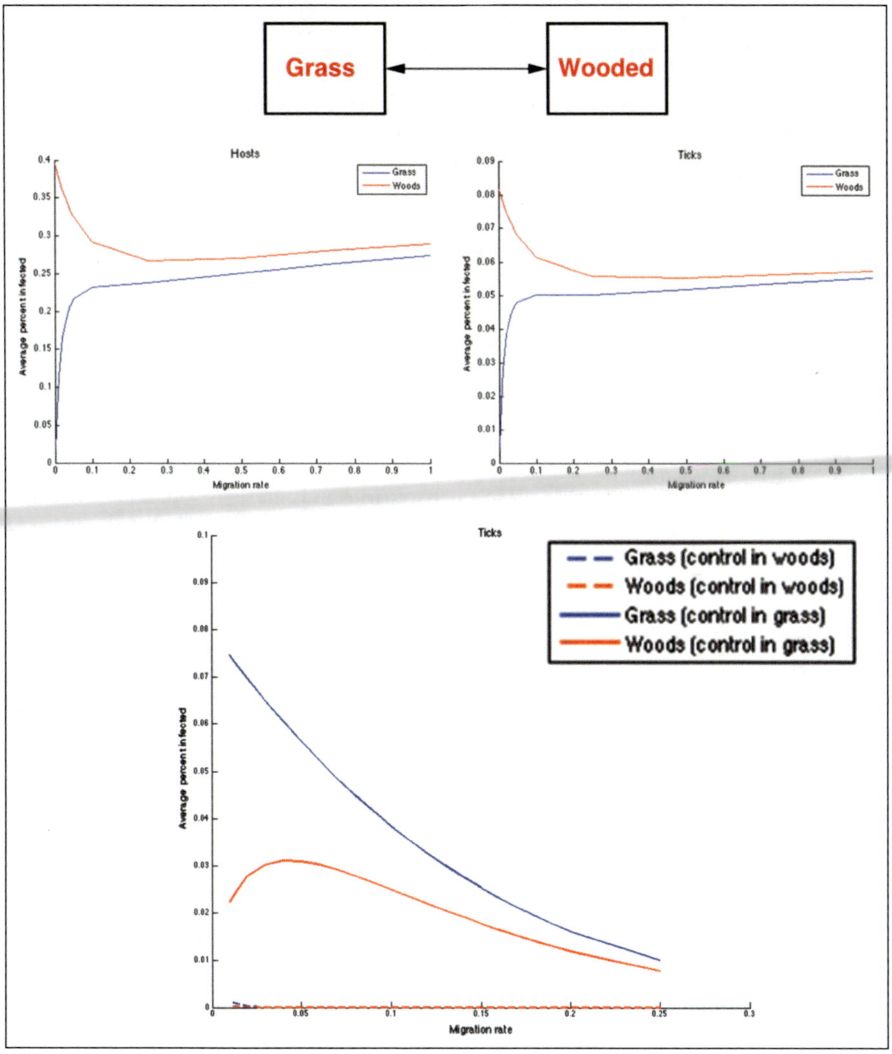

Figure 3. The effect of increased patch connectivity. In the top two graphs, the vertical axis is the average percent of infected hosts (left) and ticks (right) over the last eight years of a 30-year HME simulation. We observe that the individual effects resulting from a local ecology are lost with increased with large inter-patch migration. In the bottom graph, the vertical axis is the average percent of infected ticks over the last eight years of a 30-year HME simulation. We observe that when tick control is applied only in wooded areas, the control effort is effective regardless of migration levels. However, when the control is applied only in the grassy patch then the level of patch connectivity plays a large role in the effectiveness of the control.

The Surrounding Environment

In our remaining investigations, we consider more complex metapopulation ecologies. Future simulations will rely on a larger 5 × 5 spatial grid of patches that are connected to the nearest neighbors. Although the patches are established in a rectangular fashion, individuals move away from the center of a patch in radial directions and therefore any of the six patches that surround an interior patch would be considered equally connected to the given patch. In addition, we consider the boundaries as being reflecting boundaries, in which hosts and ticks located in edge patches are unable to cross the boundaries and thus remain in their patch rather than migrate in the boundary direction.[22] We will discuss the importance of this boundary assumption in the following section.

As noted above, the initial work on this model discussed only wooded and grassy habitat types for patches. In reality, the grassy areas would not instantly transform to wooded areas, but rather there would be some intermediate area of land with intermediate parameter values. Additionally, transition areas may be of importance to other features such as the likelihood of host-tick encounters because ticks may be more likely to use the vegetation of intermediate height to attack hosts. We simply define transition patches as having parameter values halfway between grassy and wooded patches for the host carrying capacity, K_i and the tick death rate, $\hat{b}_i$. In multiple simulations, we seek to check the importance of the inclusion of such areas in the use of the model to predict the spread of HME in various settings.

Figure 4 illustrates this point with one set of simulations. In the figure, we show side-by-side the number of infected ticks that populate a 5 × 5 spatial grid in three scenarios that are described visually in the figure. We observe there that there is predictable sensitivity to replacing grassy areas with transition zones and perhaps greater sensitivity to replacement of the wooded areas that are highly endemic. In fact, when the graphs of the percent of infected ticks in each grid are compared, there is virtually no difference between the first two scenarios and only slight variation to the third. This suggests that in applying the model to a specific scenario, transition areas could be included as grassy areas with little loss of accuracy in predictions. Inclusion of other aspects of transition zones such as increased host-finding rates will be explored in future work.

Boundary Effects

Edge effects and habitat fragmentation have long been studied to understand the impact on various processes.[16,17,22] We use a variety of scenarios for our metapopulation model to assess these effects on the predicted prevalence of the infected ticks. Recall that our initial simulations assume that our boundary is a reflecting boundary, which is a good assumption for an environment that is adjacent to a river or urban area. This assumption contributes to the high numbers of infected ticks in the corners of the graphs in Figure 4.

Continuing with the assumption of a reflecting boundary, we now consider several scenarios in which a small area of endemicity is placed in a nonendemic environment. In Figure 5, we show where endemic (yellow E) patches are located and include a graph of the percent of infected ticks in all patches each month over a time period of 30 years.

The first observation we see is that an isolated endemic patch cannot remain endemic when surrounded by nonendemic patches with sufficient migration ($m_{ij} = 0.01$) between the patches; however, a single endemic patch in the corner of our grid is sufficient to create an endemic environment for multiple surrounding patches. On the other hand, even if we double the size of the endemic area and place it near, but not on, our spatial boundary, we do not maintain an endemic HME outbreak, though we see that the outbreak takes longer to diminish. Thus, patches along a reflecting boundary will experience different dynamics than patches that are in the interior of our grid.

There are two other commonly used boundary conditions in the ecology literature. One is a toroidal, or cyclic, boundary condition in which the vertical and horizontal edges of the grid are considered to wrap around to the opposite boundary. The other is an absorbing boundary in which individuals who would migrate across the edge boundary are lost. Generally for these latter two boundary choices edge effects will be lost, though some symmetries would appear for toroidal

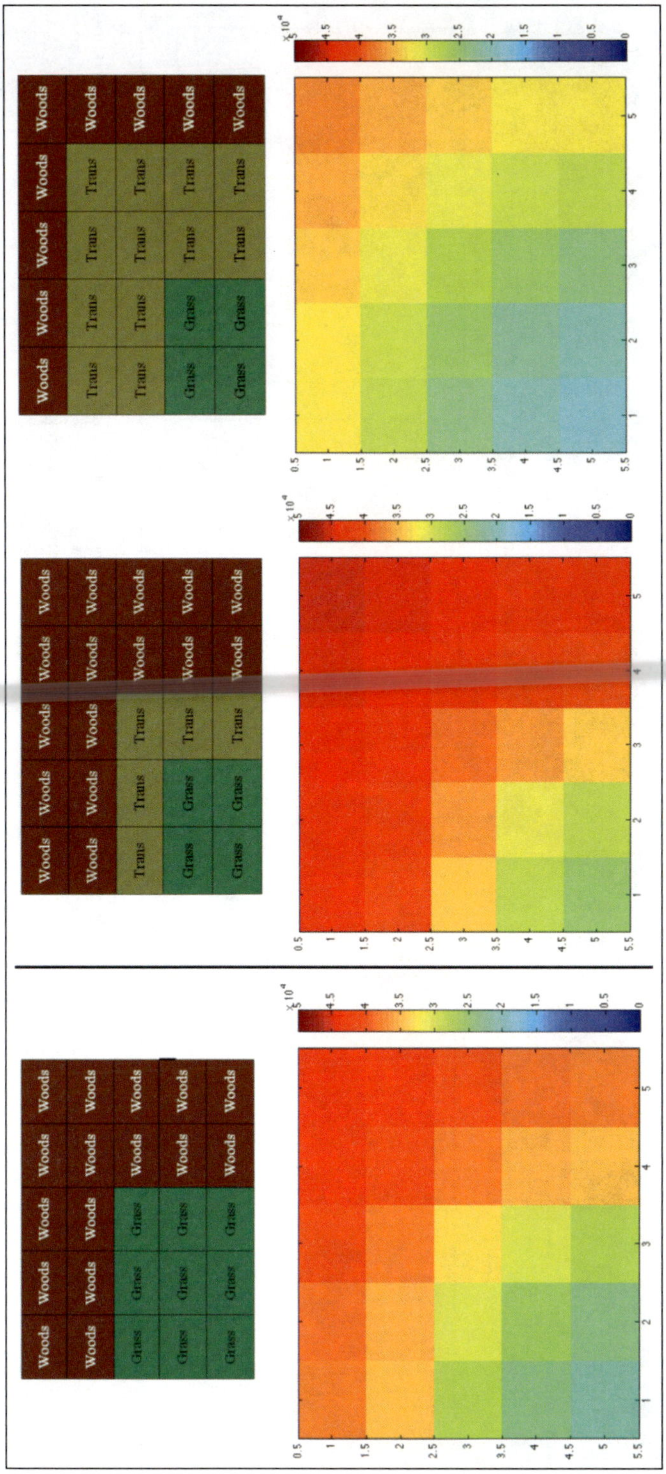

Figure 4. Exploring the effect of transition patches. The top illustrations show the patch configurations that are assumed in each simulation and the figures below show the number of infected ticks within each of the 25 patches after 1000 months with parameter sets as given in Table 1. We conclude that the inclusion of transition zones is less important than correctly categorizing wooded patches.

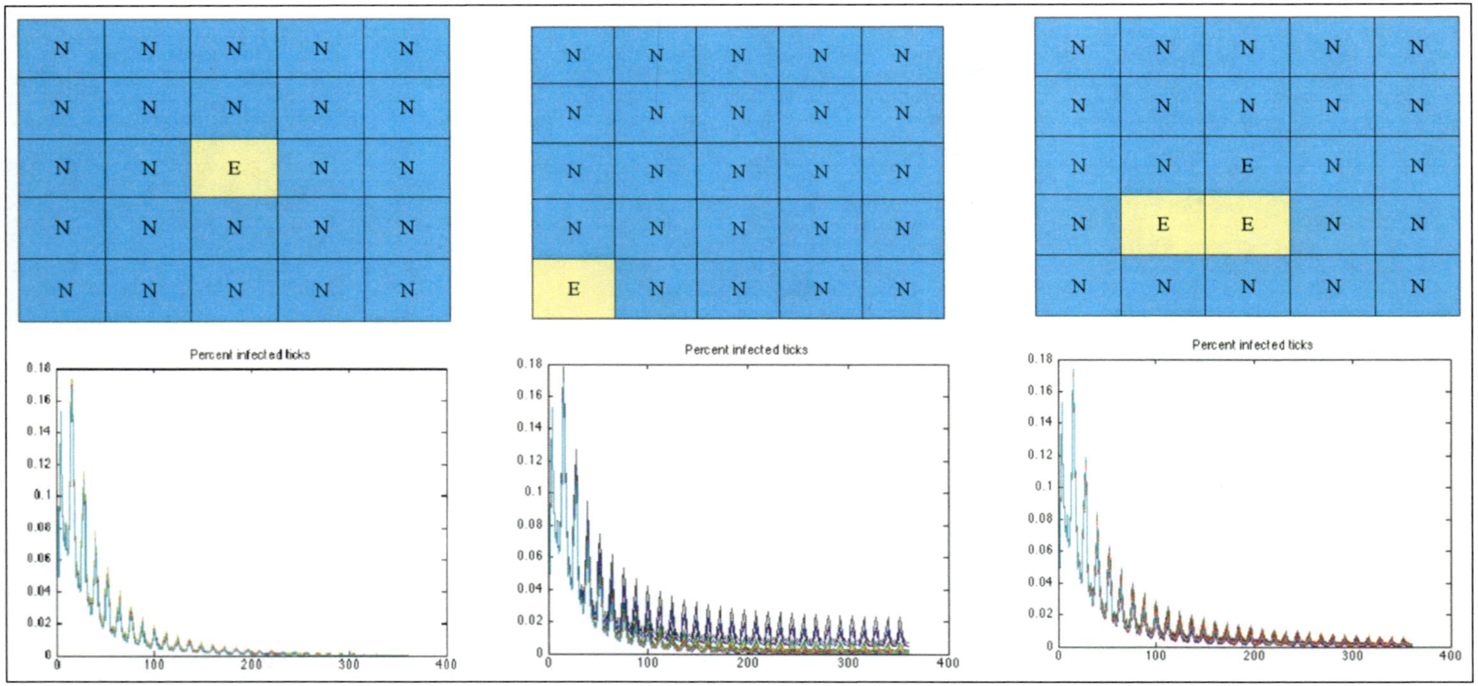

Figure 5. Reflecting boundary effects and endemicity. The simulations are run with reflecting boundaries, periodic host and death birth rates and other parameters as in Table 1. A single endemic patch (denoted by E and using the wooded patch parameters) or pair of endemic patches is surrounded by patches for which the disease is not endemic (N and using the grass patch parameters with $\hat{b}$ increased to 0.25). The percent of infected ticks in each month and in each patch is plotted. In the figure to the left, we see that the endemicity of a patch that is surrounded by nonendemic patches is quickly diffused and to the right we see that in time a pair of patches near but not on the boundary will also have endemicity diffused by migration. However, the center figure illustrates the effect of a reflecting boundary as would be seen with plots of land backing to a river or to urban development. In this case, the endemicity of the patch spreads to neighboring patches.

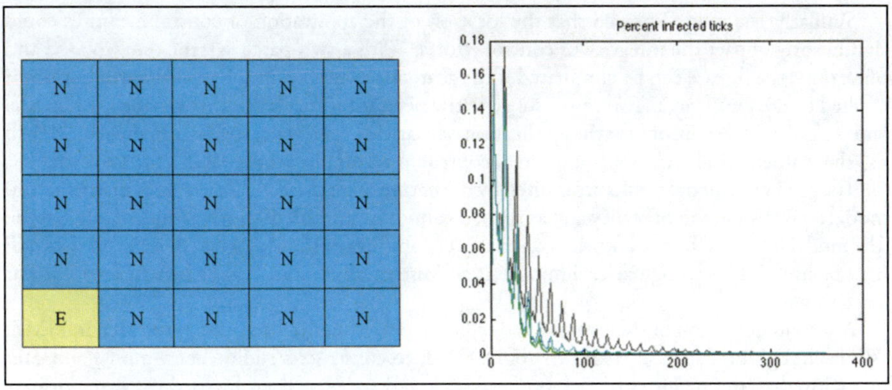

Figure 6. Toroidal boundary effects and endemicity. The simulations from Figure 6 are repeated with toroidal boundaries, periodic host and death birth rates and other parameters as in Table 1. The boundary effect observed in Figure 5 for an endemic patch along the boundary is lost when a toroidal condition is used, illustrating the importance in choosing boundary conditions that accurately reflect the actual ecology.

boundaries as an expected result of the structure.[22] As an illustration of lost boundary effects, we modify our model from our standard reflecting boundaries to a toroidal boundary condition. In Figure 6, we see that an endemic patch on the boundary of our grid will no longer remain endemic when surrounded by nonendemic patches. This result shows the importance of identifying boundary conditions. Many geographic areas may be modeled as having reflecting boundaries if there are conditions such as a river or man-made buildings and structures that would preclude the ticks and hosts from moving in that direction. On the other hand, if the studied land area is surrounded by similar ecologies then toroidal conditions would most appropriately be chosen. These model results imply that the impact of these boundaries can be significant on the predicted risk of disease in a given area.

Conclusion

Tick-borne diseases are of increasing concern for human health worldwide as the incidence of known diseases increases and new diseases are identified. The outbreaks are notoriously sporadic in space and time making them difficult to predict or control. In order to provide some insights into the complex dynamics of tick-borne diseases, we present the results from a spatially explicit, mathematical model of human monocytic ehrlichiosis.

Using a metapopulation ecology framework, the model predictions show that spatial arrangement/location plays a significant role. Within each individual patch, the values chosen for $\hat{\beta}_i$, the growth rate of ticks and $\hat{A}$, the transmission rate from hosts to ticks, were shown to have the most influence on the resulting outbreak. This indicates that favorable weather conditions that increase the tick growth rate will significantly increase the likelihood of an outbreak. Similarly, the species diversity of a given area could increase the transmission rate from hosts to ticks because many species are not susceptible to *E. chaffeensis.* This would in turn increase the likelihood of an outbreak.

Across a spatial framework, the model predictions showed that the location of a potentially endemic region determines if an outbreak will emanate out from that location or be washed out by surrounding nonendemic areas. Outbreaks are predicted to occur more commonly along areas where there is a reflecting boundary next to an endemic area. As human development continues to encroach into previously undisturbed areas, new regions that effectively have reflecting boundaries therefore create the potential for higher risk of outbreaks at the edges of development and wilderness. While transition areas were shown to be effectively modeled as grassy areas, the importance of transition into human developed areas has yet to be explored.

Similarly, the model showed that the location of the application of control measures could significantly impact the integrity of control efforts. With a two-patch system, one wooded and one grassy, the disease can be eliminated if the acaricide is applied in the wooded area but not if applied to the grassy area. This result suggests the need for future work that expands this model and seeks optimal solutions for the application of acaricide for a given spatial framework. We will use the mathematical technique of optimal control to accomplish this goal in a future work.

The predicted model results are limited by the nature of mathematical models. A mathematical model is an abstraction of reality and as such, it cannot capture all of the dynamics of any system. The model is limited by the knowledge and understandings of the biological system and thus will only be useful if we recognize its limitations and use those limitations as a benefit to point out key data gaps.

A second limitation of these results is due in turn to the limited amount of research in HME. While enough information is known about HME to construct a rudimentary model and gain many insights, the majority of tick-borne disease research has been focused on Lyme disease which has higher incidence levels. As stated previously, the advantage for working with HME is the simplifying one-host structure, but given the aggressive and seemingly indiscriminate feeding habits of *A. americanum*, additional host reservoirs may need to be added to the model system to fully capture the dynamics and risks of HME.

Our metapopulation model provides a solid basis for exploring the initial dynamics of HME across a spatial landscape. The results provide some interesting clues into the sporadic nature of outbreaks. Thus, in addition to future work in the optimal control of a potential HME outbreak, future work will also continue to expand on this model to include more information about the life history of the lone star tick as well as the impact of additional host species. As the incidence of tick-borne diseases continues to increase, insights from models such as the one presented here will provide invaluable tools for prevention of future human morbidity and mortality.

Acknowledgements

We would like to thank Louis Gross, J. Stephen Dumler and Daniel Sonenshine for their input. HDG was supported by Grant Number K25AI067791 from the National Institute Of Allergy And Infectious Diseases. The content is solely the responsibility of the authors and does not necessarily represent the official views of the National Institute of Allergy and Infectious Diseases or the National Institutes of Health.

References

1. Piesman J, Eisen L. Prevention of tick-borne diseases. Annu Rev Entomol 2008; 53:323-343.
2. Centers for Disease Control and Prevention, Summary of notifiable diseases—United States, 2005, MMWR Weekly 2007; 54(53):2-92.
3. Centers for Disease Control and Prevention, Summary of notifiable diseases—United States, 2001, MMWR Weekly 2003; 50(53):1-108.
4. Anderson BE, Dawson JE, Jones DC et al. Ehrlichia chaffeensis, a new species associated with human ehrlichiosis. J Clin Microbiol 1991; 29:2838-2842.
5. Yabsley MJ, Wimberly MC, Stallknecht DE et al. Spatial analysis of the distribution of Ehrlichia chaffeensis, causative agent of human monocytotropic ehrlichiosis, across a multi-state region. Am J Trop Med Hyg 2005; 72:840-850.
6. Mixson TR, Ginsberg HS, Campbell SR et al. Detection of Ehrlichia chaffeensis in adult and nymphal Amblyomma americanum (Acari: Ixodidae) ticks from Long Island, New York. J Med Entemol 2004; 41:1104-1110.
7. Anderson BE, Sims KG, Olson JG et al. Amblyomma americanum: a potential vector of human ehrlichiosis. Am J Trop Med Hyg 1993; 49:239-244.
8. Dumler JS, Bakken JS. Human ehrlichioses: newly recognized infections transmitted by ticks. Annu Rev Med 1998; 49:201-213.
9. Ewing SA, Dawson JE, Kocan AA et al. Experimental transmission of Ehrlichia chaffeensis (Ehrlichieae) among white-tailed deer by Amblyomma americanum (Acari: Ixodidae). J Med Entomol 1995; 34:368-374.

10. Lockhart JM, Davidson WR, Stallknecht DE et al. Isolation of Ehrlichia chaffeensis from wild white-tailed deer (Odocoileus virginianus) confirms their role as natural reservoir hosts. J Clin Microbiol 1997; 35:1681-1686.
11. Lockhart JM, Davidson WR, Dawson JE et al. Temporal association of Amblyomma americanum with the presence Ehrlichia chaffeensis reactive antibodies in white-tailed deer. J Wildl Dis 1995; 31:119-124.
12. Gaff H, Gross LJ. Analysis of a tick-borne disease model with varying population sizes in various habitats. Bull Math Biol 2007; 69:265-288.
13. Hanski I, Simberloff D. The metapopulation approach. In: Hanski I, Gilpin ME, eds. Metapopulation Biology: Ecology, Genetics and Evolution. San Diego: Academic Press, 1997:5-26.
14. Hanski I, Gilpin ME. Conceptual foundations. In: Hanski I, Gilpin ME, eds. Metapopulation Biology: Ecology, Genetics and Evolution. San Diego: Academic Press, 1997:1-4.
15. Wiens J. Metapopulation dynamics and landscape ecology. In: Hanski I, Gilpin ME, eds. Metapopulation Biology: Ecology, Genetics and Evolution. San Diego: Academic Press, 1997:43-62.
16. Allan BF, Keesing F, Ostfeld RS. Effect of forest fragmentation on Lyme disease risk. Conservation Biology 2003; 17(1):267-272.
17. Murcia C. Edge effects in fragmented forests: implications for conservation. Trends in Ecology and Evolution 1995; 10:58-62.
18. Hanski I. Metapopulation ecology, great britain: Oxford University Press, 1999.
19. Blower SM, Dowlatabadi H. Sensitivity and uncertainty analysis of complex models of disease transmission: an HIV model, as an example. Int Stat Rev 1994; 2:229-243.
20. Marino S, Hogue IB, Ray CJ et al. A methodology for performing global uncertainty and sensitivity analysis in systems biology. J Theor Biol 2008; 254(1):178-196.
21. Gaff H, Schaefer E. Results from a mathematical model for human monocytic ehrlichiosis. Proceedings of the 5th Conference on Rickettsiae and Rickettsial Diseases, a Supplement to Clinical Microbiology and Infection, to appear.
22. Comins HN, Hassell MP, May RM. The spatial dynamics of host—parasitoid systems. J Anim Ecol 1992; 61(3):735-748.

Chapter 5

Modelling Stochastic Transmission Processes in Helminth Infections

Stephen J. Cornell*

Abstract

The number of helminths within a host can only increase by the host encountering additional infectious stages, so it is important to consider not only whether a host is infected, but also the severity of its infection. Stochastic models consider explicitly the number of parasites within the host and treat infection, death and other demographic events as random processes. I discuss stochastic helminth population models of increasing degrees of complexity, starting with the infection dynamics within a single host and finishing with the full parasite lifecycle among a population of hosts. I demonstrate the mathematical techniques that can help to analyse these models and discuss the insights into parasite population biology that these methods can bring.

Introduction

The number of helminth parasites infecting a particular host matters to both the host and the parasite. Hosts with a high worm burden suffer much stronger pathology than those with a milder infection and in many cases mount an immune response that strengthens with exposure to infectious stages of the parasite. In the terminology of Anderson and May,[1] macroparasites (and helminths in particular) cannot reproduce directly within the host, so macroparasite infections develop more gradually than microparasites do because the worm burden can only increase if the host encounters additional infectious stages in its environment. Many of the processes affecting parasite demographics are density dependent—i.e., depend nonlinearly on the number of worms in the host—such as the risk that a female cannot find a mate,[2] or the modulation of infection rate by the host's immunity.[3] This means that population dynamics of helminth parasites depend on the distribution of worms among hosts and cannot be adequately modelled by considering the mean worm burden alone.

A key feature of many helminth infections is that the distribution of worm burdens is aggregated, i.e., the variance in worm burden among hosts is greater than the mean.[4] Much effort has been spent on building mathematical models to understand the causes and consequences of these patterns of infection. Aggregated distributions can be generated by heterogeneities in hosts' encounters with infectious stages, or heterogeneities in the dynamics of parasites within hosts. Some of this heterogeneity is systematic—older hosts may be better at avoiding or resisting parasites—but much heterogeneity has no obvious correlation with the host's age or other attributes.

In this chapter I shall describe stochastic models for helminth infections, where heterogeneity in host exposure and response are modelled by random processes. The most common form of randomness in these models is demographic stochasticity, so that parasite subpopulations in two different hosts may be different even when they are subject to the same underlying rates of

*Stephen J. Cornell—Institute of Integrative and Comparative Biology, University of Leeds, Leeds LS2 9JT, UK. Email: s.j.cornell@leeds.ac.uk

Modelling Parasite Transmission and Control, edited by Edwin Michael and Robert C. Spear.

infection, reproduction and death. In addition, some models have allowed the parameters determining parasite population dynamics to differ between hosts, again as a random process. Random fluctuations in environmental conditions can also play an important role in parasite dynamics,[5,6] though I shall not discuss this here. I shall describe models with increasing degrees of complexity, from the infection dynamics of a single host to parasite proliferation within a host population. I shall demonstrate mathematical techniques which can be used to approximate and understand these stochastic models, using as an example a parasite which causes increased mortality in its host. This model is an exemplar of the host-parasite interaction and has been studied in different forms by several authors.[7-12]

Infection in a Single Host

The earliest stochastic models of macroparasite transmission described the dynamics within a single host only.[13] The infection rate of the host is treated as a known quantity, which is assumed to be unaffected by the reproductive output of the worms in the host under consideration. This simplifies the analysis to the extent that explicit mathematical solutions can be obtained for models with many biological components, such as acquired immunity and parasite-induced mortality.

These models give valuable insights into how different mechanisms increase or decrease parasite aggregation and how this aggregation affects nonlinear quantities such as mating probabilities.[2] The main appeal of this technique is the possibility for detailed mathematical analysis, though as models become more biologically complicated it is sometimes necessary to use approximations. I shall illustrate the method by discussing a model that was studied by Valerie Isham.[10]

Clumped Infection and Parasite-Induced Mortality

Isham[10] studied a model for helminth infections, which was not aimed at describing any particular parasite but had two notable features:

- Hosts do not encounter infective parasite stages individually, but rather in 'clumps'. This is a reasonable assumption for many helminth infections: a biting insect may transmit many infective larvae to the unfortunate host and eggs in faeces will develop into larvae that are highly clumped in space.
- The morbidity effects of the parasite causes a higher rate of mortality in hosts with high worm burdens.

Clumped infection was already a component of the earliest stochastic parasite models[13] and is expected to lead to more aggregated parasite distributions, while density-dependent host death rate is expected to decrease aggregation.[9]

In this section, I shall reproduce Isham's analysis of the model, in order to illustrate the technique and to discuss its strengths and limitations. In the model, a host of age a is assumed to harbour m parasites. The processes by which m changes are

- Infection: $m \to m + c$ at rate $\phi(a)h_c$. This represents an infection event taking place at rate $\phi(a)$ (which may depend on host age a) and when infected the host gains a clump containing c parasites. The distribution of clump sizes is, where $\sum_{c=0}^{\infty} h_c = 1$.
- Parasite death: $m \to m - 1$ at rate $\mu_M m$, representing independent parasite death events at rate μ_M.
- Host death: The host dies at rate $\mu_H(a) + \alpha m$. This represents the host's intrinsic death rate μ_H, which may depend on its age a and an increased death rate due to the number of parasites in the host.

Under these assumptions, the following Forward equation for the joint probability $P_m(a)$ that a host reaches age a and contains m parasites can be written:

$$\frac{dp_m(a)}{dt} = -\{\mu_H(a) + \alpha m + (1 - h_0)\phi(a) + \mu_M m\} p_m(a) + \mu_M(m+1)p_{m+1}(a) + \phi(a)\sum_{c=1}^{\infty} p_{m-c}(a)h_c. \tag{1}$$

The right-hand-side term on the first line of this equation represents all the processes which change the worm burden out of state m; the terms on the second line represent processes where the worm burden changes into state m.

This equation can be analysed with the use of the probability generating function (PGF) $R(a;z) = \Sigma_{m=0}^{\infty} z^m p_m(a)$. Multiplying Eqn. (1) by z^m and summing, we note that we can simplify terms like $\Sigma_{m=0}^{\infty} mz^m p_m(a) = z\partial R(a;z)/\partial z$ to give

$$\frac{\partial R(a;z)}{\partial z} = -\{\mu_H(a) + \phi(a)[h(z) - 1]\}R(a;z) - \{(\alpha + \mu_M)z - \mu_M\}\frac{\partial R(a;z)}{\partial z}. \tag{2}$$

This equation could be solved directly, but to study age-intensity profiles we are instead interested in the probability $p(m|a)$ of having m parasites *conditional* on reaching age a. If the probability of surviving to age a is $S(a) \equiv \Sigma_{m=0}^{\infty} p_m(a) = R(a;1)$, then the standard relationship between joint and conditional probabilities gives $p_m(a) = p(m|a)S(a)$. An equation for the survival probability is obtained by substituting $z = 1$ into Eqn. (2):

$$\frac{dS(a)}{da} = -\mu_H(a)S(a) - \alpha m_M(a)S(a), \tag{3}$$

where we have written

$$\left.\frac{\partial R(a;z)}{\partial z}\right|_{z=1} = \sum_{m=0}^{\infty} mp_m(a) = S(a)\sum_{m=0}^{\infty} mp(m|a) = S(a)m_M(a),$$

m_M being the mean of m conditional on the host reaching age a.

Let $Q(a;z) = \Sigma_{m=0}^{\infty} z^m p(m|a) = R(a;z)/S(a)$ be the PGF for $p(m|a)$. Using Eqns. (2) and (3), we can derive the following equation for $Q(a;z)$:

$$\frac{\partial Q(a;z)}{\partial a} = \{\phi(a)[h(z) - 1] + \alpha m_M(a)\}Q(a;z) - \{(\alpha + \mu_M)z - \mu_M\}\frac{\partial Q(a;z)}{\partial z}. \tag{4}$$

Explicit Solution of the Model

Equation (4) can be solved explicitly by first defining $\zeta(z) = \frac{\log((\alpha + \mu_M)z - \mu_M)}{\alpha + \mu_M}$, and then making the change of variables $u = \frac{a + \zeta}{2}$, $v = \frac{a - \zeta}{2}$, $W(u, v) = \log Q(a; z)$. Eqn. (4) can then be written in the form

$$\frac{\partial W(u,v)}{\partial u} = f(u + v, u - v) + g(u + v),$$

where $f(a,\zeta(z)) = \phi(a)[h(z) - 1]$, $g(a) = \alpha m_M(a)$. This equation can be integrated directly to give

$$W(u,v) = \int_{u_0}^{u} [f(w + v, w - v) + g(w + v)]dw + F(v), \tag{5}$$

where u_0 is an arbitrary constant and $F(v)$ is an arbitrary function of its argument.

The desired solution is obtained by imposing the following two boundary conditions: (i) $Q(0, z) = p(m = 0|a = 0) = 1$, which states that the host is uninfected at birth, constrains the form for F; and (ii) $Q(a, 1) = \Sigma_{m=0}^{\infty} p(m|a) = 1$, which states that the probabilities of different worm burdens sum to unity. While there is only one arbitrary function in (5), the second boundary condition is required because $g = \alpha m_M(a)$ needs to satisfy the self-consistency condition with $m_M(a) = \partial Q(a;z)/\partial z|_{z=1}$. The final solution is

$$Q(a;z) = \exp\left[\int_{\zeta(z)-a}^{\zeta(z)} f(y + a - \zeta(z), y)dy - \int_{\zeta(1)-a}^{\zeta(1)} f(y + a - \zeta(1), y)dy\right],$$

where f is a general function and for the specific form $f(a,\zeta(z)) = \phi(a)[h(z) - 1]$ we get

$$Q(a;z) = \exp\left[\int_{z}^{\theta(a;z)} \frac{\phi(a + \zeta(x) - \zeta(z))[1 - h(x)]}{(\alpha + \mu_M)x - \mu_M}dx + \int_{\theta(a;1)}^{1} \frac{\phi(a + \zeta(x) - \zeta(1))[1 - h(x)]}{(\alpha + \mu_M)x - \mu_M}dx\right], \quad (6)$$

where

$$\theta(a;z) = \frac{\mu_M + \left[(\alpha + \mu_M)z - \mu_M\right]e^{-(\alpha+\mu_M)a}}{\alpha + \mu_M}.$$

Note that the equivalent solution in ref. 10 was derived by assuming that $\phi(a)$ is independent of a, though as shown here a solution can be obtained without this assumption.

When Can This Type of Model Be Solved Explicitly?

We are now in a position to discuss the type of models which can be solved by this method.

- Eqn. (2) is a first-order partial differential equation because both host and parasite death are linear functions of m. If some processes took place at a rate proportional to m^2, then there would be terms of the type $\partial^2 P / \partial z^2$ and there is a much smaller chance that an explicit solution to the PDE would be possible. The problem would become completely intractable if rates had a still more complex dependence (e.g., exponential) on m. This severely limits the degree of biological realism of models for which explicit solutions are available.
- The component of the host's death rate that is independent of m does not bias survivorship as a function of m and as a result the expression (6) does not depend on $\mu_H(a)$. This means that the results can be applied to organisms with any type of age-death relationship.
- Even though an explicit solution is possible for this model, the compactness of the solution belies its usefulness. In principle, the value of $p(m|a)$ can be obtained by expanding Eqn. (6) as a series in powers of z, though as noted by Isham this does not give very convenient expressions. Moreover, the expression is still more cumbersome when the infection rate $\phi(a)$ depends on age a. Nevertheless, some low moments of the distribution can be obtained in a useful form, as discussed in the next section.

Effect of Clumping and Parasite-Induced Mortality on Aggregation

The mean $m_M(a) = \Sigma_{m=0}^{\infty} mp(m|a)$ and variance $V_M(a) = \{\Sigma_{m=0}^{\infty} m^2 p(m|a)\} - m_M^2(a)$ of the worm burden, conditional on the host reaching age a, can be straightforwardly obtained from derivatives of $Q(a;z)$ evaluated at $z = 1$. For the case ϕ = constant, using Eqn. (6) we have

$$m_M(a) = \frac{\phi}{\alpha}\left[1 - h\left(\theta(a;1)\right)\right]$$

$$V_M(a) = \frac{\phi}{\alpha}h'(1) - \frac{\mu_M}{\alpha}m_M(a) - \frac{\phi}{\alpha}h'\left(\theta(a;1)\right)e^{-(\alpha+\mu_M)a}$$

The mean $m_M(a)$ increases with age a but approaches a finite limit as a increases. The variance to mean ratio $I_M(a) = V_M(a)/m_M(a)$ is a measure of the degree of aggregation in parasite burdens, being unity for a Poisson distribution. As discussed in ref. 10, I_M is unity when infecting clumps contain only one parasite, but increases when infecting clumps are bigger. Parasite-induced mortality has a more complicated effect on I_M, however. The limiting value of $I_M(a)$ for large a can either decrease or increase with α/μ_M, depending on the details of the distribution of clump sizes.

The model was extended by Herbert and Isham[14] to allow for more than one life history stage of the parasite and also for nonMarkovian processes in life history progression (i.e., the time spent in different stages was not assumed to be exponentially distributed). The influence of mortality on parasite aggregation was further explored by Barbour and Pugliese,[15] who assumed nonclumped infection but considered both density-dependent parasite mortality and a range of scenarios for density-dependent parasite-induced host mortality.

Moment Closure

As discussed above in 'Explicit solution of the model', for many models it is not possible to obtain an exact solution for the distribution of worm burdens. However, it is possible to obtain an approximate solution by using a technique called 'moment closure'. While this is not necessary for the model described above, it is instructive to see how the approximation works in this case.

The mean m_M and the variance V_M of the worm burden can be expressed as derivatives of $Q(a;z)$ at $z = 1$, so from Eqn. (4) these quantities satisfy

$$\begin{aligned} \frac{dm_M}{da} &= \phi(a)h'(1) - \alpha V_M(a) - \mu_M m_M(a) \\ \frac{dV_M}{da} &= \phi(a)\left[h''(1) + h'(1)\right] + \mu_M m_M(a) - 2\mu_M V_M(a) \\ &\quad + \alpha\left\{3m_M(a)V_M(a) + m_M^3(a) - T_M(a)\right\}, \end{aligned} \tag{7}$$

where $T_M(a) = \Sigma m^3 p(m|a)$ is the expectation of the cube of the worm burden. We find in general that the rate of change of the n'th moment of the distribution depends on moments of order higher than n. This hierarchy of equations cannot be solved, because the solution to any equation depends on the solution of an equation that has not yet been solved.

The technique of moment closure gets around this problem by approximating a higher-order moment in terms of a combination of lower-order ones. For instance, Isham[10] used the normal approximation for the third moment, in which case the term multiplying α on the right-hand side of Eqn. (7) (the third order cumulant) is zero. The above equations are now 'closed' as they depend only on m_M and V_M and can be solved. Reference 10 gives an explicit solution as a function of a and compares the results of the moment closure to the exact solution.

Note that the moment equations form an unclosed hierarchy, even in a case such as this where an explicit exact solution for the moments themselves is available. However, the moment closure approximation can give approximate analytical or numerical solutions for a much wider class of problems. In particular, it is not restricted to processes which depend linearly on the internal variables as discussed in 'When can this type of model be solved explicitly?'.

Acquired Immunity

Immunity to many helminth infections is thought to be acquired progressively in response to exposure to infection challenge. Deterministic age-structured models have been used to investigate the importance of the host's immune response in shaping patterns of helminth infection as a function of age.[3,16] Immunity has also been investigated in single-host stochastic models.

Chan and Isham[17] studied a model similar to the one described above, with the additional assumption that the host has a stochastic integer-valued 'immunity' variable I which increased at a rate proportional to the current infection level. In order to aid tractability, they assumed that the infection rate was modulated by a linearly decreasing function of I. This leads to a first order PDE for the PGF of the number of parasites and immunity variable, for the reasons discussed in 'Explicit solution of the model'. This form for the infection rate is actually negative when I is high enough, but that was not found to lead to mathematical problems for the parameters studied by the authors. It was not possible to obtain explicit solutions for the PGF, because of the particular nonlinear coefficients of the first-order partial derivatives, but the equations for the first and second moments of the distribution formed a closed hierarchy so could be solved without requiring a closure assumption.

Host Heterogeneity

A further cause of parasite aggregation is intrinsic differences between hosts. Hosts may differ in the rate at which they encounter parasites or in their ability to mount an immune response. The host represents the parasite's habitat, so any differences in hosts, e.g., size and/or age, may affect their development, reproduction and mortality.

If mathematical solutions for the distribution of worm burdens is possible for a host with given parameters, then results for a heterogeneous host population can be obtained by averaging these results (either analytically or numerically) over the host parameters.[18] Isham[10] noted that analytical results obtained in this may be unwieldy and not helpful intuitively. It is also possible to incorporate host heterogeneity at the level of moment equations. Chan and Isham[17] followed this approach in a model of schistosomiasis and discovered that, while the equations for the first and second moments formed a closed system for a single host, the moment hierarchy was not closed when host heterogeneity was taken into account. Two types of host heterogeneity were considered—either differences in parasite encounter rates, or differences in immune response. The former was found to have a strong effect on parasite aggregation, but the latter had only a very weak effect. Further work using the same model[19] showed that stochastic variability may obscure the expected age-intensity relationship generated by acquired immunity.

Mating Probability and Population Genetics

One of the advantages of exact solutions of stochastic single-host infection-death models is that the behaviour of nonlinear functions of the worm burden may be investigated explicitly. Cornell et al[20] investigated a model of early-season dynamics of nematodiasis in sheep. The study focused on the reproductive output of female parasites and did not consider nonlinear processes which might affect the dynamics at high parasite density. Helminths are generally assumed to be promiscuous, but at low density there is a risk that a female may find no male to mate with in the same host. The resulting depression in average fecundity is known as the mating probability.[2] Mating probabilities can be estimated by assuming a given distribution of worms among hosts, such as Poisson or negative binomial[2] and when calculated in this way they decrease to zero at low density. However, Cornell et al found that clumped infection led to the mating probability being greater than zero in this limit. This shows the importance of considering the mechanisms that generate aggregation when modelling parasite population dynamics.

Cornell et al[20] also considered the population genetics of the parasites' offspring and found that if infecting clumps have a multinomial distribution of genotypes that follow Hardy-Weinberg equilibrium then the offspring show small deviations from Hardy-Weinberg equilibrium. However, if the infective clumps were more heterogeneous, then the offspring distributions showed strong deviations from Hardy-Weinberg equilibrium and in particular there was a strong promotion of rare homozygotes.

Competition between Parasite Species

Many hosts will harbour more than a single parasite species and each species may strongly affect the dynamics of other parasite species within the host.[21] Bottomley et al[22] have extended the single-host framework to allow for more than one parasite species. They assumed that the density of adult worms affected the mortality of larval stages of either species and solved a stochastic model approximately by moment closure. When interactions between individuals of the same species were antagonistic, worm burdens were found to be underdispersed when interactions between species were weaker than interactions within species and overdispersed otherwise. However, when interactions within species were mutualistic, the variance to mean ratio was found to be close to unity.

Infection among Multiple Hosts

The models considered above in 'Infection in a single host' assume that the infection rate is independent of the parasite dynamics within the host. In the absence of any feedback from the parasite's reproduction to the abundance of infective stages of the parasite, this method is not able to investigate the evolution of the force of infection over time, or the potential of control strategies to eradicate the parasite. In order to model the ecology of the parasite, it is necessary to consider the dynamics of the parasite stages that live outside the definitive host. This makes the model considerably more complicated, because there may be many hosts of different ages and infection status. The sub-populations of parasites in each of the hosts become coupled and the host themselves may also undergo complicated population dynamics.

Nevertheless, in many cases there are suitable approximations which simplify the models and make it possible to analyse their behaviour. In the present section I shall describe these techniques and again illustrate them with a specific example.

Hybrid Models

The problem of multiple hosts is simplified if one assumes that the host population is large and that the infectious stages generated by the parasite in one host subpopulation is equally likely to infect any of the other hosts (whether they are transmitted directly or via a vector organism). Each host now only sees an average force of infection (though the infection process could still be in the form of clumps), so a host's infection rate is statistically independent of its worm burden. This means that the single host models described in 'Infection in a single host' can be used to describe the dynamics within each host, while the abundance of the infectious stage is a deterministic quantity related to the mean reproductive output of the parasites infecting a host. This type of hybrid model is discussed in detail by Nåsell.[23] Hybrid models are often cited as a motivation for studying single-host stochastic models, though in fact few such studies do actually relate the infection process to the mean reproductive output of the parasite.

Hybrid models assume that the dynamics of the host and/or the parasite's infectious stages are deterministic, but allow full stochastic dynamics within the host. A moment closure assumption is often used to obtain an approximate solution of the within-host dynamics. I shall illustrate this method using a host-parasite model that was studied by Anderson and May[7] and later by Kretzschmar and Adler.[24] This model is very similar to the single-host model studied by Isham[10] and discussed above in 'Clumped infection and parasite-induced mortality', though without clumped infection and where the infection rate is related to the parasite population size. Using the parameter definitions in Table 1, the model takes the following form:

- Assume there are $n_m(t)$ hosts with m parasites at time t, so the number of hosts in the system is $H(t) = \Sigma_{m=0}^{\infty} n_m(t)$ and the total number of parasites is $P(t) = \Sigma_{m=0}^{\infty} m n_m(t)$.
- Each host is infected at a rate $\phi(t)$. Under the assumption that infective stages develop rapidly outside the host and are much shorter lived that the adult parasites in the hosts, the force of infection is related directly to the current parasite population in the form $\phi(t) = \lambda \frac{P(t)}{H_0 + H(t)}$ (see ref. 24).
- Parasites die at the per capita rate μ_M.
- Hosts are born at per capita rate β and a host with m parasites dies at a per capita rate $\mu_H + \alpha m$. When a host dies, its parasites die too.

Given these processes, the equations for the rate of change of n_m are:

$$\frac{dn_0}{dt} = -(\mu_H + \phi)n_0 + \mu_M n_1 + \beta H(t) \tag{8}$$

$$\frac{dn_m}{dt} = -\left[\mu_H + \phi + m(\alpha + \mu_M)\right]n_m + \mu_M(m+1)n_{m+1} + \phi n_{m-1}. \tag{9}$$

Note the similarity between these equations and Eqn. (1) for the case $h(c) = \delta_{c,1}$.

Table 1. Definition of model parameters used throughout this chapter

Symbol	Definition
m	Number of parasites in host
a	Host age
t	Time
c	Number of infecting parasites in a clump
h_c	Probability of c parasites in a clump
$p_m(a)$	Probability that a host reaches age a and has m parasites
$p(m\|a)$	Probability that a host has m parasites given that it has age a
$R(a;z)$	Probability generating function of $p_m(a)$
$Q(a;z)$	Probability generating function of $p(m\|a)$
P	Mean number of parasites per host
H	Mean number of hosts
β	Host birth rate
μ_H	Host death rate in absence of parasites
α	Additional host death rate per parasite
μ_M	Parasite death rate
ϕ	Host infection rate
λ	Asymptotic infection rate per parasite
H_0	Host infection efficiency
$n_m(t)$	Number of hosts with m parasites

Deterministic Models with Parasite Aggregation

Some of the earliest models of helminth infection dynamics were deterministic, but nevertheless allowed for the fact that the distribution of parasites between hosts was aggregated. Anderson and May[7] used this approach (see also ref. 8) to discuss processes that stabilise or destabilise host-helminth interactions. An equation for dH/dt is obtained by summing Eqns. (8,9) over m:

$$\frac{dH}{dt} = (\beta - \mu_H)H - \alpha P,$$

and an equation for P is obtained by multiplying Eqns. (8,9) by m and summing:

$$\frac{dP}{dt} = \frac{\lambda PH}{H_0 + H} - (\mu_H + \mu_P)P - \alpha\sum_{m=0}^{\infty} m^2 n_m(t).$$

The term $\Sigma_{m=0}^{\infty} m^2 n_m(t)$ represents the sum of square numbers of parasites in hosts, which depends on the details of the distribution of parasites among hosts. Anderson and May[7] related this to the mean number of parasites per host by assuming a negative binomial distribution with mean $\frac{P}{H}$ and aggregation parameter k. The generating function for this distribution is $\Sigma_{m=0}^{\infty} z^m n_m = H(t)[1+\frac{P}{Hk}(1-z)]^{-k}$ and the sum of square numbers of parasites is $\Sigma_{m=0}^{\infty} m^2 n_m = P + \frac{P^2(k+1)}{Hk}$. Since the second moment is approximated in terms of the first by assuming a particular distribution, this is actually a form of moment closure.

Under this approximation, the system is reduced to two ordinary differential equations which can be analysed with standard methods. As shown in reference 7, a positive value of k leads to a stable equilibrium between host and parasite, while a negative value of k (corresponding to a conventional binomial distribution, which is underdispersed) destabilises the equilibrium. They concluded that parasite aggregation can stabilise host-parasite interactions.

While this approach is not able to quantify the degree of aggregation which is expected to arise from a given stochastic model, it make it possible to explore the consequences of aggregation on nonlinear processes such as mating probabilities[2] and the evolution of drug resistance.[25] It has recently been used to discuss in some detail the importance of different density dependent processes on transmission,[26] control[27] and drug resistance[28] in helminth infections.

Moment Closure for the Variance

Rather than assuming a fixed degree of aggregation of parasites among hosts, Kretzschmar and Adler[24] instead assumed that the distribution of parasites will itself evolve due to the stochastic dynamics. They first wrote an equation for $Q = \Sigma_{m=0}^{\infty} m^2 n_m$ by multiplying (8,9) by m^2 and summing:

$$\frac{dQ}{dt} = -(2\mu_M + \mu_H)Q + (2\phi + \mu_M)P + \phi H - \alpha\sum_{m=0}^{\infty} m^3 n_m.$$

To approximate the final term, they again assumed a negative binomial distribution of worms between hosts with mean $\frac{P}{H}$ and shape parameter k so that $\Sigma_{m=0}^{\infty} m^3 n_m = \frac{(k+1)(k+2)P^3}{k^2H^2} + 3Q - 2P$. However, rather than assume a fixed aggregation parameter k they constrained it by stipulating that the first and second moments of n_m be P and Q. This means that $k = \frac{P^2}{H(Q-P)-P^2}$, so finally we have

$$\sum m^3 n_m = \frac{P(Q-P)}{H}\left(1 + \frac{H(Q-P)}{P^2}\right) + 3Q - 2P$$

and the equations for H, P and Q form a closed hierarchy (note that Kretzschmar and Adler[24] used a slightly different formalism in terms of the variance to mean ratio, but their presentation is equivalent to that shown here).

While the formalism is more complicated than that of Anderson and May[7] (three differential equations rather than two), it can still be analysed with standard methods. Kretzschmar and Adler[24] also studied the case where parasites were distributed according to a Neyman Type-A distribution and found parameter values where assuming one class of distribution predicted a stable host-parasite equilibrium and the other predicted instability. More specifically, they found that aggregated distributions would only produce a stable equilibrium if the variance to mean relationship increased with increasing mean worm burden. Their study underlines the sensitivity of the predictions of moment closure models to the type of closure that is assumed.

Rosà and Pugliese[11] extended Kretzschmar and Adler's model to incorporate clumped infection and free-living stages of the parasite, as well as host heterogeneity. They found the latter to be a much more strongly stabilising effect than clumped infection. Bottomley et al[29] used a related hybrid approximation (though assuming normal rather than negative binomial parasite distributions) to study the coexistence of parasite species within a constant population of hosts, deriving equations related to the Lotka-Volterra equations.

Fully Stochastic Models

If the host population is not so large as to be essentially infinite, then the abundance of infective stages is no longer statistically independent of the worm burden of any particular host and the 'hybrid model' approximation cannot be applied. Moreover, stochastic dynamics of the host population might also need to be taken into account. When this is the case, analytical approximations are typically unavailable and results are usually obtained by individual-based computer simulations. These are relatively straightforward to program: one takes account of each host's parasite burden, as well as other attributes such as host age and immune status as necessary. The free-living stages may be modelled as explicit populations, or as a 'force of infection' variable (which is appropriate when the numbers of infectious stages are extremely large). Parasite infection and death events can be simulated as stochastic processes, where the time to the next event is an exponentially distributed pseudorandom variable and the type of event is determined pseudorandomly by weighting each possibility by its rate.[9] Alternatively, a discrete time step can be assumed and the number of births, deaths, etc. per timestep can be generated as pseudorandom numbers with appropriate distributions (see e.g., ref. 30).

Explicit stochastic simulation models of this type are commonly used as tools to inform control strategies, such as the LYMFASIM program to model lymphatic filariasis[31] and SCHISTOSIM for schistosomiasis.[32] It is not possible to analyse such models in the same way as analytical models, so optimal control strategies can only be found by trial and error. However, this might not be a serious handicap given modern computing power. Simulation models are capable of handling intricate, biologically realistic processes where an analytical approach would be insurmountably difficult.

While inevitably more complicated, fully stochastic models can nevertheless sometimes be understood by comparing their behaviour to single-host or hybrid models. Since hybrid models assume that the host population size is infinite, the fully stochastic model will be similar to a hybrid model when the host population is large enough. It should be stressed that, in this context, 'population size' means the number of other host that can be infected by the offspring of the parasite population in one host; in this sense, an infinite population where each individual can only infect nearby neighbours has a very small effective population size. The hybrid approximation assumes that the abundance of infective stages is effectively a deterministic variable, but when the host population size (and therefore the adult population size) is finite the abundance of infective stages will fluctuate. These fluctuations will be larger larger when the host population is smaller. This idea that host population size 'tunes' the fluctuations in infection rate has been explored by Cornell et al.[33,34] There follows a discussion of some effects of finite host population sizes on helminth population dynamics.

Mating Probability

As discussed in the section 'Mating probability and population genetics', worm fecundity is depressed at low density if the worms reproduce sexually. As a result, there is an Allee effect[35]—i.e., the parasite population will tend to die out if the density of infectious stages is below a certain threshold. In an infinite host population, where the dynamics of the force of infection are deterministic, the parasite will always go extinct if it starts below this threshold. However, the force of infection can fluctuate stochastically above this threshold when the host population is finite. This means that epidemics may be more likely in small populations than in large ones if the initial force of infection is low.[33] If infection is clumped, however, the mating probability is increased and the Allee effect may be removed.[34]

Population Genetics and Antihelminthic Resistance

When the host population is small, the offspring of the parasites within one particular host are more likely to reinfect the same host than if host population is large. As a result, smaller host populations increase inbreeding in the parasite population, which increases homozygosity and therefore will promote antihelminthic resistance if this is a recessive trait.[36] If infection is clumped and clumps represent collections of larvae whose parents all live in the same host, then these hosts will be more genetically homogeneous than the parasite population as a whole.[37] As discussed in 'Mating probability and population genetics', this can also lead to inbreeding and thus significantly enhance the production of rare homozygous offspring.

Stability and Population Cycles

Rosà et al[12] used individual based simulations to extend their study of host-parasite interactions (as discussed in 'Moment closure for the variance') beyond the hybrid approximation used in reference 11. They found that, when host populations were of the order of 100 or more, the mean population sizes were similar to the predictions of the hybrid model. They also found that the processes which stabilised population dynamics in the deterministic approximation would also tend to stabilise stochastic populations. However, they found that the populations would often exhibit sustained cycles even when the deterministic approximation predicted damped cycles.

Conclusion

In this chapter I have focused on the mathematical techniques which can be used to analyse stochastic models of helminth population dynamics. In fact, stochastic population modelling is increasingly performed by computer simulation, a trend which will no doubt continue. The strength of simulation models is that they can be made arbitrarily flexible and complicated. For example, spatial variation in the force of infection can straightforwardly be incorporated into a simulation model. Even when an analytical approximation would be appropriate (e.g., a hybrid approximation if there are many hosts), it may give few advantages over a simulation model if one only needs numerical results for specific parameter values and indeed the mathematical method may be slower at producing results. Stochastic simulations are very useful for making predictions provided that the parameters are known. However, it can be very tricky to fit these parameters to data and the noise in simulation output can give the illusion of biological realism. As with any model, it is important that simulation models are validated against the real world before trusting their predictions.

Nevertheless, I believe it would be a mistake to dismiss mathematical analysis as impractical or unnecessary. Mathematical analysis can elucidate robust principles and generalities which are opaque from black-box simulation models. An example of the sort of insight which can be obtained from analysis is the condition for aggregating processes to stabilise host-parasite interactions (see sections 'Deterministic models with parasite aggregation' and 'Moment closure for the variance').

Immunity and host heterogeneity are two phenomena which are important determinants of patterns of infection, yet which are difficult to measure directly. In many cases, acquired immunity is thought to increase with the host's cumulative exposure to infection challenge.[3] However, there are many qualitatively similar mathematical relationships between the determinants of immunity and its effects, while in a model one has to make a specific choice. Very few of these possibilities have been explored and—while they can give good qualitative[3,16] and even quantitative[38] agreement with data—the 'correct' or 'best' formalisms are unknown. There is further ambiguity when formulating stochastic models of immunity—for example, in ref. 18 immunity is assumed to be integer valued and the size of this 'unit' determines the strength of stochasticity in immunity. Intrinsic host heterogeneity and demographic stochasticity are additional confounding factors[19] and bring additional parameters which need to be fitted to data. These uncertainties reduce the power of models to make quantitative predictions.

Analytical methods have another advantage over individual-based simulation models, beyond their ability to bring insight to a problem rather than just results. Helminth parasites can live in very large populations (tens of thousands of gastrointestinal nematodes can live in one sheep) and host populations can also be very large. A brute-force simulation, where every worm's birth and death is accounted for, could be beyond currently available computer power. However, mathematically derived approximations can provide tools to make simulations feasible for very large systems. For example, a diffusion approximation for the number of parasites can allow many birth and death events at once but still model the demographic noise correctly.[39] Similarly, mathematical expansions about the hybrid approximation can take account of a large but finite number of interacting hosts[33,40] and even model spatially explicit populations of hosts.[41] Tools of this sort are under active development and promise to help us apply stochastic models with greater biological realism to understand and manage parasite populations.

References

1. Anderson RM, May RM. Infectious Diseases of Humans. Oxford: Oxford University Press, 1991.
2. May RM. Togetherness among schistosomes: its effects on the dynamics of the infection. Math Biosci 1977; 35:301-343.
3. Woolhouse MEJ. A theoretical framework for the immunoepidemiology of helminth infection. Parasite Imunol 1992; 14:563-578.
4. Shaw DJ, Dobson AP. Patterns of macroparasite abundance and aggregation in wildlife populations: a quantitative review. Parasitology 1995; 111:S111-S133.
5. Marion G, Renshaw E, Gibson G. Stochastic modelling of environmental variation for biological populations. Theor Popul Biol 2000; 57:197-217.

6. Morgan ER, Medley GF, Torgerson PR et al. Parasite transmission in a migratory multiple host system. Ecol Modell 2007; 200:511-520.
7. Anderson RM, May RM. Regulation and stability of host-parasite population interactions: I. regulatory processes. J Anim Ecol 1978; 47:219-247.
8. May RM, Anderson RM. Regulation and stability of host-parasite population interactions: II. destabilizing processes. J Anim Ecol 1978; 47:249-267.
9. Anderson RM, Gordon DM. Processes influencing the distribution of parasite numbers within host populations special emphasis on parasite-induced mortalities. Parasitology 1982; 85:373-398.
10. Isham VS. Stochastic models of host-macroparasite interaction. Ann Appl Probab 1995; 5:197-210.
11. Rosà R, Pugliese A. Aggregation, stability and oscillations in different models for host—macroparasite interactions. Theor Popul Biol 2002; 61:319-334.
12. Rosà R, Pugliese A, Villani A et al. Individual-based vs. deterministic models for macroparasites: host cycles and extinction. Theor Popul Biol 2003; 63:295-307.
13. Tallis GM, Leyton M. Stochastic models of populations of helminthic parasites in the definitive host. Math Biosci 1969; 4:39-48.
14. Herbert J, Isham VS. Stochastic host-parasite interaction models. J Math Biol 2000; 40:343-371.
15. Barbour AD, Pugliese A. On the variance-to-mean ratio in models of parasite distributions. Adv Appl Probab 2000; 32:701-719.
16. Hudson PJ, Dobson AP. Macroparasites: observed patterns in naturally fluctuating animal populations. In: Grenfell BT, Dobson AP, eds. Ecology Of Infectious Diseases In Natural Populations. Cambridge: Cambridge University Press, 1995.
17. Chan MS, Isham VS. A stochastic model of schistosomiasis immuno-epidemiology. Math Biosci 1998; 151:179-198.
18. Grenfell BT, Wilson K, Isham VS et al. Modelling patterns of parasite aggregation in natural populations: trichostrongylid nematode-ruminant interactions as a case study. Parasitology 1995; 111:S135-S151.
19. Chan MS, Mutapi F, Woolhouse MEJ et al. Stochastic simulations and the detection of immunity to schistosome infections. Parasitology 2000; 120:161-169.
20. Cornell SJ, Isham VS, Grenfell BT. Drug resistant parasites and aggregated infection—early-season dynamics. J Math Biol 2000; 41:341-360.
21. Lello J, Fenton A, Stevenson IR et al. Competition and mutualism among the gut helminths of a mammalian host. Nature 2004; 428:840-844.
22. Bottomley C, Isham V, Basanez MG. Population biology of multispecies helminth infection: interspecific interactions and parasite distribution. Parasitology 2005; 131:417-433.
23. Nåsell I. Hybrid Models of Tropical Infections. Berlin: Springer-Verlag, 1985.
24. Kretzschmar M, Adler FR. Aggregated distributions in models for patchy populations. Theor Popul Biol 1993; 43:1-30.
25. Anderson RM, May RM, Gupta S. Nonlinear phenomena in host-parasite interactions. Parasitology 1989; 99:S59-S79.
26. Churcher TS, Ferguson NM, Basanez MG. Density dependence and overdispersion in the transmission of helminth parasites. Parasitology 2005; 131:121-132.
27. Churcher TS, Filipe JAN, Basanez MG. Density dependence and the control of helminth parasites. J Anim Ecol 2006; 75:1313-1320.
28. Churcher TS, Basanez MG. Density dependence and the spread of anthelmintic resistance. Evolution 2008; 62:528-537.
29. Bottomley C, Isham V, Basanez MG. Population biology of multispecies helminth infection: Competition and coexistence. J Theor Biol 2007; 244:81-95.
30. Saul A. Computer model of the maintenance and selection of genetic heterogeneity in polygamous helminths. Parasitology 1995; 111:531-536.
31. Plaisier AP, Subramanian S, Das PK et al. The lymfasim simulation program for modeling lymphatic filariasis and its control. Methods Inf Med 1998; 37:97-108.
32. Vlas SJ, Van Oortmarssen GJ, Gryseels B et al. Schistosim: a microsimulation model for the epidemiology and control of schistosomiasis. Parasitology 1996; 55:170-175.
33. Cornell SJ, Isham VS. Ultimate extinction of the promiscuous bisexual Galton-Watson metapopulation. Aust N Z J Stat 2004; 46:87-98.
34. Cornell SJ, Isham VS, Grenfell BT. Stochastic and spatial dynamics of nematode parasites in farmed ruminants. Proc R Soc Lond B Biol Sci 2004; 271:1243-1250.
35. Allee WC. Animal Aggregations: a study in general sociology. Chicago: University of Chicago Press, 1931.
36. Cornell SJ, Grenfell BT. Spatiotemporal infection dynamics and the evolution of drug resistance in nematode parasites of ruminants. In: Poulin R, Morand S, Skorping A, eds., Evolutionary Biology Of Host-Parasite Relationships: Theory Meets Reality. Elsevier, 2000.

37. Cornell SJ, Isham VS, Smith G et al. Spatial parasite transmission, drug resistance and the spread of rare genes. Proc Natl Acad Sci USA 2003; 100:7401-7405.
38. Cornell SJ, Bjornstad ON, Cattadori IM et al. Seasonality, cohort-dependence and the development of immunity in a natural host-nematode system. Proc R Soc Lond B Biol Sci 2008; 275:511-518.
39. Marion G, Renshaw E, Gibson G. Stochastic effects in nematode infections of ruminants. IMA J Math Appl Med Biol 1998; 15:97-116.
40. Ross JV. A stochastic metapopulation model accounting for habitat dynamics. J Math Biol 2006; 52:788-806.
41. Ovaskainen O, Cornell SJ. Space and stochasticity in population dynamics. Proc Natl Acad Sci USA 2006; 103:12781-12786.

Chapter 6

Modelling Environmentally-Mediated Infectious Diseases of Humans:
Transmission Dynamics of Schistosomiasis in China

Justin Remais*

Abstract

Macroparasites of humans are sensitive to a variety of environmental variables, including temperature, rainfall and hydrology, yet current comprehension of these relationships is limited. Given the incomplete mechanistic understanding of environment-disease interactions, mathematical models that describe them have seldom included the effects of time-varying environmental processes on transmission dynamics and where they have been included, simple generic, periodic functions are usually used. Few examples exist where seasonal forcing functions describe the actual processes underlying the environmental drivers of disease dynamics. Transmission of human schistosomes, which involves multiple environmental stages, offers a model for applying our understanding of the environmental determinants of the viability, longevity, infectivity and mobility of these stages to controlling disease in diverse environments. Here, a mathematical model of schistosomiasis transmission is presented which incorporates the effects of environmental variables on transmission. Model dynamics are explored and several key extensions to the model are proposed.

Introduction

A common feature of many of the most debilitating macroparasites of humans is their dependence on environmental life-stages subject to dynamic climactic, ecological, hydrological and other conditions. This phase can be wholly environmental, where for example infected humans excrete parasite eggs in feces and others are exposed via contaminated food or, as in the case of hookworm, where contact with contaminated soil can result in penetration of the parasite through the intact skin. Alternatively, the environmental phase may consist of time spent in an intermediate host, such as a snail or fish, itself subject to heterogeneous environments. Transmission of human schistosomes involves environmental phases of both types and thus understanding the environmental determinants of the viability, longevity, infectivity and mobility of these phases is key to conceptualizing disease transmission and ultimately controlling disease in diverse environments.

Schistosomes enter the environment as eggs that hatch in water into a free-swimming miracidium that seeks a snail of the appropriate species to infect. Asexual reproduction in the snail produces cercariae, another free-swimming aquatic stage with a lifespan on the order of a day, which penetrate the intact skin of a definitive host and mature into adult worms. The worms sexually pair and the female lays copious numbers of eggs that are the source of

*Justin Remais— Rollins School of Public Health, Department of Environmental and Occupational Health, Emory University. Atlanta, Georgia. USA. Email: justin.remais@emory.edu

Modelling Parasite Transmission and Control, edited by Edwin Michael and Robert C. Spear.

pathogenic response in the host. Some of these eggs find their way into the feces or urine, are excreted and the cycle begins again. The intermediate host, a freshwater snail and the two free-living aquatic stages are known to be subject to environmental stresses such as temperature[1] and shear forces present in the water column.[2] For *S. japonicum*, the species that causes schistosomiasis in east and southeast Asia, transmission is further complicated by the fact that a variety of mammals can serve as the definitive host,[3] including rodents, dogs, cats, pigs and water buffalo, the latter of which is particularly important to sustaining transmission in the lower Yangtze environment.[4]

In China, considerable progress has been made since the 1950s controlling transmission of *S. japonicum* in humans and domestic animals. From a total of 433 endemic counties in 1959, the disease has been eliminated from 260 counties leaving approximately 800,000 infected people and another 60 million at risk.[5,6] However, these represent only a small fraction of the worldwide total of schistosomiasis cases that the World Health Organization estimates at 200 million, 85% of which are in Africa.[7] Most of these infections are suffered by poor people, particularly children and most are preventable and treatable, although effective vaccines remain a hope for the future.

Where schistosomiasis transmission has been eliminated, targeted environmental modifications have often played an important role.[8] Conversely, major environmental changes such as water development projects have often led to a sustained elevating effect on schistosomiasis prevalence.[9] The underlying mechanisms shaping this relationship are poorly understood. An expansion in the preferred habitat of intermediate host snails is often implicated in these prevalence increases, yet few data exist to fortify this claim.[10] In China, recent evidence points to the influence of changing water levels on intermediate host populations.[11] Yet a clear mechanistic understanding of the processes that lead to increased disease is lacking and therefore opportunities to mitigate the disease impact of water projects using engineering or design principles is limited.[12]

Schistosomes are not alone among disease systems where mechanisms bridging environmental factors and epidemiological parameters have been poorly characterized. For example, although it has been well established that meningococcal meningitis in western Africa exhibits seasonal patterns, the particular causes remain uncertain and could range from low humidity to wind speed.[13] Similarly, multiple drivers have been proposed for the seasonal nature of cholera, including rainfall, temperature and planktonic blooms. Yet the specific roles of these drivers have not been resolved and well-established dynamic features, such as the second cholera peak experienced in endemic regions in south Asia, have gone largely unexplained.[14]

Given the limited mechanistic understanding of environment-disease interactions, mathematical models that describe them have seldom included the effects of time-varying environmental processes on transmission dynamics. Where they have been included, seasonality is commonly incorporated phenomenologically, using mathematical functions that are periodic in time and therefore describe in a generic way the seasonal variation in a parameter—a simple sinusoidal function is a common example. Few examples exist where seasonal forcing functions describe the actual processes underlying the environmental drivers of disease dynamics.[15] Because models that incorporate seasonality are sensitive to which parameters are externally forced as well as the shape of their forcing, there is a pressing need to identify the actual mechanisms at play. These mechanisms can include seasonal behaviors of definitive hosts, environmental forcing of vectors and intermediate hosts, sensitivities of parasite survival in the environment and annual variation in host births and deaths.

Understanding the mechanisms that tie environmental change to changes in disease dynamics is crucial for the development of comprehensive control strategies that may be more sustainable and cost effective in the long run. For the case of West Nile virus, for example, simulation studies have suggested that concentrating pesticide spraying efforts during the spring, when most transmission occurs among birds, could be more effective than the current practice of spraying in response to human cases in the late summer and early fall when mosquito numbers are already in decline.[16] Ultimately, models and management practices that incorporate the timing of key events such as intermediate host reproduction and parasite development are essential to developing more successful control strategies. Understanding these mechanisms is also vital for estimating the long-term impact of impending climate change at global and regional scales on environmentally mediated diseases, whereas current projections are, to a large extent, empirically-based. Indeed, it has been argued in the case of malaria, for example, that models which are mechanistic, based on plausible underlying drivers of the system and basic biology, rather than empirical relationships, are more useful for predicting and responding to, the influence of climate change.[17]

Table 1 summarizes the evolution from simple deterministic models, to complex spatially-explicit, individually-based models appropriate for studying re-emergent scenarios in the Sichuan environment, where connectivity and environmental heterogeneity structure the dynamics of transmission. Iterative evaluation of alternative models in the light of field data is the essence of the modelling process in the application presented here. Below I summarize a model of schistosomiasis transmission in western China which aims to incorporate mechanistic environment-parasite relationships, in the hopes of understanding the local determinants of transmission and its control in endemic settings. Spatial extensions to the model are discussed and an alternative, stochastic framework is proposed for application to re-emergent disease.

Modelling Schistosome Transmission

The use of mathematical models in the study of schistosomiasis dates back to the 1960s, when a four-parameter model was first proposed and used to explain the dynamics of endemic disease.[18,19] Since then, a number of models have been developed and used to explore the biological and epidemiological characteristics of schistosome species and their hosts, with a majority of them focused on *S. mansoni* and *S. haematobium*[20-28] and a few on *S. japonicum*.[29,30] This literature has three notable characteristics, it is explanatory rather than predictive, it is focused on phenomenological and, thereby, generalizable aspects of disease transmission and, for the most part, it has relied on analytical rather than computational methods of analysis. Koopman[31] has written of the successes and limitations of these models in general and where they fit into a more comprehensive mathematical approach. Thus far, these models have had a very limited impact on field studies and control programs.[27,32] One reason for this is the difficulty in adapting models to site-specific conditions, such as local climactic factors and intermediate host dynamics.

To date, we have used a model[33] of schistosomiasis transmission for our work in China with tactical rather than strategic objectives. Our focus is on site-specific transmission and the issue of selecting from the limited array of feasible control modalities that are effective and sustainable in a particular village. This is because Chinese experience, as well as our recent investigations, has clearly shown considerable variability in the prevalence and intensity of human infection in villages with similar agriculture but that are geographically proximate.[34] Hence, we regard the model as a platform for the synthesis of general knowledge of the mechanisms of disease transmission, quantitative estimates of biological parameter values and the local factors influencing transmission. To that end, the model has been extended to incorporate additional phenomena and additional data. Here, I build on the underlying model structure and parametrization described elsewhere,[33] incorporating the influence of additional environmental phenomena.

The Model

The structure of the delay-differential equation model is shown schematically in Figure 1. Three state variables are tracked in the model, worm burden in each risk group, the density of susceptible snails in each environment and the density of infected snails in that environment. Here risk group refers to occupational subgroups known in this region to exhibit pronounced differences in the timing, intensity and location of water contact and corresponding infection levels, including farmers, students and others, the latter including domestic workers, teachers,

Table 1. The evolution of modelling frameworks for Schistosoma japonicum *transmission in Sichuan, China*

	Deterministic		Stochastic		
	Single-Population Model	**Multi-Risk Group Model**	**Multi-Risk Group Model**	**Individual-Based Model**	**Spatially-Explicit Individual-Based Model**
Heterogeneous populations	No	Somewhat	Somewhat	Yes	Yes
Assumption-free worm aggregation process	No	No	No	Yes	Yes
Stochastic parasite introduction	No	No	Yes	Yes	Yes
Spatial heterogeneity[a]	No	No	No	No	Yes
Model complexity	Simple	Moderate	Moderate	Complex	Complex
Suitable for studying control in endemic areas	Fair	Good	Not appropriate	Not appropriate	Not appropriate
Suitable for studying emergence	Fair	Fair	Good	Good	Good

[a]While both deterministic and stochastic models have the potential to be spatially-explicit, stochastic models of individuals interacting in space provide additional fidelity.

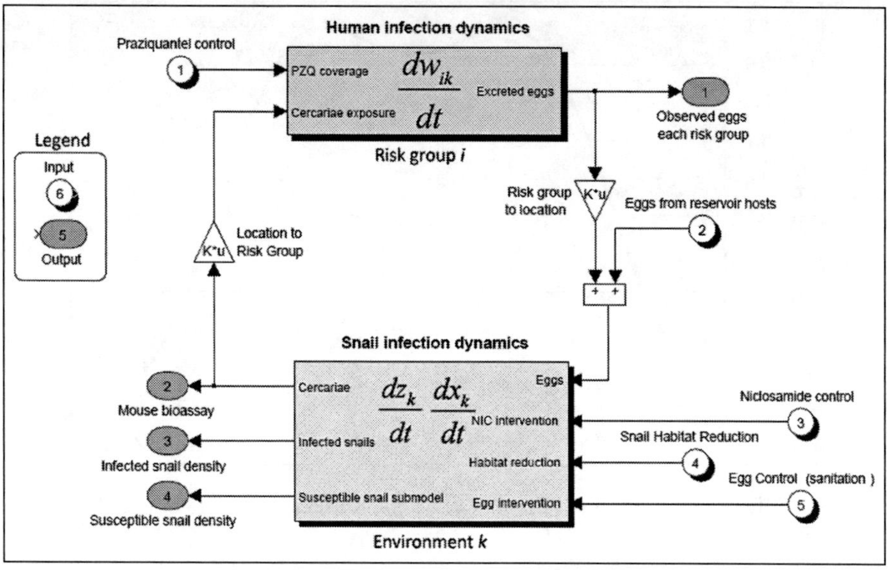

Figure 1. Diagram illustrating the structure of a multi-risk group model incorporating field data as both inputs and outputs. The host population is divided into *i* groups and the environment partitioned into *k* environments.

etc.[33] Environment refers to the land in which a risk group lives and farms. Hence, for each occupational group, *i*, living in environment *k*, the mean worm burden is given by the solution of the state equation:

$$\frac{dw_{ik}}{dt} = \alpha e^{-\mu_w \tau_w} S_i(t-\tau_w)\gamma_{ik} C_k(t-\tau_w) f(w_{ik}) - \mu_w w_{ik}(t) \tag{1.1}$$

where

$S_i(t)$ is the water exposure index of occupation group *i* and reflects the seasonal variation in water contact;

$e^{-\mu_w \tau_w}$ is the fraction of worms surviving the development time in humans;

α is the number of parasites acquired per cercaria per m^2 skin surface contact;

$f(w_{ik})$ is the density dependent worm establishment function which describes a process in which the likelihood of developing into an adult worm is assumed to be reduced when the worm burden is high due to a 'crowding effect', to concomitant immunity, or both;

μ_w is the worm mortality rate; and

$C_k(t-\tau_w)$ is the mean spatial density of cercariae in irrigation system *k* at time $t-\tau_w$. The time delay is due to the fact that the rate of change in the number of adult worms at time *t* is due to exposure to cercariae at time $t-\tau_w$ where τ_w is the worm development period in human hosts.

Modelling Cercariae-Environment Interactions

Cercariae are the free-living aquatic stage of the parasite which can infect humans and other mammals. They are negatively geotropic and positively phototropic, thus cercariae accumulate at the surface of water where they seek an appropriate mammalian host. They are highly susceptible to environmental stressors, including desiccation, turbulence in the water column, water temperature, aquatic chemistry and light.[2,35-39] Water temperature and flow are key determinants

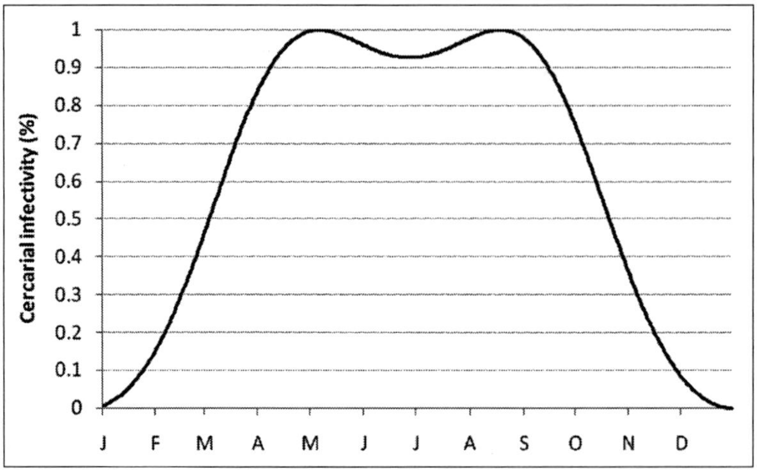

Figure 2. Daily cercarial infectivity for one year (2004) in Shian 5 as determined by water temperature following Upatham et al.[35]

of cercarial viability and thus $C_k(t)$ is dependent on the infected snail population as modified by these environmental factors:

$$C_k(t) = I_c(T_1)\frac{r_c(t)}{A_s}\sigma A_b z_k(t) \tag{1.2}$$

where

$I_c(T_1)$ is the temperature-dependent infectivity of cercariae, described below;
T_1 is the surface water temperature, measured directly using an automated logger, or estimated from air temperature using a model described below; and
A_s is the nominal surface water area of the village irrigation system;
$r_c(t)$ the precipitation-and/or irrigation-dependent modulation of the average daily cercarial production $[\sigma A_b z(t)]$ which enters the aquatic environment, defined briefly below and in detail elsewhere;[40]
σ is the cercarial production per infected snail per day;
A_b is the area of snail habitat;
$z_k(t)$ is the infected snail density.

Temperature-Dependent Cercarial Activity

Cercarial activity, including host-seeking, surface seeking, host penetration and survival are known to be temperature sensitive. Experiments that examine the influence of temperature on successful penetration and establishment in animal hosts reveal the combined effect of temperature on multiple activities.[35] Cercariae exposed to temperatures between 15 and 30 degrees C show the highest worm recovery rates from mouse hosts. Above and below this range, recovery rates decrease, resulting in the annual infectivity cycle depicted in Figure 2 using temperature data for the Shian 5 study village in 2004. This relationship is incorporated in the model as $I_c(T_1)$, the temperature-dependent infectivity of cercariae, serving as one source of seasonal limitation of transmission in the framework presented here.

Flow-Dependent Cercarial Activity

Cercarial production is modulated by the availability of water in channels, $r_c(t)$, which can be predicted from precipitation and temperature using a conceptual rainfall-runoff model, IHACRES,[41,42] described elsewhere[40] and modeled following the simple binary formulation:

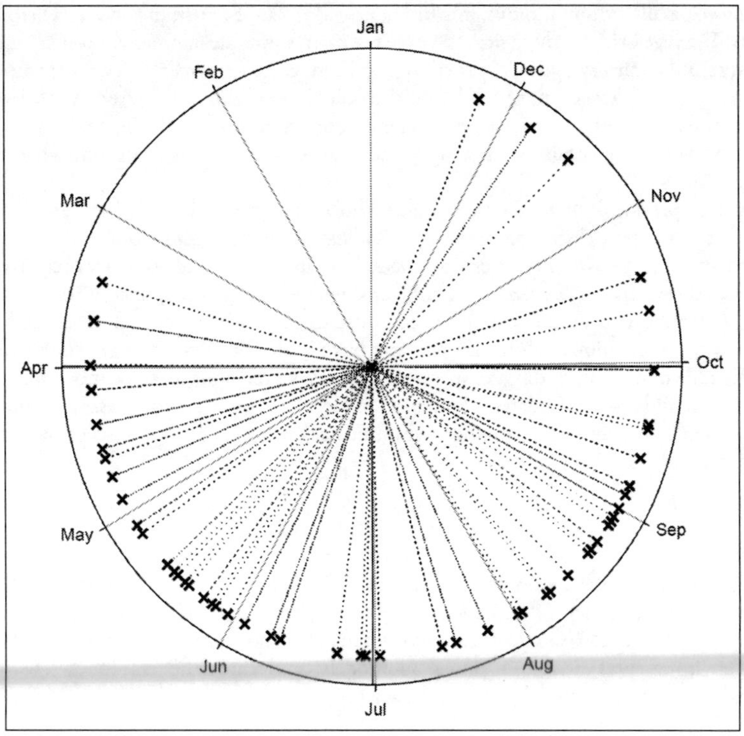

Figure 3. Polar plot of annual (2003) daily water flow classifications as predicted by the IHACRES hydrological model. Symbols (x) represent "transmission days," days with sufficient channel flow to allow for egg hatching and the coincidence of water contact and cercariae. Reprinted with permission from: Remais J, Liang S, Spear RC, Environ Sci Technol 2008; 42(7):2643-2649. © 2008 American Chemical Society.

$$r_c(t) = \begin{cases} 0 & q_t < \Omega_c \\ 1 & q_t \geq \Omega_c \end{cases} \tag{1.3}$$

where q_t is the normalized, IHACRES-predicted channel discharge at time step t; and Ω_c is the discharge threshold for cercarial release. If flow falls below the threshold for cercarial release, then $r_c(t) = 0$, effectively prohibiting cercarial penetration of hosts. When the flow threshold is met or exceeded, $r_c(t) = 1$ and transmission proceeds unimpeded. Thus, during and after rain events, when flowing water is available, cercarial dispersion and penetration can occur. This formulation is consistent with the ecology of *Oncomelanian* snails, which reside above the waterline but are submerged and shed cercariae when channel flows rise.[43] A sample classification of daily $r_c(t)$ in one study site for the year 2003 is given in Figure 3.

Modelling Snail-Environment Interactions

Models of schistosome intermediate hosts have typically explored a limited number of functional forms and environmental variables, such as Woolhouse and Chandiwana[44] and Woolhouse,[45] who selected simple nonlinear models relating water temperature and *B. globosus* recruitment and linear models relating mortality and water temperature. Woolhouse and Chandiwana[46] adapted their previous model[44] for flowing water environments, adding the effect of high rainfall. In contrast to the intermediate hosts of African schistosomes, modelling of the *Oncomelania hupensis* host of *S. japonicum* is rare.

O. hupensis snails are amphibious, inhabiting irrigation canals, riparian zones and littoral environments. The vegetation in these sites serves to maintain a suitable microenvironment, including temperature and humidity, as well as providing food and refuge resources. Juveniles are submerged during early stages of development, while adults are often found above the water line on vegetation and on shaded moist soil. Adults persist under environmental stress by closing their shell opening with a maneuverable operculum, allowing for aestivation and making them somewhat resistant to dry conditions.[47,48]

Liang et al[33] previously used a temperature-dependent recruitment model coupled with constant annual mortality to model seasonal abundance fluctuations of *O. hupensis*, but no direct measurements of recruitment, mortality or environmental variables were made to construct this model. Others have shown that *O. hupensis* is highly sensitive to seasonal weather conditions including flooding, temperature and humidity.[5,43] In response to these sensitivities, another study[49] used a mark-recapture technique to directly measure birth and mortality processes under changing environmental conditions, finding temperature and heavy precipitation to be most influential in determining abundance. A validated population model for *O. hupensis* was presented, suitable for predicting snail abundance in changing environments. In this model, the susceptible snail state equation is defined as:

$$\frac{dx_k}{dt} = g\left(t-t_x, \bar{E}_{t-t_x}\right)x_k(t-t_x) - h\left(t, \bar{E}_t\right)x_k(t) \tag{1.4}$$

where population gains in environment k are accomplished by the recruitment term $g\left(t-t_x, \bar{E}_{t-t_x}\right)$, lagged by a temperature-dependent development time, t_x, required to reach mature size (estimated for *O. hupensis* elsewhere[50]) and losses are accounted for by mortality term $h(t, \bar{E}_t)$, where $\bar{E}_t$ is a vector of environmental variables at time t. Submodels g and h are defined as follows, with model fitting and parameters described in detail elsewhere:[49]

$$g\left(t-t_x, \bar{E}_{t-t_x}\right) = \beta_1 e^{-0.5\left[\log\left(\frac{T(t-t_x)}{\beta_2}\right)\Big/\beta_3\right]^2} + \beta_4 e^{-0.5\left[\frac{R(t-t_x)-\beta_5}{\beta_6}\right]^2} \tag{1.5}$$

where $T(t-t_x)$ is air temperature and $R(t-t_x)$ is a count of rain events >15 mm per month at time $t-t_x$ and β_{1-6} are fit parameters; and

$$h\left(t, \bar{E}_t\right) = \delta_1 + 4\delta_2 \frac{e^{\left[-\frac{T(t)-\delta_3}{\delta_4}\right]}}{\left(1+e^{\left[-\frac{T(t)-\delta_3}{\delta_4}\right]}\right)^2} + \delta_5 R(t) \tag{1.6}$$

where $T(t)$ is air temperature, $R(t)$ is a count of rain events >15 mm per month and δ_{1-5} are fit parameters.

Fits of submodels g and h to environmental data are shown in Figure 4. Notice that the susceptible snail state equation is not dependent on other, endogenous transmission model state variables. Consequently, the susceptible snail model can be calibrated independent of the full transmission model, thus economizing the computation required for calibration, described further in the chapter by Spear and Hubbard in this volume.

State variable $z_k(t)$, the density of infected snails in environment k, is given by the solution to:

$$\frac{dz_k}{dt} = \rho e^{-\mu_z \tau_z} \xi(T_1) M_k(t-\tau_z) x_k(t-\tau_z) - \mu_z z_k \tag{1.7}$$

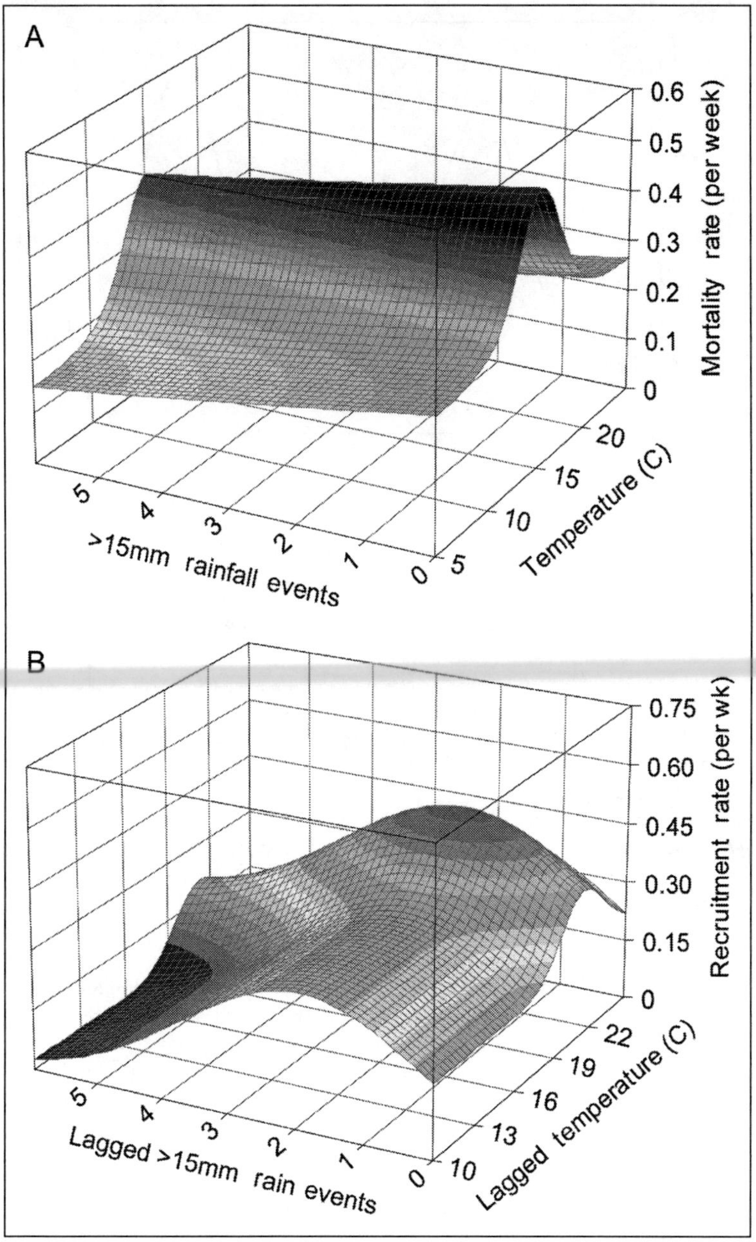

Figure 4. A) Relationship between instantaneous per capita recruitment rate for *O. hupensis robertsoni* and mean air temperature and mean number of rainfall events >15 mm (month^{-1}). Climate data are lagged by t_s as discussed elsewhere.[49] B) Relationship between instantaneous per capita mortality rate for *O. hupensis robertsoni* and mean air temperature and mean number of rainfall events >15 mm (month^{-1}). Reprinted with permission from: Remais J, Hubbard A, Wu Z, Spear R. J Appl Ecol 2007; 44(4):781-791. © 2007 Blackwell Publishing Ltd.

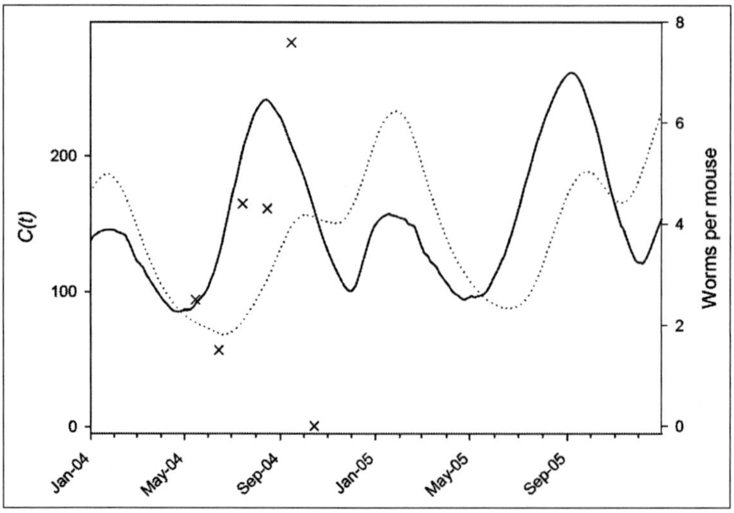

Figure 5. Mean of 1000 simulated predictions of daily cercarial output in a simulated village (Shian 5) generated using the temperature- and precipitation-driven snail model (–) of Remais et al[49] and the model (····) used by Liang et al[33] Symbols (x) are mean worms recovered from mice (n = 94) deployed in 2001 in Shian 5 as reported by Spear et al.[52]

where

$x_k(t)$ is the density of susceptible snails in environment k;
ρ is the fraction of those miracidia which successfully infect snails;
ξ is a parameter representing the degree of spatial convergence of the distribution of snail hosts and miracidia;
μ_z is the mortality rate of infected snails; and
$M_k(t)$ is the mean density of miracidia in the irrigation system in environment k, derived from hatched eggs, a process described below.

The implications of an environmentally-driven snail model are shown in Figure 5 using a transmission model previously calibrated for Shian 5, described in detail elsewhere[51] and in the chapter by Spear and Hubbard in this volume. The spring snail population peak generated by the model leads to significant infected snail numbers earlier in the year when compared to the model used by Liang.[33] As a consequence, the onset of peak cercarial release into waterways is shifted back by more than a month, from late September to mid-August, a prediction that is in line with available cercarial concentration data from field studies in Shian 5.[52] Comparisons to field data of this sort can highlight how environmentally-driven intermediate host models can bring model performance into better agreement with real world observations.

Modelling Ova-Environment and Miracidia-Environment Interactions

Total egg production from all risk groups is modeled as:

$$E_k(t) = \frac{1}{2} h \sum_i g_i n_i w_{i,k} \Phi(w_{i,k}, \kappa_{w_i}) \tag{1.8}$$

where

h is eggs per gram stool (EPG) per worm pair based on Hubbard et al;[53]
g_i is the average stool production of a member of the ith group;
n_i is the number of people in ith group whose stool is used as fertilizer;
$\Phi(W_{i,k}, \kappa_{w_i})$ is worm mating probability following May[54] and described elsewhere.[33]
The factor $\frac{1}{2}$ converts mean worm burden to worm pairs.

Before hatching into miracidia, excreted eggs are subject to environmental stress. They are resilient, however and can persist for days on fields before being washed into irrigation channels by a precipitation event.[40] A composite parameter representing on-field inactivation of eggs can be calculated from literature values of egg resilience and a simple first-order inactivation process can be used to express viable eggs, $E^{*}(t)$, as a function of the sum of decaying eggs contributed since the last flow event:[40]

$$E_k^{*}(t) = \sum_{T-\tau_\varepsilon}^{T} E_k(t) e^{-\varepsilon_d (T-t)} \tag{1.9}$$

where

$E_k^{*}(t)$ is the sum of viable eggs shed by infected humans in environment k since the last flow event;

$E(t)$ represents eggs excreted into environment, defined above;

ε_d is the decay constant governing inactivation of eggs lying dormant on fields between flow events;

$T - \tau_E$ is the time since last flow event.

Miracidia are short-lived, free-swimming and are drawn to light, accumulating near the surface of water where they seek an appropriate snail host. They are sensitive to water temperature and aquatic chemistry, with the former exerting a pronounced influence on viability.[1,55-57] Experimental data of the influence of temperature on miracidial infectivity have shown optimal activity between 15 and 30 degrees C,[58] a relationship incorporated into the model of the net effective density of miracidia in environment k, $M_k(t)$, a function of viable eggs in the environment, $E_k^{*}(t)$:

$$M_k(t) = I_m(T_1) \frac{r_e(t)}{A_s} \beta E_k^{*}(t) \tag{1.10}$$

where

$I_m(T_1)$ is the surface water temperature dependent miracidial infectivity to snails analogous to $I_c(T_1)$ discussed above for cercariae;

A_s is the nominal surface water area of the village irrigation system;

$r_e(t)$ is the precipitation-and/or irrigation-dependent modulation of the average daily miracidial production $[\beta E(t)]$ which enters the aquatic environment, defined briefly below and in detail elsewhere;[40]

β is the fraction of the total daily egg production of infected villagers returned into the environment as fertilizer, adjusted for the presence of sanitation.

Flow events provide opportunities for viable eggs to hatch and I therefore define $r_e(t)$ analogous to the cercarial equation at time t as:

$$r_e(t) = \begin{cases} 0 & q_t < \Omega_e \\ 1 & q_t \geq \Omega_e \end{cases} \tag{1.11}$$

where q_t is the normalized, IHACRES-predicted discharge at time step t, as above. Here, if water flow falls below the threshold for egg hatching, Ω_e, $r_e(t) = 0$ and eggs lie dormant. When the flow threshold is met or exceeded, $r_e(t) = 1$ and viable eggs on fields are washed into the irrigation system, where they hatch and can infect snails.

Model Parameters

The model was structured and parameterized to allow the use of as much of the field data as can be feasibly collected with the methods available in rural China. This includes environmental data (described below), cross-sectional data on snail population density, seasonally varying water contact patterns by group and survey data on the intensity of human infection. Some of these data are inputs to the model and some are used for parameter estimation. The issue of parameter

Table 2. Parameter ranges and environmental inputs for transmission model following Liang et al.[51] The distribution for all parameters is uniform except for log-uniform distribution for α, ρ and γ_s. Data estimates marked * are available in Table 4 in Liang et al.[51]

Parameters	Interpretation and Unit	Ranges	References
Biological			
τ_w	Development time of worms in human host (day)	20-40	59
μ_w	Worm natural mortality (/day)	0.000183-0.0014	59
h	Eggs excreted (/worm pair/gram feces)	0.768-2.72	53
μ_z	Patent and latent snail death rate (/day)	0.0063-0.033	60
σ	Cercarial production (/sporocyst/day)	20-50	61,62
h_{PZQ}	Efficacy of praziquantel	0.8-0.95	63,64
D_{D1}	Degree-days for sporocyst development	1550-1950	65
T_{D1}	Threshold temperature for sporocyst development (°C)	12-15	62
α	Schistosome acquired (/cercaria/m^2 contact)	0.0001-0.5	-
ρ	Intermediate host infection (/miracidium/m^2 surface water)	0.000001-0.0005	-
β_{1-6}	Intermediate host recruitment parameters	See ref	49
δ_{1-5}	Intermediate host mortality parameters	See ref	49
Site-specific			
w_{0i}	Initial worm burden in the *i*th group	Data estimate*	Local data
Z_0	Initial density of infected snails	Data estimate*	Local data
$S_i(t)$	Water contact index	Data estimate*	Local data
x_0	Initial density of susceptible snails	Data estimate*	Local data
κ_{0i}	Initial worm aggregation parameter	Data estimate*	Local data
x_0	Initial mean snail density	17-35	66
γ_i	Spatial index for the distribution and interaction between exposure and cercariae for ith group	Data estimate*	Local data
ξ	Spatial index for the distribution and interaction between snails and miracidia	1	-
Inputs			
T_1	Water temperature (°C)	No constraint	Local data
T_2	Snail microenvironment temperature (°C)	No constraint	Local data
C_{chemo}	Chemotherapy coverage	Data estimate*	Local data
P	Rainfall (mm/day)	No constraint	Local data
$r(t)$	Precipitation-driven channel flow (binary)	Data estimate	40

estimation is complex but central to our approach. Table 2 lists parameter values for the model and their literature sources. When used to study interventions (discussed in detail in the chapter by Seto and Carlton in this volume), a fundamental challenge is to reduce the residual uncertainty in model output, or its behavior more broadly, after as much of the local data as possible has been utilized to narrow the posterior distributions of the parameter values (see chapter by Spear and Hubbard in this volume).

To that end, we have conducted a variety of field studies to better understand the importance of certain elements of the model, or to obtain parameter estimates relevant to the biology of the snail or parasite specific to the region in which we work. Examples are the value of the parameter describing the production of parasite eggs per mated worm pair per gram of stool,[53] the importance of rainfall in determining infected snail densities and the concentration of cercariae in irrigation water.[52] There is no question that the modelling approach, with the ultimate objective of designing effective intervention strategies to meet public health objectives, has led us to seek quantitative estimates of factors controlling disease transmission that have not been of great interest to Asian parasitologists since the work of Pesigan.[67]

Environmental Data

Modelling the environmental drivers of seasonality requires an accurate dataset of environmental variables, acquired by measurement where possible and prediction where not. As in all environmental monitoring, strict quality assurance measures need be taken, including instrument calibration/certification, statistically valid sampling designs, reference sites and data verification.[68,69] In the work described here, air temperature, barometric pressure and relative humidity are collected relatively easily throughout the study region using continuous loggers (Hobo Onset H21-002) sampling at 12 minute intervals, validated with regional data available from the National Climatic Data Center.[70] Likewise, water temperature and water column height (stage) are logged at the same interval using similar equipment (Hobo Onset U20-001-01, U22-01) in a representative sample of irrigation channels. To estimate flow ($m^3\ s^{-1}$) from stage (m) in these channels, flow measurements must be made at multiple flow volumes in order to construct a simple rating curve. Daily precipitation is collected using a combination of tipping gauges (Hobo Onset RG3-M) and manually read rain gauges. Where data were missing due to equipment or staff error (typically accounting for $\ll 1$ percent of data points in the study), data were obtained from the NOAA weather station located at the Xichang municipal airport (World Meteorological Organization ID 56571), approximately 13 km from the study sites. Where water temperature was not directly measured, it was estimated from air temperature using a standard, simple linear model:[71]

$$T_w(t) = \alpha + \beta T_a(t) \tag{1.12}$$

where T_w = water temperature, T_a = air temperature and α and β are fit parameters. Time lags were excluded from the model as the observed lags (<4 hours) were much shorter than the averaging period (1 day), as is typical for temperature predictions in shallow channels.[72]

Model Dynamics

The model described above, when parameterized as described in the chapter by Spear and Hubbard, generates predictions of the sort depicted in Figure 6. Mean worm burdens for the three risk groups in Shian 5 are summarized for 1000 simulations over the five year period 2001-2005. The impact of two chemotherapies, modeled as described in the chapter by Seto and Carlton in this volume, is shown in the Figure. One characteristic that can be explored using this simulation environment is the time-to-return for worm burdens after chemotherapy. As is evident in the plot, worm burden returns to precontrol levels in the farmer group in less than 3 years, while the student and other group require considerably more time to rebound, owing to their differing exposure profiles.

The seasonal rise in worm burden following the second simulated chemotherapy can be seen in Figure 7 which plots the acquisition (or loss by mortality) of worms in the three risk groups. Note that the acquisition of new worms in the Figure represents exposures to cercariae that took place as many as 6 weeks prior. Notable is the bimodal farmer pattern which results from exposures during the spring planting season. While similar activities occur during the late winter and early spring during the harvest of the winter crop, cercarial shedding and infectivity is limited in this period due to

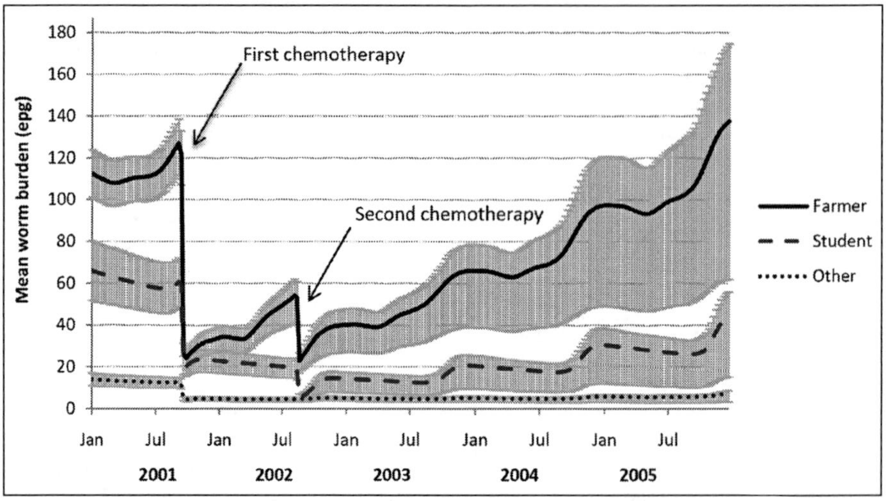

Figure 6. A 5-year prediction of 1000 simulated time profiles (line: mean; envelope: 25th and 75th percentiles) of mean worm burden for each of the three risk groups in Shian 5 following two chemotherapies (coverage based on field data).

low temperatures and limited precipitation. During the spring planting, however, temperatures just exceed the limits for cercarial activity and spring rains provide opportunities for the coincidence of cercariae and water contact. The timing and nature of interventions can be selected based on these seasonal patterns, as described in the chapter by Seto and Carleton in this volume.

All three risk groups experience a 'shut-down' of worm acquisition in the late fall related to the coincidence of temperature decreases, lower snail numbers and reduced water contact activities. The time-varying $I(T)$, $r(t)$ and $s(t)$ terms are at their minimum values in the late fall through winter and 'turn on' again in the spring, remaining 'on' through the transmission season. Of great interest is to explore the effects of these gating functions[73] on the transmission process and their sensitivity

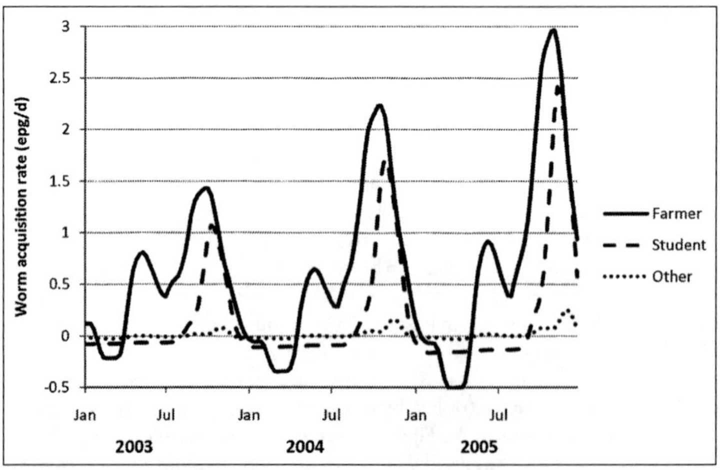

Figure 7. The derivative of worm burdens following the second chemotherapy in Figure 6, representing the rate of worm acquisition or loss (expressed as epg/d).

to local and regional environmental changes. Furthermore, efforts are needed to extend the model to account not just for environmental variables, but for their spatial distribution.

Modelling Spatial Connectivity

Human schistosomes are model organisms for the study of and response to, the spread of disease in space and time, as their transport through the environment takes place along discrete pathways. Parasites are carried in advective flows along canals and streams as both larvae and ova. Within intermediate snail hosts, parasites are conveyed among and between aquatic and riparian habitats and as adult worms, human and animal hosts serve as the transport mode. With respect to the *S. japonicum* parasite, I term these flows parasite diffusion, using the phrase to encompass all diffusive pathways along which parasites are transported into new and existing locales. The presence of suitable pathways can affect the probability of emergence of transmission, the level of worm burden within a community once transmission is established and how transmission spreads to neighboring areas. What is more, the degree to which an endemic or emergent community is connected can have important implications for the efficacy and sustainability of various control strategies.

In a preliminary exploration of parasite diffusion, the travel time of the free-swimming forms of the parasite or snail larvae due to advective transport in typical irrigation systems was estimated,[74] showing empirically that there is significant transport of viable parasite larvae within irrigation channels and that transport of larval stages occurs over considerable distance, with viable organisms detectable as far as 400 m from source snails.[38] Using these key transport parameters in a follow-up project, the impact of larval transport on endemic disease transmission was assessed using a spatial-temporal model of networked villages,[75] showing that diffusion of larvae via the surface water pathway, on its own, influences not just the intensity of transmission in a village, but also the effectiveness of standard interventions. Such a model allows us to better understand a number of phenomena specific to the endemic situation, such as which villages serve as "sinks" in the network, villages where worm burden can accumulate because they lie at the bottom of a watershed of numerous connected upstream villages.

The implications of a connected landscape have been explored extensively in ecology, where metapopulation models[76] describe the effect of migration between connected patches on population conservation. Likewise, environmental and social connections can promote the persistence of schistosomiasis and challenge efforts to control transmission. While hydrological connectivity is relatively straightforward to characterize, social connectivity is considerably more difficult to measure and express mathematically. The indirect transmission of schistosomiasis differs from recent epidemiological modelling of social connectivity and contact networks for communicable disease spread.[77-83] Within the context of the endemic transmission situation, small-scale human mobility can spread parasites from village to village. This effect may be small though in comparison to other social behaviors, such as the renting or selling of water buffalo which, if infected, can potentially release much larger numbers of eggs into the environment. While these factors might be modeled much like hydrological connectivity over the small scale via inter-village flows, they differ from the hydrological situation in that these processes can occur over much larger spatial scales and much less predictably. While difficult to estimate precisely, we look to future field data to inform the probabilities that define the movements of heterogeneous hosts; much theoretical and empirical work is needed in this area.

Extending the Modelling Framework

Our current research attention has shifted from endemic disease to disease re-emergence, a phenomenon we have documented in the mountainous region of Sichuan Province.[84] For studying re-emergence, human infection may be more appropriately modeled by risk groups with stochastic parasite establishment in a heterogeneous environment. The simplest form would be a stochastic compartmental model, where the risk group structure would be maintained with each compartment being comprised of a number of identical individuals. However, the

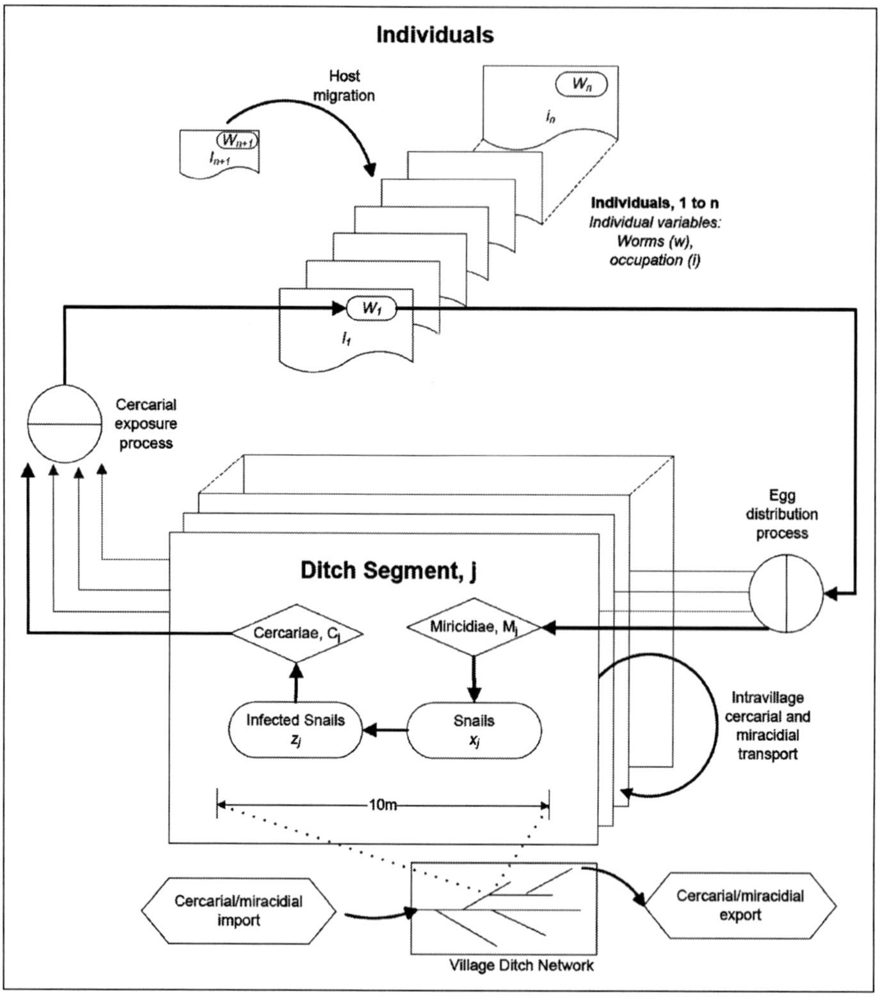

Figure 8. A diagrammatic conceptualization of the re-emergence model. The frame of reference of the flowchart is the interaction of one individual with the ditch environment. Stochastic introduction processes include host migration and import of cercariae and miracidiae, while cercarial exposure, egg distribution and hydrologic transport processes are implemented on a spatially-explicit, segmented mapping of the ditch environment.

static representation of the aggregation of parasites in humans in the deterministic model, even within our risk groups, does not translate easily to the re-emergent situation where, initially, the population is parasite-free. Hence, an individually-based model is preferred.[31] Individual-based micro-simulation models have been utilized in studying schistosomiasis transmission in endemic settings, but without our emphasis on environmental factors.[85] In ecology there has also been considerable interest in individually-based models.[86] There, the analog to our discrete population of humans is a population of animals in a heterogeneous, but continuous environment. There are particularly interesting approaches being explored which might allow us to naturally utilize our GIS data base and GPS-based maps in an individually-based model.[87]

Figure 8 schematically represents the elements of a new model in which the stochastic introduction of parasites is implemented by means of migration of infected hosts and advective parasite transport. Heterogeneous ditch environments, then, serve as platforms wherein eggs are released from infected individuals who, along with uninfected individuals, traverse the waterway environments and are potentially infected by contact with cercarial contaminated water. Stochastic implementations of the egg release and cercarial exposure processes are particularly suitable for a system where transmission is strongly conditioned by both environmental and behavioral factors that defy deterministic formulation. Moreover the stochastic model proposed here provides the structure to capture the potentially large influence of chance events that have been recognized to govern early epidemic dynamics, even in relatively large populations.[88-90]

Conclusion

The wide array of processes discussed herein can be conceptualized at various spatial and temporal scales, evaluated for their relative abilities to capture relevant transmission dynamics, including seasonal dynamics, and incorporate available field data. Measuring the potential drivers of seasonality may be relatively straightforward in the case of climate, but measuring and formalizing patterns of human behavior, particularly in a spatially explicit context, remain formidable challenges.[91] Indeed, iterative evaluation of alternative models in the light of field data is the essence of the modelling process in our application. Ultimately, quantifying and synthesizing the interaction between environmental and social determinants in transmission models offers great promise for developing novel modes of control in diverse environments.

References

1. Anderson RM, Mercer JG, Wilson RA et al. Transmission of Schistosoma-mansoni from man to snail—experimental studies of miracidial survival and infectivity in relation to larval age, water temperature, host size and host age. Parasitology 1982; 85:339-360.
2. Upatham E. Effect of a waterfall on the infectivity of St. Lucian Schistosoma mansoni cercariae. Trans R Soc Trop Med Hyg 1973; 67(6):884-885.
3. Riley S, Carabin H, Ne et al. Multi-Host transmission dynamics of Schistosoma japonicum in Samar Province, the Philippines. PLoS Medicine 2008; 5(1):e18.
4. Davis GM, Wu WP, Chen HG et al. A baseline study of importance of bovines for human Schistosoma japonicum infections around Poyang Lake, China: Villages studied and snail sampling strategy. Am J Trop Med Hyg 2002; 66(4):359-371.
5. Zhou XN, Wang TP, Wang LY et al. The current status of schistosomiasis epidemics in China. Zhonghua Liu Xing Bing Xue Za Zhi 2004; 25(7):555-558.
6. Wang RB, Wang TP, Wang LY et al. Study on the re-emerging situation of schistosomiasis epidemics in areas already under control and interruption. Zhonghua Liu Xing Bing Xue Za Zhi 2004; 25(7):564-567.
7. W.H.O. Prevention and control of schistosomiasis and soil-transmitted helminthiasis: Report of a W.H.O expert commitee, 2001.
8. Sleigh A, Li X, Jackson S et al. Eradication of schistosomiasis in Guangxi, China. Part 1: Setting, strategies, operations and outcomes, 1953-92. Bull World Health Organ 1998; 76(4):361-372.
9. W.H.O. Parasitic diseases in water resources development: The need for intersectoral negotiation. Geneva: World Health Organization, 1993.
10. Steinmann P, Keiser J, Bos R et al. Schistosomiasis and water resources development: systematic review, meta-analysis and estimates of people at risk. The Lancet Infectious Diseases 2006; 6(7):411-425.
11. Seto EYW, Wu W, Liu H-Y et al. Impact of changing water levels and weather on oncomelania hupensis hupensis populations, the snail host of Schistosoma japonicum, downstream of the three Gorges dam. EcoHealth 2008; in press.
12. Jobin W. Dams and Disease: Ecological Design and Health Impacts of Large Dams, Canals and Irrigation Systems. London: Routledge, 1999.
13. Pascual M, Dobson A. Seasonal patterns of infectious diseases. PLoS Med 2005; 2(1):e5.
14. Pascual M, Bouma MJ, Dobson AP. Cholera and climate: Revisiting the quantitative evidence. Microbes Infect 2002; 4(2):237-245.
15. Kendall BE, Briggs CJ, Murdoch WW et al. Why do populations cycle? A synthesis of statistical and mechanistic modeling approaches. Ecology 1999; 80:1789-1805.

16. Altizer S, Dobson A, Hosseini P. Seasonality and the dynamics of infectious diseases. Ecological Letters 2006; 9(4):467-484.
17. Thomas CJ, Hay SI. Global climate change and malaria—Authors' reply. The Lancet Infectious Diseases 2005; 5(5):259-260.
18. Macdonald G. The dynamics of helminth infections, with special reference to schistosomiasis. Transactions of the Royal Society of Tropical Medicine and Hygiene 1965; 59(5):489-506.
19. Hairston NG. An analysis of age-prevalence data by catalytic models. A contribution to the study of bilharziasis. Bull World Health Organ 1965; 33(2):163-175.
20. Anderson RM, May RM. Helminth infections of humans—mathematical-models, population-dynamics and control. Advances in Parasitology 1985; 24:1-101.
21. Barbour AD. Modeling the transmission of schistosomiasis: An introductory view. American Journal of Tropical Medicine and Hygiene 1996; 55(5):135-143.
22. Chan MS, Guyatt HL, Bundy DAP et al. The Development of an age-structured model for schistosomiasis transmission dynamics and control and its validation for Schistosoma-mansoni. Epidemiology and Infection 1995; 115(2):325-344.
23. Chan MS, Bundy DAP. Modelling the dynamic effects of community chemotherapy on patterns of morbidity due to Schistosoma mansoni. Transactions of the Royal Society of Tropical Medicine and Hygiene 1997; 91(2):216-220.
24. Feng Z, Li CC, Milner FA. Schistosomiasis models with density dependence and age of infection in snail dynamics. Math Biosci 2002; 177-178:271-286.
25. Gryseels B. Uncertainties in the epidemiology and control of schistosomiasis. American Journal of Tropical Medicine and Hygiene 1996; 55(5 Suppl):103-108.
26. Woolhouse ME, Hasibeder G, Chandiwana SK. On estimating the basic reproduction number for Schistosoma haematobium. Tropical Medicine and International Health 1996; 1(4):456-463.
27. Woolhouse MEJ. On the application of mathematical models of schistosome transmission dynamics .1. natural transmission. Acta Tropica 1991; 49(4):241-270.
28. Woolhouse MEJ. Mathematical models of transmission dynamics and control of schistosomiasis. American Journal of Tropical Medicine and Hygiene 1996; 55(5):144-148.
29. Yu JM, Yuan HC, J YQ et al. A transmission model for schistosomiasis japonica in lake Marchlands region. Chinese Journal of Public Health 1998; 17(6):347-350.
30. Williams GM, Sleigh AC, Li Y et al. Mathematical modelling of schistosomiasis japonica: Comparison of control strategies in the People's Republic of China. Acta Tropica 2002; 82(2):253-262.
31. Koopman JS, Jacquez G, Chick SE. New data and tools for integrating discrete and continuous population modeling strategies. Ann N Y Acad Sci 2001; 954:268-294.
32. Chan MS. The consequences of uncertainty for the prediction of the effects of schistosomiasis control programmes. Epidemiology and Infection 1996; 117(3):537-550.
33. Liang S, Maszle D, Spear RC. A quantitative framework for a multi-group model of schistosomiasis japonicum transmission dynamics and control in Sichuan, China. Acta Trop 2002; 82(2):263-277.
34. Spear RC, Seto E, Liang S et al. Factors influencing the transmission of Schistosoma japonicum in the mountains of Sichuan Province of China. Am J Trop Med Hyg 2004; 70(1):48-56.
35. Upatham ES, Kruatrachue M, Khunborivan V. Effects of physicochemical factors on the infection of mice with Schistosoma japonicum and S. mekongi cercariae. Southeast Asian J Trop Med Public Health 1984; 15(2):254-260.
36. Radke MG, Ritchie LS, Rowan WB. Effects of water velocities on worm burdens of animals exposed to Schistosoma mansoni cercariae released under laboratory and field conditions. Exp Parasitol 1961; 11:323-331.
37. Webbe G. The effect of water velocities on the infection of animals exposed to Schistosoma mansoni cercariae. Ann Trop Med Parasitol 1966; 60(1):78-84.
38. Lowe D, Xi J, Meng X et al. Transport of Schistosoma japonicum cercariae and the feasibility of niclosamide for cercariae control. Parasitol Int 2005; 54(1):83-89.
39. Jewsbury J. Effects of water velocity on snails and cercariae. Parasitology Today 1985; 1(4):116-117.
40. Remais J, Liang S, Spear RC. Coupling hydrologic and infectious disease models to explain regional differences in schistosomiasis transmission in Southwestern China. Environ Sci Technol 2008; 42(7):2643-2649.
41. Jakeman AJ, Hornberger GM. How much complexity is warranted in a rainfall-runoff model. Water Resources Research 1993; 29(8):2637-2649.
42. Jakeman AJ, Littlewood IG, Whitehead PG. Computation of the Instantaneous unit-hydrograph and identifiable component flows with application to 2 small upland catchments. Journal of Hydrology 1990; 117(1-4):275-300.
43. Pesigan TP, Farooq M, Hairston NG. Studies on Schistosoma japonicum infection in the Philippines. 2. The Molluscan Host. Bulletin of World Health Organization 1958; 18:481-578.

44. Woolhouse MEJ, Chandiwana SK. Population biology of the freshwater snail bulinus-globosus in the Zimbabwe highveld. Journal of Applied Ecology 1990; 27(1):41-59.
45. Woolhouse MEJ. Population biology of the freshwater snail biomphalaria pfeifferi in the Zimbabwe highveld. Journal of Applied Ecology 1992; 29(3):687-694.
46. Woolhouse ME, Chandiwana SK. Population dynamics model for bulinus globosus, intermediate host for Schistosoma haematobium, in river habitats. Acta Trop 1990; 47(3):151-160.
47. Rozendaal JA. Vector control. Methods for use by individuals and communities. Geneva: World Health Organization, 1997.
48. Davis GM, Wilke T, Zhang Y et al. Snail-Schistosoma, Paragonimus interactions in China: Population ecology, genetic diversity, coevolution and emerging diseases. Malacologia 1999; 41(2):355-377.
49. Remais J, Hubbard A, Wu Z et al. Weather-driven dynamics of an intermediate host: mechanistic and statistical population modelling of oncomelania hupensis. Journal of Applied Ecology 2007; 44(4):781-791.
50. Hong Q, Zhou X, Sun L et al. Imapct of global warming on transmission of schistosomiasis in China. IV. Accumulated temperature for development of generations of oncomelania hupensis in natural environment. Chinese Journal of Schistosomiasis Control 2003; 15(4):269-271.
51. Liang S, Spear RC, Seto E et al. A multi-group model of Schistosoma japonicum transmission dynamics and control: Model calibration and control prediction. Trop Med Int Health 2005; 10(3):263-278.
52. Spear RC, Zhong B, Mao Y et al. Spatial and temporal variability in schistosome cercarial density detected by mouse bioassays in village irrigation ditches in Sichuan, China. Am J Trop Med Hyg 2004; 71(5):554-557.
53. Hubbard A, Liang S, Maszle D et al. Estimating the distribution of worm burden and egg excretion of Schistosoma japonicum by risk group in Sichuan province, China. Parasitology 2002; 125:221-231.
54. May RM. Togetherness among schistosomes—effects on dynamics of infection. Mathematical Biosciences 1977; 35(3-4):301-343.
55. Chernin E. Some host-finding attributes of Schistosoma-mansoni Miracidia. American Journal of Tropical Medicine and Hygiene 1974; 23(3):320-327.
56. Upatham ES. The effect of water temperature on the penetration and development of St. Lucian Schistosoma mansoni miracidia in local biomphalaria glabrata. Southeast Asian J Trop Med Public Health 1973; 4(3):367-370.
57. Donnelly FA, Appleton CC, Schutte CH. The influence of salinity on the ova and miracidia of three species of Schistosoma. Int J Parasitol 1984; 14(2):113-120.
58. Shao BR, Xu X. Artificial infection of schistosome on oncomelania. Chinese Medical Journal 1956; 42:357-372.
59. Anderson RM, May RM. Infectious Diseases Of Humans: Dynamics And Control. Oxford; New York: Oxford University Press, 1991.
60. Mac Kenzie WR, Schell WL, Blair KA et al. Massive outbreak of waterborne cryptosporidium infection in Milwaukee, Wisconsin: Recurrence of illness and risk of secondary transmission. Clinical Infectious Diseases 1995; (21):57-62.
61. Qian BZ, Qian J, Xu DM et al. The population dynamics of cercariae of Schistosoma japonicum in oncomelania hupensis. Southeast Asian J Trop Med Public Health 1997; 28(2):296-302.
62. Pesigan TP, Hairston NG, Jauregui JJ et al. Studies on Schistosoma japonicum infection in the Philippines. 2. The molluscan host. Bulletin of World Health Organization 1958; (18):481-578.
63. Stelma FF. Immuno-epidemiology, morbidity and chemotherapy in a community recently exposed to Schistosoma mansoni infection. A study in northern Senegal. (PhD Thesis), Rijksuniversiteit te Leiden, 1997.
64. Liang YS, Coles GC, Doenhoff MJ. Short communication: Detection of praziquantel resistance in schistosomes. Tropical Medicine and International Health 2000; 5(1):72-72.
65. Sun LP, Zhou XN, Hong QB et al. The preliminary study on the growing degree day (GDD) of Schistosoma japonicum development in the intermediate snail host, oncomelania hupensis. Chinese Journal of Zoonoses 2001; 17(4):80-82.
66. Zhao WX, Gu XG, Xu FS et al. An ecological observation of oncomelania hupensis robertsoni in Xichang, Daliang Mountains, Sichuan. Sichuan Journal of Zoology 1995; 14(3):119-121.
67. Pesigan TP, Farooq M, Hairston NG et al. Studies on Schistosoma japonicum infection in the Philippines. 1. General considerations and epidemiology. Bull World Health Organ 1958; 18(3):345-455.
68. US EPA. Guidance on choosing a sampling design for environmental data collection. Washington, DC: Government Printing Office, 2002.
69. US EPA. Guidance on environmental data verification and data validation. Washington, DC: Government Printing Office, 2002.
70. NOAA. Global hourly surface data: National oceanic and atmospheric administration, national climatic data center, 2008.
71. Stefan HG, Preud'homme EB. Stream temperature estimation from air temperature. Water Resources Bulletin 1993; 29(1):27-45.

72. Erickson TR, Stefan HG. Linear air/water temperature correlations for streams during open water periods. Journal of Hydrologic Engineering 2000; 5(3):317-321.
73. Liang S, Seto E, Remais J et al. Environmental effects on transmission and control of parasitic diseases exemplified by schistosomiasis in Western China. Proc Natl Acad Sci USA 2007; 104(17):7110-7115.
74. Maszle DR, Whitehead PG, Johnson RC et al. Hydrological studies of schistosomiasis transport in Sichuan Province, China. Sci Total Environ 1998; 216(3):193-203.
75. Xu B, Gong P, Seto E et al. A spatial-temporal model for assessing the effects of intervillage connectivity in schistosomiasis transmission. Annals of the Association of American Geographers 2006; 96(1):31-46, in press.
76. Hanski I. Habitat connectivity, habitat continuity and metapopulations in dynamic landscapes. Oikos 1999; 87(2):209-219.
77. Halloran ME, Longini IM Jr. Using validation sets for outcomes and exposure to infection in vaccine field studies. Am J Epidemiol 2001; 154(5):391-398.
78. Koopman JS, Chick SE, Simon CP et al. Stochastic effects on endemic infection levels of disseminating versus local contacts. Math Biosci 2002; 180:49-71.
79. Newman ME. Spread of epidemic disease on networks. Phys Rev E Stat Nonlin Soft Matter Phys 2002; 66(1 Pt 2):016128.
80. Sander LM, Warren CP, Sokolov IM et al. Percolation on heterogeneous networks as a model for epidemics. Math Biosci 2002; 180:293-305.
81. Grais RF, Ellis JH, Kress A et al. Modeling the spread of annual influenza epidemics in the US.: the potential role of air travel. Health Care Manag Sci 2004; 7(2):127-134.
82. Rvachev LA, Longini IM. A mathematical-model for the global spread of influenza. Mathematical Biosciences 1985; 75(1):1-1.
83. Sattenspiel L, Herring DA. Simulating the effect of quarantine on the spread of the 1918-19 flu in Central Canada. Bull Math Biol 2003; 65(1):1-26.
84. Liang S, Yang C, Zhong B et al. Re-emerging schistosomiasis in hilly and mountainous areas of Sichuan, China. Bull World Health Organ 2006; 84(2):139-144.
85. Vlas SJ, Van Oortmarssen GJ, Gryseels B et al. Schistosim: A microsimulation model for the epidemiology and control of schistosomiasis. Am J Trop Med Hyg 1996; 55(5 Suppl):170-175.
86. Grimm V. Ten years of individual-based modelling in ecology: What have we learned and what could we learn in the future? Ecological Modelling 1999; 115(2-3):129-148.
87. Bian L. The representation of the environment in the context of individual-based modeling. Ecological Modelling 2003; 159(2-3):279-296.
88. Lloyd-Smith JO, Galvani AP, Getz WM. Curtailing transmission of severe acute respiratory syndrome within a community and its hospital. Proc R Soc Lond B Biol Sci 2003; 270(1528):1979-1989.
89. Hufnagel L, Brockmann D, Geisel T. Forecast and control of epidemics in a globalized world. Proc Natl Acad Sci USA 2004; 101(42):15124-15129.
90. Riley S, Fraser C, Donnelly CA et al. Transmission dynamics of the etiological agent of SARS in Hong Kong: Impact of public health interventions. Science 2003; 300(5627):1961-1966.
91. Grassly NC, Fraser C. Seasonal infectious disease epidemiology. Proc Biol Sci 2006; 273(1600):2541-2550.

Chapter 7

Parameter Estimation and Site-Specific Calibration of Disease Transmission Models

Robert C. Spear* and A. Hubbard

Abstract

The use of mathematical models for developing management options for controlling infectious diseases at a local scale requires that the structure and parameters of the model reflect the realities of transmission at that scale. Data available to inform local models are generally sparse and come from diverse sources and in diverse formats. These characteristics of the data and the complex structure of transmission models, result in many different parameter sets which mimic the local behavior of the system to within the resolution of field data, even for a model of fixed structure. A Bayesian approach is described, at both a practical and a theoretical level, which involves the assignment of prior parameter distributions and the definition of a semi-quantitative goodness of fit criteria which are essentially priors on the observable outputs. Monte Carlo simulations are used to generate samples from the posterior parameter space. This space is generally much more constrained than the prior space, but with a highly complex multivariate structure induced by the mathematical model. In applying the approach to a model of schistosomiasis transmission in a village in southwestern China, calibration of the model was found to be sensitive to the effective reproductive number, R_{eff}. This finding has implications both for computation time for the Monte Carlo analysis and for the specification of field data to efficiently calibrate the model for transmission control.

Introduction

The theme of this book, the use of mathematical transmission models for the management and control of parasitic disease, carries the implicit assumption that the parameters of the model, as well as its structure, are directly applicable to the environment in which transmission is to be managed. While models have been used to great effect to elucidate the social and biological phenomena central to the patterns of disease on large spatial and temporal scales,[1] this has rarely been true for developing control tactics in local environments where the devil is in the details. The challenge is to use local data to tailor the model to reflect local conditions to a degree adequate for informing control tactics at that scale. In this chapter we propose a particular approach that was developed in the context of environmental models, mainly in hydrology and recognizes that the relevant data for model calibration are diverse and come in many formats and with varying degrees of confidence in their accuracy and precision.[2]

In this volume it can be seen that mechanistically-based models are particularly well suited for the analysis of the transmission dynamics of infectious diseases. A major reason for their

*Corresponding Author: Robert C. Spear—Center for Occupational and Environmental Health, School of Public Health, University of California, Berkeley CA. Email: spear@berkeley.edu

Modelling Parasite Transmission and Control, edited by Edwin Michael and Robert C. Spear.

popularity is that the structure of the model reflects known processes and mechanisms that underlie the observable features of disease transmission. However, mathematical models often become complex as they expand to incorporate more mechanistic detail and the default is often to include more detail rather than less.[3] It is generally a major challenge to match the appropriate level of detail of the model to the purposes of a particular study.

The complexity of mechanistically-based models is in contrast to purely statistical models which are aimed at efficiently summarizing relationships between variables in a particular set of data by utilizing more general and mathematically simpler structures. Both statistical and mechanistic mathematical models include sets of parameters whose values must be specified in order to use the model for subsequent forecasts or analyses. Parameter estimation is central to the statistical modelling process in the sense of finding parameters that result in best fits to the specific data set about which the analysis revolves. In contrast, there is often prior information on the values of at least a subset of the parameters of mechanistically-based models independent from the particular application at hand. Further, some of the parameters of mechanistic models can often be estimated by independent experiments conducted before or during the modelling project. Hence, a mechanistically-based model can be thought of as a platform for the merging three types of information on the processes being modeled:

- The set of causal hypotheses that describe the current qualitative understanding of how the different processes and variables are inter-related (the structure of the model);
- Quantitative information that may be available from the literature or from independent laboratory or field experiments (prior information on the model parameters);
- The data specific to the particular problem under analysis (generally measurements dependent on a subset of the state variables of the system).[4]

The challenge is to merge these three types of information with the objective of identifying effective management options to diminish disease transmission. This goal immediately places some constraints on the spatial and temporal scale of the processes under analysis as well as the nature and format of data available to inform the model both qualitatively and quantitatively. For example, there is an important subset of disease transmission models that implicitly depend on, although seldom acknowledge, environmental factors like air temperature or water availability. Hence, an important determinant of spatial scale concerns the relationship between the human population subject to infection and the spatial homogeneity or lack thereof in these key environmental variables. In some cases the land of a single village may be the relevant environment and in others a set of villages may be more appropriate, for example, if they lie within a connected and relatively homogeneous patch of vector habitat. Temporal scales range from days and weeks in the case of disease outbreaks to months and years for tracking the prevalence or intensity of endemic disease.

Ideally the model structure and scale are well matched to the data set available for analysis. This may not be the case in that data collection is often dictated by political boundaries and logistical considerations rather than the realities of the transmission processes. The underlying issue is one of connectivity which is easily understood by noting that it is not effective management to suppress transmission in one village if its neighbors continue to harbor ample infected people and/or vectors and the pathways exist to reintroduce them into the controlled village at the first opportunity. In any case, the nature and quantity of the problem-specific data strongly condition the approach to parameter estimation/calibration. Therefore, we now sketch the characteristics of data we postulate to be typically available for use in the management of environmentally-mediated infectious diseases.

Local Data

The local data relevant to understanding the transmission dynamics of infectious diseases in human populations can be roughly divided into two categories which we shall term survey data and monitoring data. Monitoring data are defined as those data that are collected routinely at locations within the geographical area of interest and at regular time intervals. Environmental temperature and rainfall amounts are the most commonly available data of this sort. Survey data are those acquired over particular time periods of intensive investigation or, alternately, episodically as resources become available. In the developing world, prevalence or intensity of infection in the human population or the abundance of vectors or intermediate hosts are examples of survey data whereas, for reportable diseases in the developed world, such data may be classified as monitoring data if it is reliably summarized in weekly or monthly reports.

From a modelling perspective, the essential difference between monitoring and survey data is that the former are amenable to time series methods of summarization and analysis of which a great many are available.[5,6] However, when addressing the issues of management and control of disease transmission in the developing world, time series data are seldom available on the key variables and processes of interest beyond the weather. For the most part, what is known of these variables is available only via survey data. Hence, time series models may be useful submodels of larger mechanistic simulation models of the overall processes of interest, but the data to be "fit," i.e., the outcome data that define the likelihood, are generally survey data. Moreover, survey data are often from surveys of different sorts collected at different times and places within the region of interest making it is very difficult to generalize about the nature, structure, or amount of such data. Hence, approaches to parameter estimation that depend on specifically structured data, e.g., monitoring data, are of limited utility in disease management analyses typical of the developing world.

A second general characteristic of the data available to inform model parameterization and calibration relates to transferability, or lack thereof, from village to village or region to region. It is useful to distinguish between two data types in this context, site-specific data and what will be called biological data, that is, data specific to the disease, the infectious agent, its vectors, or intermediate hosts that can be assumed to apply to host and/or vector populations that reside or range over extended geographical areas. For example, the average number of eggs produced per day by an adult female schistosome within a human host should not vary locally, but might over longer distances or between genetically isolated parasite populations. The importance of the distinction between site-specific and biological parameters is that estimates of biological parameters obtained from one site can, in principal, be transferred to a second area within the same ecological zone whereas this is not often the case for site-specific parameters related to local agriculture or water availability.

Neither the biological nor the site-specific parameters of a well-developed disease transmission model will ever be known precisely. Even with the most mathematically tractable models and copious time series data, the best one can hope to achieve are a set of parameter estimates but also an estimate of the residual uncertainty in their values. The attractive possibility, however, is that as the model is applied to new situations the residual uncertainty in biological parameter values, as reflected in a statistical distribution that characterizes the current information about a parameter, will continually narrow. This possibility calls out for a Bayesian approach to the parameter estimation problem. For a number of years, we have utilized a particular approach that is well suited to the nature and amount of data available in disease transmission studies. As originally formulated the approach was admittedly ad hoc, but understandable and practical. It has been widely applied and extended, mainly in the field of hydrology.[7] More recently we have recognized that this approach can be subsumed under a more general framework termed Bayesian Melding by Poole and Raftery.[8] Here the procedure will be described in its practical context and later summarized more generally using their terminology.

In short, the procedure begins with the selection of a specific model structure and the specification of statistical distributions describing the uncertainty in each constant model parameter including, perhaps, those of time variable functions like the temporal pattern of exposure as driven by agricultural activity. The model structure and parameter distributions describe an ensemble of models which are then studied via Monte Carlo methods. The third element of the approach is the specification of a set of criteria to assess the performance of each model realization against the field data, that is, a measure of the goodness-of-fit. However, here goodness-of-fit is defined quite generally and can be based on any observable feature of the model's output, for example the timing between events, or by requiring that a state variable lies in some plausible range at certain times.

Because field data are often imprecise or approximate, it is sometimes unwise to overemphasize fits to point estimates. In earlier studies, this concern led to a binary criterion of fit, pass/not-pass, where passing implies that a particular simulation run was judged to mimic the essential features of the field data.[2,9] In this variant of the method, each parameter vector associated with passing or not-passing simulations is saved for subsequent statistical analysis. The Bayesian nature of the approach can be seen by recognizing that the passing parameter vectors are samples from a multivariate posterior distribution which is generally very much more constrained than the original parameter distributions, the priors that were used in the initial Monte Carlo runs.

The multivariate posterior distribution of passing parameters is the concrete expression of the fact that many different parameter combinations can produce model behavior consistent with field observations. This is not an artifact of the pass/fail classification, but it implies that, in general, these complex models are not invertible, that is, it is generally not possible to back-calculate a single optimum point in parameter space that produces the best fit, at least in any traditional sense. Beven has written extensively about this characteristic of complex environmental models, extending it to incorporate different model structures and termed it equifinality.[10] Particularly relevant to applications of complex models to disease control is the observation that "one implication of rejecting the concept of an optimal parameter set and accepting the concept of equifinality is that the uncertainty associated with the use of models in prediction might be wider than has hitherto been considered..."[7]

A Calibration Example

To exemplify the calibration process we propose, the model of *S. japonicum* transmission presented in the chapter by Remais is utilized in the context of exploring site-specific control options as discussed in the chapter by Seto and Carlton. Structurally, the model is comprised of a set of nonlinear differential-difference equations which track mean village worm burden in three groups of villagers; farmers, students and others who live in the village but who have limited water contact often due to non-agricultural employment outside the village.[11] The rationale for dividing the population into three groups relates to their differences in water contact, both in timing and degree. The fourth state variable is the mean village density of infected snails. The delay terms are the development time of the parasite in the human host and the prepatent interval in snails. As recounted in Chapter 12, the model is in the tradition that began with MacDonald,[12] but adapted to the epidemiological and environmental determinants of transmission in the hilly and mountainous areas of China. In this setting there is very limited evidence of acquired immunity as a function of age or infection history and very similar patterns in disease prevalence and intensity between adult men and women within a village.[13]

Including the degree-day model of snail development, the model has 10 biological parameters, all of which are constant and 15 site-specific parameters of which 9 are constant and 6 are time-varying. Examples of time varying parameters are the average water contact rate, $S_i(t)$, of each group of villagers over the season and the village average density of susceptible snails, $X(t)$. The latter is treated as a time variable parameter rather than a state variable because, even in highly endemic conditions, the prevalence of infection in snails does not exceed 2%. The density of susceptible snails, however, is generated by an independent submodel which is driven by temperature and rainfall data as described in Chapter 6.

The next step in the process is to specify plausible ranges for the biological parameters from the literature as described in a paper on the calibration of an earlier version of the model.[14] As discussed in that paper, there is no magic formula for assigning the prior parameter distributions. Some are reasonably well defined in the literature and others can only be estimated from informed guesses. However, there are technical reasons elaborated below that motivate the assumption that the individual parameter priors are uniformly distributed at least for the initial stages of the analysis that are addressed here. (Other distributional assumptions will be seen to require the full Bayesian melding procedure to properly estimate the posterior distribution as discussed below.) Similarly, survey data inform the initial ranges to be assigned to the site-specific parameters, for example, the area of snail habitat in a particular village.

Table 1. Parameter and initial condition limits

Parameter	Definition and Units	Limits
Biological		
τ_w	Development delay of worms in humans (days)	20-40
μ_w	Worm mortality rate (/day)	0.000183-0.0014
μ_z	Infected snail mortality rate (/day)	0.0063-0.033
h	Parasite egg excreted (/worm pair/gm feces)	0.768-2.72
σ	Cercarial production (/sporocyst/day)	0.1-5
h_{PZQ}	Efficacy of Praziquantel	0.8-0.95
D_{D1}	Degree-days for sporocyst development	1550-1950
T_{D1}	Threshold temp for sporocyst development °C	12-15
α	Schistosome acquisition (/cercaria/m² contact)	0.0001-0.5
ρ	Snail infection (/miracidium/m² surface water)	0.000001-0.0005
Site specific		
ω_{0i}	Initial worm burden in i-th group	90-140, 60-110, 30-60
z_0	Initial density of infected snails	0.401-0.773
κ_{0i}	Worm aggregation parameter in i-th group	0.4-1.1, 0.15-0.72, 0.5-0.8
A_h	Area of snail habitat (m²)	8004
A_s	Area of village surface water (m²)	2208
Time-varying		
$\tau_z(t)$	Parasite development in snail (days)	Submodel
$T_1(t)$	Air temperature (°C)	Monitoring data
$P(t)$	Rainfall (mm/day)	Monitoring data
$X(t)$	Uninfected snail density (snails/m²)	Submodel
$S_i(t)$	Daily water contact i-th group (m²/day)	Survey data
$r(t)$	Water availability (fraction of full)	Set to unity

In general, the ranges for the site-specific parameters are narrower than is the case for the biological parameters. However, the price paid is that collection of the necessary site-specific data is often labor intensive. Again, the ranges stipulated for both types of parameters are regarded as the limits of uniform distributions. Table 1 contains the ranges used in the main model and identifies the time variable parameters used in this example. More complete listing are given in Liang et al[15] and in Chapter 6.

The particular village for the calibration example is Chaunxing Xinlong 7 which lies in hilly terrain in Xichang County of southwestern Sichuan Province. The village is hydrologically isolated from its neighbors and can be treated as a single unit, at least from the perspective of determinants of its endemic level. The field data on which the criteria for passing simulations are defined are based on cross-sectional infection surveys in 2000 and 2002, as detailed in Liang et al[14] and supplemented by a more recent data set from 2003-2004.[15] These are survey data as defined above and are comprised of village-average infected snail densities, generally in the spring of the year, average egg excretion data among farmers and students in the winter and estimates of village average cercarial density from mouse bioassays once per month from June through August of 2001.[13]

Table 2 contains the ranges of the passing criteria. Local daily air temperature and rainfall data are used as inputs to each simulation for the uninfected snail population submodel, the temperature dependent infectivity parameters, $I_c(T)$, $I_m(T)$ and for the degree-day calculation on which the developmental delay of the parasite in the snail is based.

The simulations include environmental interventions in the 2003-04 period, notably focused snail control and the introduction of a number of household anerobic digesters that accept both human and pig wastes and subsequently provide biogas for cooking. These units also inactivate schistosome eggs with high efficiency, thereby decreasing the egg input into the environment, although the digested material is still used as fertilizer. Details are given in Liang et al.[15]

Because the timing of the collection of the various types of survey data is seldom coincident, the initial conditions of the state variables also need to be assigned and these, too, are typically generated by sampling from prior distributions the ranges of which are also included in Table 1. This can be tricky in models which contain developmental delays, such as the current example. A delay τ necessitates an initial function, rather than a single initial value, in the interval $(-\tau, 0)$. Further complications arise from a temperature-variable delay as in the present case. However, these are not complications introduced by this particular approach to parameter estimation, but are generic to any simulation application.

With the model structure given (Chapter 6), the prior parameter distributions and the distributions of the initial conditions specified (Table 1) and the pass/fail criterion defined (Table 2), Monte Carlo simulation trials can begin. In each simulation the parameters and initial conditions are randomly drawn from the uniform prior distributions, the state equations integrated over the specified interval and the calibration conditions checked at the appropriate times. The run is classed as a pass or a fail, the parameter vector stored for subsequent analysis and the process repeated.

Table 2. Passing criteria

Dates	Risk Group	Lower Limit	Upper Limit
Human infection		***epg***	
15-30 Nov 2000	Farmer	70	200
	Student	20	120
19-24 Oct 2002	Farmer	11	37
	Student	1	8
28 Nov-10 Dec 2004	Farmer	0.6	21
	Student	0	12
Infected snail density		***Snails/m²***	
20-30 Jun 2000		0.43	0.85
23 Jun-1 Jul 2001		0.12	0.38
15 June-5 July 2003		0.041	0.218
Mouse bioassay		***Worms/mouse***	
30-31 July 2001		0	2
30-31 Aug 2001		0	5
26-27 Sept 2001		0	1
30-31 Oct 2001		0	1
30 Jun-1 July 2003		0	1

As has been recounted elsewhere, the fraction of simulations which meet the passing criteria is generally very small, 1% would be excellent, but not to be expected in the initial runs.[4] The difficulty of achieving reasonable numbers of passes for subsequent analysis is generally regarded as the principal weakness of the approach. As a result, it is seldom feasible to use statistical analyses that rely on numbers of parameter vectors for passing simulations in excess of a few thousand, at least for complex models. In any case, there is a good reason to run the model on a fast computer with efficient simulation software. In the present case, after initial debugging, the model was run on a Dell Optiplex 755 (Intel Core 2 Duo E6850 3.00 GHz CPU) using Matlab® version R2007a. Over the 2000-2004 calibration period, 306 passing simulations were obtained in 90,100 trials which ran overnight.

It is possible that the structure of the model, or the prior parameter ranges, are inconsistent with simultaneously meeting all calibration criteria. The result is no passing simulations, or perhaps very few. Hence, it is wise to save from each simulation run a vector indicating whether each of the components of the pass criterion was met (1) or not met (0). For a passing run, all elements of this vector are unity and, for failing runs, at least one entry will be zero. In the current example, a failing simulation produced the criterion vector keyed in order to Table 2:

$$[1, 1, 0, 1, 1, 1, 0, 1, 1, 1, 1, 1, 1, 1]$$

where the two failing sub-criteria are the mean egg excretion from farmers in October 2002 and the village average infected snail density June 2000. Calculating the correlation between each of the elements of this criterion vector is often instructive. A large negative correlation may point to either prior distributions with incompatible ranges or a structural deficiency in the model. In either case, it is an issue to be addressed before going into production runs.

The Posterior Parameter Space

An underlying reason for the complexity of the posterior parameter space is that the calibration criteria are generally dependent on functions of the basic model parameters rather than dependent on their values individually. For example, if an important rate in the model is dependent on the product of two or more parameters, as in equation 1.1 of Chapter 6, passes can occur over the entire range of each if the others assume a value yielding a product in the correct range. Similarly, for linear differential equation models of the form $dx/dt = Ax$, where A is a constant matrix, the eigenvalues of the characteristic equation, which determine the dynamic behavior of the system, are complicated nonlinear functions of the elements of the A matrix. Hence, if the calibration criteria are sensitive to the eigenvalues of A, the passing space of the basic parameters is geometrically complex and almost certain to be a small fraction of the volume of the space defined by their independent prior distributions. For these complex models CPU time per simulation is generally sufficiently costly to motivate the search for ways to increase the pass rate.

Recent work in our group aimed at understanding the role of the site-specific parameters in reducing or eliminating transmission suggests a strategy for increasing the pass rate that may be useful for many infectious disease analyses where the basic reproductive number R_0 is of central interest.[15] The example of the eigenvalues of the A matrix discussed above motivates the speculation that the calibration criterion may depend on R_0. If so, even an approximate analytical expression of its dependence on model parameters might provide some insight into the complicated shape of the pass region and thereby suggest a means of increasing the pass rate. In the present case, a simplified model which leads to such an approximate R_0 is obtained by merging the three sub-populations in the village into one, assuming neither of the model's nonlinearities is operational and that the time variable delay is fixed at some median value. This leads to the simplified model:

$$\frac{dw}{dt} = a_{11}\alpha_{11}(t-\tau_w)z(t-\tau_w) - \mu_w w(t)$$

$$\frac{dz}{dt} = a_{21}\alpha_{21}(t-\tau_z)w(t-\tau_z) - \mu_z z(t)$$

where $w(t)$ is the mean worm burden in the human population, $z(t)$ the average infected snail density in the village environment, τ_w the developmental delay of the parasite in humans, τ_z the developmental delay of the parasite in snails and the $\alpha(t-\tau)$ parameters are the time-variable exposure or environmental parameters each normalized by dividing by their annual maximum values.

The simplified model then allows the approximation $R_0 = P_s P_b$, where P_s is a function of site-specific parameters and P_b of biological parameters. Woolhouse, in his very useful discussion of modelling and the control of schistosome transmission dynamics, used a similar approach and a two parameter description of R_0, but focusing on the human-to-snail and snail-to-human portions of the transmission cycle.[16] In the present case, the parameter aggregations are:

$$P_s = \frac{S\gamma A_b X \xi \beta n g_0}{\mu_z A_s^2} \quad \text{and} \quad P_b = \frac{\alpha\sigma\rho h e^{-\tau_w \mu_w}}{\mu_w}$$

There are four parameters included in P_s which are not listed in Table 1. γ and ξ relate to spatial inhomogeneites in the distribution of infected snails and of human water contact over the village irrigation system. As noted above, the parameters S and X are the annual maxima of the time variable water contact and uninfected snail density, respectively and arise from the normalization of those parameters in the simplified model. Because of this normalization and the fact that interventions affecting $X(t)$ and the parameter β were implemented in the 2003-2004 period, the value of R computed from these relations for any given simulation run can be expected to overestimate the postcontrol or effective reproductive number, R_{eff}. That is, it is likely that declining worm burdens post 2004 will occur for values of R based on the foregoing approximations above 1. This is essentially a scaling issue which does not alter the approach to achieving an increased pass rate. However, the value of 1 looses its traditional relevance. To underscore that fact, we redefine $R' = P_s P$ and below refer to R' rather than R_0 or R_{eff}.

Since P_s and P_b are known functions of their respective parameter sets, each can be computed for each simulation run together with the value of R'. The question is if this value, derived from the approximate model, is useful in discriminating between passes and fails in the full model. As noted earlier, an initial run over the full 2000-2004 calibration period yielded 306 passes from 90,100 simulations for a pass rate of 0.34%. Figure 1 shows the cumulative distribution functions of R', P_b and P_s separately for the 306 passes and 500 randomly selected fails. The passing and failing distributions differ for all three with all passing distributions shifted to the left. That is, parameters associated with simulations which meet the goodness-of-fit criteria combine to yield lower values of all three of these parameter aggregations, but R' and P_b are shifted to a much greater extent than P_s. Indeed, the shift in R' is largely due to the influence of P_b. None of the values of R' for the passing simulations exceeded a value of 4.0 where over 75% of the fails did so. That is, the calibration criterion are quite sensitive to R' in this particular case. It appears that for practical purposes $R' < 4.0$ is a necessary, but not sufficient, condition, for passing simulations. This suggests using the value of R' as a prefilter to increase the pass rate in subsequent runs. After each random draw of the parameters vector from the prior distribution, R' is calculated and the simulation conducted only if $R' < 4.0$. A full calibration run, from 2000-2004 utilizing the prefilter, led to 353 passes in 24,700 simulations which is about a 4-fold increase in the pass rate to 1.43%. However, there is obviously a great deal of complexity remaining to be explored in the remaining portion of the space where $R' < 4.0$ which accounts for the 24,200 failing simulations. Recall that the approximate model underlying the R' calculation does not take account of delay parameters τ_w and τ_z, nor the parasite aggregation parameter k among other potential candidates potentially important to meeting the calibration criteria.

Although we have focused on the importance of parameter combinations, it is sometimes instructive to look at the univariate marginal distributions of individual parameters under passes and fails. Figure 2 shows these distributions for four parameters of the P_b aggregation, α, σ, ρ and h. (Definitions in Table 1.) For passing runs, these distributions show that somewhat higher values occur for α and σ, markedly lower values for ρ including a cutoff at about 6×10^{-6} and that preferred values of h lie in the middle of the prior range. The most useful information in this

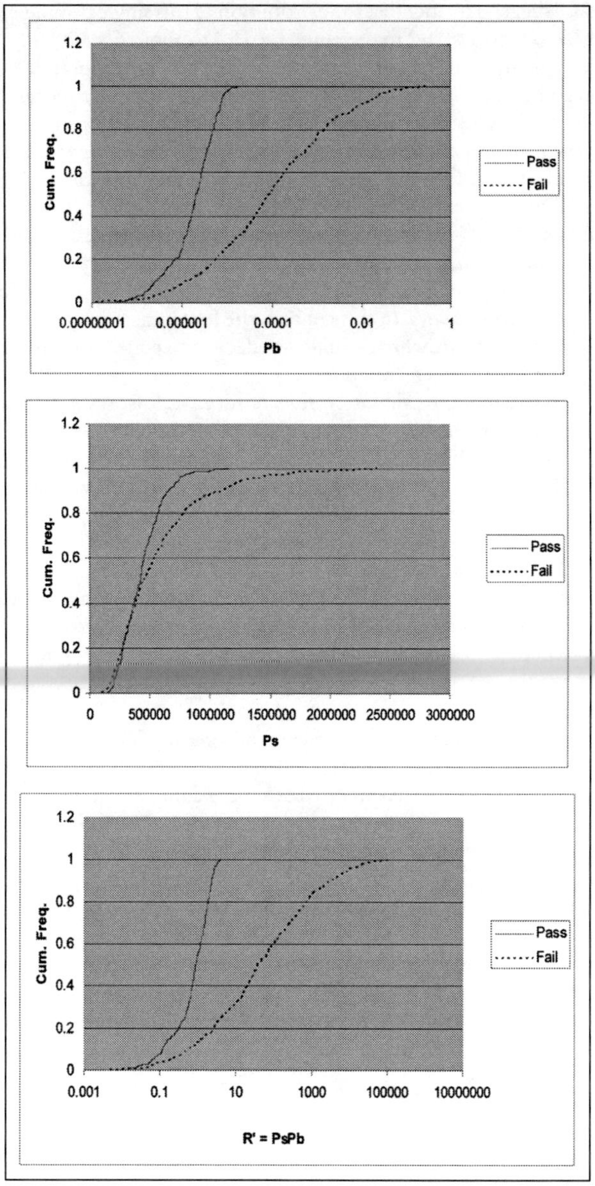

Figure 1. Cumulative distribution functions of the biological parameters, P_b, the site-specific parameters, P_s and their product, R', for 306 simulations which pass and 500 simulations which fail the calibration criteria.

example relates to ρ and the issue of parameter transportability. Recall that the original motivation for separating the biological parameters, P_b, from the site specific parameters, P_s, arose from the assumption that the biological parameters were transportable from one village to another in this region. With that in mind, the difference in the distribution of P_b under passes and fails, shown in Figure 1b, suggests that we extend the prefilter idea to constrain P_b to be less than 4×10^{-5} in future applications of the model in the Xichang region. The univariate marginal distribution of ρ, shown in Figure 2C, suggests further that the distribution of this single parameter be altered from its prior range of 1×10^{-6} to 5×10^{-4} to the range 1×10^{-6} to about 6×10^{-6}.

There are clearly many other possible and potentially useful explorations of the posterior space, but the foregoing illustrate the challenge and its complexity. More generally, however, the R_0 strategy, both for both increasing the proportion of passing simulations and understanding the characteristics of the posterior parameter space, even in an approximate way, is a strategy of that might be of general utility in informing disease transmission modelling for management purposes. In such applications one is generally focused on the goal of assuring that the zero equilibrium state is stable and that the range of attraction of the zero state is as large as possible in the face of considerable uncertainty in many parameter values. Hence, preliminary studies using models to determine the nature and extent of calibration data sensitive to R_0 should be useful in setting data collection priorities for field surveys. In this context, the foregoing example leads to the speculation that there might be field data which is not cost effective to collect with respect to estimating R_0 as a prelude to studying management options.

Assuming that, in the end, the calibration runs result in at least several hundred passing parameter vectors, these vectors can be used directly to explore the effects of intervention strategies as recounted in Liang et al.[14] In the present example, suppose that it was feasible to reduce the extent of the snail habitat by 20%, perhaps by concreting a portion of the village's irrigation ditches. The issue is then to determine if this intervention would lead to a high probability of decreasing transmission with time. Suppose this intervention was to be implemented right after the end of the calibration runs in 2004 and the forecast was to run until 2009. The baseline run from 2004-09 would use the passing vectors unaltered. The habitat reduction run would allow the comparison by using the same vectors with the value of A_b in each of the passing vectors decreased to 80% of its original value. Other control options could be studied in the same fashion.

Bayesian Melding

As noted earlier, the pass/fail procedure for incorporating field data into the mechanistic modelling framework is a special case of what is called Bayesian melding.[8] We will first summarize the approach and then indicate how it can provide a more complete description of the posterior parameter space where the prior distributions are not uniform. The following exposition below draws heavily on an earlier description of the approach.[17]

In Bayesian melding, two priors on the output are compared. One prior is based on literature or field data as to what is reasonable output. These have been referred to above as the calibration criteria and are summarized in Table 2 for the example used herein. The other prior on the output is induced by running the model on prior information on input parameters, e.g., Table 1 in the example. These two output priors are "melded" together and inverted to the input parameter space, thereby refining the estimate of the input parameters. In detail and using Poole and Raftery's terminology,[8] let,

$$M:\theta \rightarrow \phi, \quad \theta \in \Theta \subseteq \Re^{n}, \quad \phi \in \Phi \subseteq \Re^{p},$$

where M is the deterministic model that relates a n-vector of input parameters, θ, to a p-vector of outputs:

$$\phi = M(\theta)$$

From the current example, one component of θ might be, h, the number of eggs/worm pair/gm feces and a component of ϕ might be $w_{ik}(t)$, the worm burden in the ith village, kth risk group at time t. Define the posterior, joint model of inputs and outputs to be

$$\pi(\theta,\phi) \propto \begin{Bmatrix} p(\theta, M(\theta)) \text{ if } \phi = M(\theta) \\ 0 \text{ otherwise} \end{Bmatrix}$$

where $p(\theta, M(\theta))$ is the premodel joint distribution. One can think of $p(\theta, M(\theta))$ as containing the statistical information and relationships among the parameters and state variables before

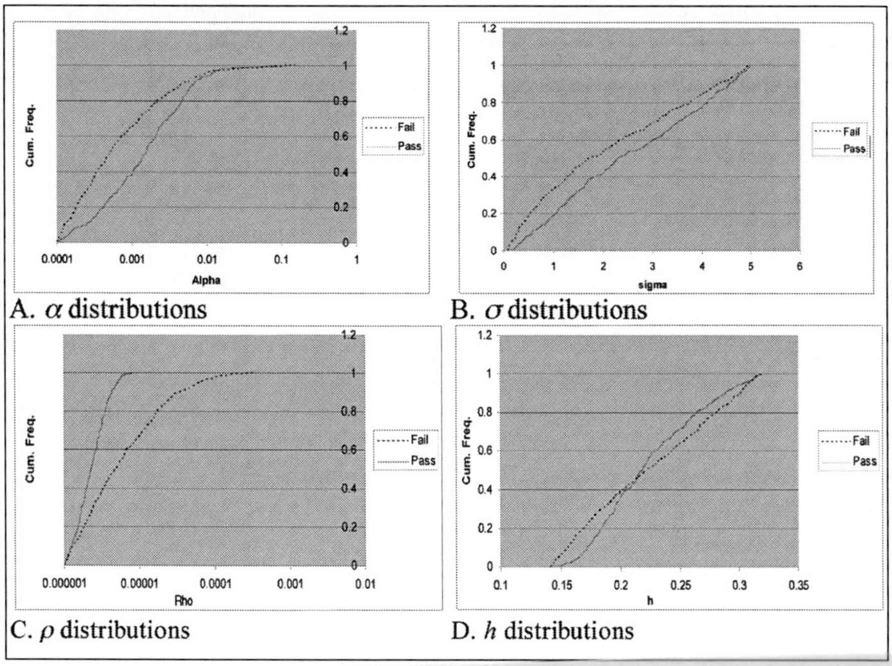

Figure 2. Univariate marginal distributions of α, σ, ρ and *h* for 306 passing and 500 failing simulations.

considering how these inputs determine the outputs, ϕ. The postmodel joint distribution, $\pi(\theta, \phi)$, however, only puts mass on input/output combinations consistent with the model, so it is a rescaled version of $p(\theta, M(\theta))$ with the mass of the impossible input/output combinations set to 0. In the schistosomiasis model, for instance, if the prior $p(\theta, M(\theta))$ allows positive probability on the combination $h = a$ and $w_{ik}(t) = b$, whereas the structure of the model suggest such a combination could never exist (although each values are acceptable in different combinations), then the posterior puts mass 0 on (a, b). The interesting distribution with respect to the estimation of input parameters is the marginal posterior of the inputs, or,

$$\pi(\theta) \propto p(\theta, M(\theta)).$$

As discussed above, we will ignore the possibility of having sufficient data to construct meaningful likelihood on the outputs. Thus, one can typically write the prior as,

$$p(\theta, \phi) \propto q_1(\theta) q_2(\phi)$$

where q_1 and q_2 are the prior distributions for θ and ϕ, respectively. If there is sufficient data on one of the outputs, so that a likelihood could be defined (say, the distribution of egg counts given the mean per village, group is negative binomial), say $L(\phi) \equiv p(D_\phi|\phi)$, where is the data on the outputs, now we have

$$p(\theta, \phi) \propto q_1(\theta) q_2(\phi) L(\phi).$$

As discussed by Poole and Raftery, adding the likelihood here removes the problem of Borel's paradox, where the posterior distribution on the outputs is no longer dependent on the arbitrary parameterization of the model, M. Though we have not defined a likelihood in our example, optimally and not surprisingly for many reasons, one would have sufficient data on the outputs in order to use something more akin to a typical Bayesian procedure.

Because $\phi = M(\theta)$, the model, M and the prior q_1, induce another, independent prior on ϕ, say $q_1^*(\phi)$. If one had a single prior on the outputs, there are existing Bayesian synthesis methods to define a posterior on inputs. Bayesian melding is a method for reconciling these two priors on the output. Specifically, Poole and Raftery suggest logarithmic pooling, or ,

$$\tilde{q}^{[\phi]}(\phi) \propto q_1^*(\phi)^{\alpha} q_2(\phi)^{1-\alpha}$$

resulting in a weighted (on the log-scale) average of the prior on the outputs induced by the prior on the inputs, $q_1^*(\phi)$ and the specified prior on the outputs, $q_2(\phi)$, resulting in a single prior, $\tilde{q}^{[\phi]}(\phi)$. The choice of α is discussed in their paper. This prior on the outputs is then inverted to get a "melded" prior on the inputs, or $\tilde{q}^{[\theta]}(\theta) = \tilde{q}^{[\phi]}(\phi)$, for $\theta = M^{-1}(\phi)$. This leads to a posterior on the inputs, $\pi^{[\theta]}(\theta) \propto \tilde{q}^{[\theta]}(\theta) L(\theta)$, which itself leads to standard Markov Chain Monte Carlo approaches to estimating posterior distribution on the input parameters, θ.

The distillation of the above technical discussion for our purposes, that is without a formal likelihood, is that Bayesian melding takes existing information on the input parameters in the form of a prior distribution, $q_1(\theta)$ and new information/expert opinion on the outputs, $q_2(\phi)$ and constructs a new, refined information on the parameter inputs, $\pi^{[\theta]}(\theta)$. Note that the pass/no pass procedure simply places a uniform prior on a subspace of Φ^S, $q_2(\phi) \propto I(\phi \in \Phi^S)$ where Φ^S is the acceptance region of Φ. This implies that $\tilde{q}^{[\phi]}(\phi) \propto q_1^*(\phi)$ if $q_2(\phi) > 0$ and $= 0$ otherwise. Thus, the choice of α does not affect the pass/no pass procedure and a random sample from the implied $\pi(\theta, \phi) \propto q_1(\theta)\tilde{q}^{[\phi]}(\phi)$ if $\phi = M(\theta)$, 0 otherwise, can be generated as follows: 1) generate a random θ from $q_1(\theta)$, 2) calculate $\phi = M(\theta)$, 3) keep (θ, ϕ) if $q_2(\phi) > 0$ (a pass).

As suggested, most procedures that attempt to estimate the marginal posterior, $\pi^{[\theta]}(\theta)$, do so, not by finding the distribution directly, but by methods which generate random samples from the underlying distribution of interest, one example being the pass/no pass method as discussed above. The result is repeated random draws from $\pi^{[\theta]}(\theta)$. So, defining the posterior distribution is an exercise in multivariate density estimation. Tree-based density estimation is a nonparametric multivariate density estimation technique particularly well-suited to estimation of high-dimensional data. In addition, newly developed nonlinear principal components methods permit the discovery of strong nonlinear relationships among the parameters that conventional principal components methods would not discover. Depending on the model and the prior information on the inputs and outputs, our experience suggest that strong nonlinear relationships (sometimes deterministic nonlinear relationships suggesting that an optimal set of parameters based on $\pi^{[\theta]}$ (θ) is not identifiable) among the posterior distribution of the input parameters is common in these types of disease models. Thus, examination of these posterior distributions can help to reparameterize the model in ways that avoid identifiability problems.

Conclusion

The fundamental fact underlying the approach to parameter estimation presented in this chapter is that there are almost certain to be many parameter combinations that are equally consistent with the local data available to calibrate the model. Moreover, when used in a forecasting mode, these combinations can lead to different qualitative performance of the model. For example, the effective reproductive rate for some may be greater than one and less than one for others. So, the search for effective control strategies seeks the feasible control option that will maximize the probability of declining transmission over time.

It is in the context of future forecasting that the full Bayesian melding concept becomes attractive. If likelihoods can be estimated for the prior distributions, then the posterior distribution can support more informed and hopefully more reliable, inference on the outcomes of the various intervention options. This point further illustrates that, if the objective is to design effective local control strategies, there is a very intimate relationship between laboratory-generated data, field data and the modelling process. It can be argued that low esteem in which modelling is held by many epidemiologists and infectious disease experts arises from the rarity of data-informed modelling studies of the sort we advocate. Conversely, modelers can argue that much of the available laboratory and field data is of marginal relevance to control programs. There is ample opportunity for progress at the intersection.

Acknowledgements

We are grateful to Ishaan Swarup who carried out the simulations discussed in this chapter. The research on which the chapter is based was supported by the US National Institute of Allergy and Infections Disease (R01-AI50612) the US National Science Foundation/NIH (EID 0622743).

References

1. Anderson RM, May RM. Infectious diseases of humans: Dynamics and Control: Oxford Science Publications, 1991.
2. Spear R, Hornberger G. Eutrophication in Peel Inlet: II. Identification of critical uncertainties via generalized sensitivity analysis. Water Research 1980; 14:43-49.
3. Beck M, Ravetz J, Mulkey L et al. On the problem of model validation for predictive exposure assessments. Stochas Hydrol Hydraulics 1997; 11:229-254.
4. Spear RC. Large simulation models: Calibration, uniqueness and goodness of fit. Environmental Modeling and Software 1998; 12:219-228.
5. Box G, Jenkins G, Reinsel G. Time Series Analysis: Forecasting And Control. 4th ed: John Wiley, 2008.
6. Chatfield C. The Analysis Of Time Series: An Introduction. 6th ed: CRC Press, 2004.
7. Beven K, Freer J. Equafinality, data assimilation and uncertainty estimation in mechanistic modeling of complex environmental systems using the GLUE methodology. J Hydrology 2001; 249:11-29.
8. Poole D, Raftery AE. Inference for deterministic simulation models: The Bayesian melding approach. J Am Stat Assn 2000; 95:1244-1255.
9. Grieb TM, Shang N, Spear RC et al. Examination of model uncertainty and parameter interaction in the global carbon cycling model. Environ Intl 1999; 25:787-803.
10. Beven K. Prophecy, reality and uncertainty in distributed hydrological modeling. Adv Water Resources 1993; 16:41-51.
11. Liang S, Maszle DM, Spear RC. A quantitative framework for a multi-group model of schistosomiasis japonicum transmission dynamics and control in Sichuan, China. Acta Tropica 2002; 82:263-277.
12. Macdonald G. The dynamics of helminth infections, with special reference to schistosomes. Trans R Soc Trop Med Hyg 1965; 59(5):489-506.
13. Spear R, Seto E, Liang S et al. Factors influencing the transmission of Schistosoma japonicum in the mountains of Sichuan province. Am J Trop Med Hyg 2004; 70(10):48-56.
14. Liang S, Spear RC, Seto E et al. A multi-group model of Schistosoma japonicum transmission dynamics and control: model calibration and control prediction. Trop Med Intl Health 2005; 10:263-278.
15. Liang S, Seto E, Remais J et al. Environmental effects on parasitic disease transmission exemplified by schistosomiasis in western China. PNAS 2007; 104:7110-7115.
16. Woolhouse MEJ. On the application of mathematical models of schistosome transmission dynamics. II. Control. Acta Tropica 1992; 50:189-204.
17. Spear RC, Hubbard A, Liang S et al. The use of disease transmission models for public health decision-making: Towards an approach for designing intervention strategies for schistosomiasis japonicum. Environ Hlth Perspectives 2002; 110:907-915.

Chapter 8

Modelling Malaria Population Structure and Its Implications for Control

Caroline O. Buckee* and Sunetra Gupta

Abstract

Mathematical models of malaria transmission have been used to inform the design of malaria control programs since the mid 20th century, and many of these models have provided useful insights into the complexity of the disease. Among developing countries, however and particularly in sub-Saharan Africa, malaria remains a major cause of morbidity and mortality. One of the main difficulties in controlling the most virulent human malaria parasite, *Plasmodium falciparum*, is its genetic diversity, which confounds attempts to design an effective vaccine. The population structure of *P. falciparum* remains poorly understood but plays a key role in determining epidemiological patterns of disease and the development of immunity. We discuss the seminal model of malaria transmission developed by Ross and MacDonald, and the modifications that have been made since to include more realism. We show that age profiles of disease and serological data support a theoretical model in which the parasite population is diverse and structured into several antigenic types and highlight the implications of this structure for controlling malaria. Lastly, we discuss the current sequence data on parasite antigen genes that are important for the aquisition of immunity, and the results of a new analysis of *P. falciparum* population structure at the genomic level.

Introduction

Malaria remains a major global public health problem, causing about 500 million cases and over a million deaths each year.[1] Although the disease has been eliminated in many parts of the developed world due to intense chloroquine use and mosquito control programs during the 1950s and 1960s, the logistical difficulties associated with the implementation of these programs in many developing countries, as well as the emergence of resistance to DDT in mosquitoes and to chloroquine in the parasite, led to a discontinuation of the WHO's Global Malaria Eradication Program in 1972. The breakdown of malaria control led to a resurgence of malaria cases worldwide and severe epidemics in areas such as Madagascar and Sri Lanka (reviewed recently in ref. 2). Renewed interest in controlling the disease in the past decade has led to significant funding for basic research and control efforts, however. The design of appropriate control programs relies on an understanding of the complex dynamical interactions between vector ecology, the lifecycle of malaria parasite and the pathology of the disease in humans, which remain poorly understood. The complexity of these interactions leads to nonlinear relationships between important epidemiological parameters

*Corresponding Author: Caroline O. Buckee—Department of Zoology, University of Oxford, South Parks Road, Oxford, U.K., OX1 3PS. Email: caroline.buckee@zoo.ox.ac.uk.

Modelling Parasite Transmission and Control, edited by Edwin Michael and Robert C. Spear.

and the use of mathematical models can provide an important tool for understanding the potential effects of control programs on malaria prevalence and disease.

Ronald Ross formulated the seminal model of malaria transmission between humans and mosquitoes in 1911, and it was subsequently modified by MacDonald in the 1950's. Based on a simple framework which tracks the proportions of human and mosquito populations that are susceptible and infected, these models emphasized the relative importance of different aspects of the transmission system in determining malaria prevalence. They were used to successfully inform the design of control programs in many countries, by providing support for the intense efforts to kill adult mosquitoes by the global eradication program, for example. The Ross-Macdonald model is deficient, however, in capturing certain essential features of malaria transmission, such as naturally acquired immunity. Numerous models have been developed to add "realism" to the original framework, but the fact remains that we still lack a thorough understanding of the relationship between infection and immunity in the host and its relation to the biology of the malaria parasite. One of the major reasons for this is the genetic and phenotypic complexity of the parasite itself. Figure 1 illustrates the lifecycle of the most virulent human malaria parasite, *Plasmodium falciparum*, highlighting its variety of morphological forms. Disease symptoms in humans are associated with the blood stage of the parasite, when rounds of asexual replication occur within red blood cells leading to recurring fever, anemia and occasionally coma and death. It remains unclear why some

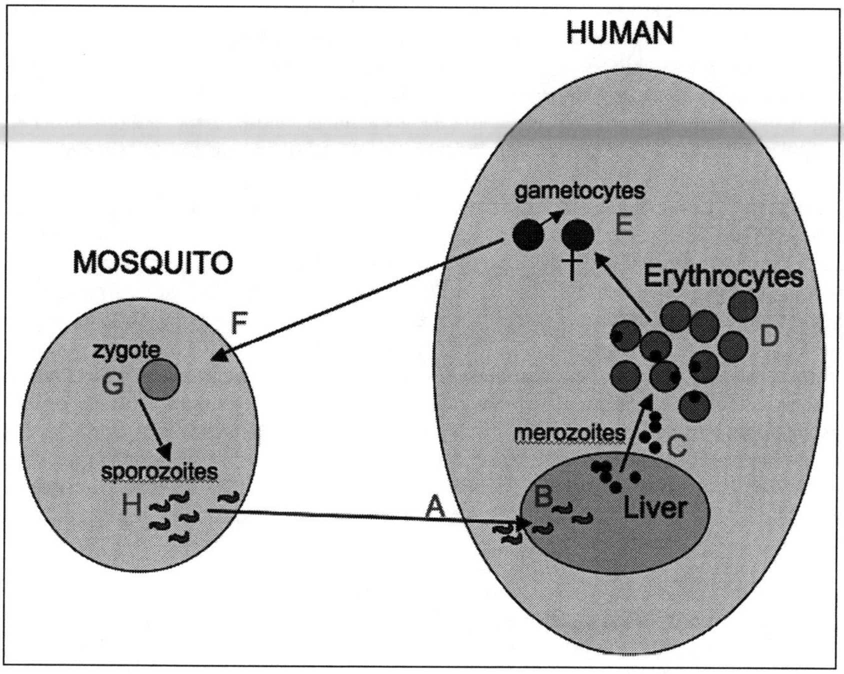

Figure 1. The lifecycle of *Plasmodium falciparum*. A) Sporozoites are injected into the human host when an adult female mosquito takes a blood meal. They travel rapidly to the liver (B), where they invade liver cells and undergo rounds of asexual reproduction and produce hundreds of merozoites. Upon release into the bloodstream (C), merozoites invade red blood cells and begin cycles of replication, lysis of the cell and repeated invasion (D). It is during these rounds of replication that disease symptoms manifest themselves. Some merozoites differentiate into male and female gametocytes (E), the sexual forms of the parasite responsible for transmission. Mosquitoes take up gametocytes upon feeding (F). Within the mosquito gut gametocytes unite and form zygotes (G), mature, eventually cross the gut barrier and travel as newly formed sporozoites (H) to the salivary glands for transmission back to the human host.

individuals suffer these complications and others don't. Although a range of host factors such as genetic background and immune status are undoubtedly important in determining disease outcome, heterogeneity within the parasite population also appears to play a major role.

By the late 19th century, distinct species of the malaria parasite *Plasmodium*, as well as different 'broods' or 'colonies' within these species, had been identified as producing different kinds of fever and other variable disease phenotypes such as quinine resistance and virulence (reviewed in ref. 3). This heterogeneity within the parasite population was also apparent in age profiles of infection and disease, which suggested that naturally acquired immunity builds up gradually and requires exposure to many different parasites. Furthermore, serological studies have shown that immunological responses to parasite isolates seem to be primarily isolate-specific.[2-4] We will discuss this evidence that the targets of immunity are polymorphic, focusing on mathematical models exploring the mechanisms behind and the implications of a parasite population made up of independently transmitted parasite types.[5,6] The epidemiological parameters of transmission are significantly affected by this assumption and this must be taken into account when designing control problems.

Sequencing projects have now given us insight into the structure of the parasite population at the genomic level. Phylogenetic analysis of these sequences is often impossible for malaria antigens, however, due to high rates of homologous and nonhomologous recombination. New techniques for understanding patterns of genetic diversity are being developed that do not rely on multiple alignments or particular evolutionary models. We discuss results from these approaches, which support the notion that there may be restricted antigenic types that could be targeted by vaccination. A solid link between genetic and phenotypic structure is needed to understand how this gene sequence data relates to expression in the host and disease phenotype, however. Mathematical models will continue to play an important role in uncovering this link and in our understanding of the relationship between the population structure of *Plasmodium falciparum* and malaria epidemiology and control.

Adding Realism to the Basic Framework of the Ross-MacDonald Models

The seminal models of malaria transmission formulated by Ross and MacDonald have remained the framework for the vast majority of theoretical models describing malaria epidemiology since they were developed during the first half of the 20th century. Although these models capture many defining features of malaria epidemiology, there are several important assumptions affecting their utility in guiding public health programs. First, mosquito biting rates in the model are assumed to be homogeneous, despite evidence that mosquitoes have systematic preferences for certain hosts.[7] Second, it is assumed that humans do not acquire any immunity to the parasite, which we will show is incompatible with both serological evidence and age profiles of disease and infection. Third, human, mosquito and parasite populations are assumed to be homogeneous. This chapter will review all these assumptions, but will focus on recent attempts to understand parasite population structure and the implications of these studies for estimates of the basic reproduction number and the design of vaccines.

Heterogeneity in Biting Rates and Susceptibility

Many models have been developed which add various levels of complexity to the basic Ross-MacDonald framework. One of the first issues to be addressed was the assumption of homogeneous biting rates by mosquitoes. Studies measuring the proportion of "freshly fed" mosquitoes containing human blood have shown that host selection by mosquitoes is not random,[7] as assumed in the original model. In fact, mosquitoes appear to actively select hosts harboring gametocytes, the stage of the parasite lifecycle responsible for transmission.[8] By dividing the human population into subpopulations with different biting rates, Dye and Hasibeder [9] showed that the assumption of uniform biting leads to an underestimation of R_0, the basic reproductive rate of the disease, even without assuming that mosquitoes preferentially select infected hosts. Heterogeneity in biting rates also leads to changes in the estimates for overall prevalence of infection, which under assumptions of homogeneous biting rates is overestimated in highly endemic areas and underestimated in low transmission regions.[9] It also causes large differences between subpopulations of hosts with respect to their contribution to transmission,[10] such that control programs targeting hosts that are bitten frequently will be most successful. Furthermore, current estimates of how heterogeneous biting rates relate to the prevalence of infection in the field show that the interaction between the two may be variable and this heterogeneity must also be taken into account when modelling the effects of control programs on malaria transmission.[11,12]

Extensions of Dye and Hasibeder model, in which host populations were also subdivided with respect to susceptibility and duration of infection, explored the effects of genetic heterogeneity in the host population on malaria transmission. Variability between hosts due to innate genetic differences in blood group or sickle cell type, for example, was modeled by assuming that two types of host exist with high or low susceptibility, long or short duration of infection. Calculation of R_0 then required the inclusion of the variance in these parameters and their covariance with respect to each other (reviewed in ref. 17).[13] In this case R_0 may be increased or decreased, depending on whether the covariance between parameters is positive or negative. The prevalence of infection will always be increased, however, when there is significant variance in host susceptibility or the length of infection. Again, the inclusion of heterogeneity in these models highlighted the often counterintuitive effects of changes to a highly complex system and the importance of mathematical models in our understanding of these nonlinearities.

Superinfection

It has long been recognized that the population structure of the malaria parasite may be better represented as multiple "broods" than a single entity and that different "broods" may coinfect one host (reviewed in ref. 3). MacDonald criticized the original Ross model in which hosts cannot be reinfected until they have cleared current infections and suggested a model of superinfection in which "broods" co-infected humans independently of each other. Dietz[14] incorporated superinfection into a model of malaria transmission by dividing the host population into different classes depending on the number of infections they harbor. He explored a range of assumptions about the density-dependent constraints on multiple infections, ranging from models in which infections occur completely independently of each other, to those in which only one infection is possible. With complete density-dependence (i.e., the original model) the prevalence predicted was much lower than in models that include superinfection. As in previous studies, models allowing superinfection showed that for higher R_0 values, changes in transmission due to control programs have little effect on prevalence. These models also highlighted the importance of the type of density-dependence occurring both in the human host and in the mosquito, which both remain poorly understood, for the estimation of R_0.

Incorporating Immunity

One of the most obvious problems with the Ross-MacDonald model is the absence of acquired immunity in humans. The relationship between infection and immunity to the malaria parasite is complex and poorly understood, although its inclusion in mathematical models is vital for an understanding of the potential effects of vaccination. Models by Aron and May[15-17] incorporated observations of semi-immune adults living in endemic areas by adding an immune class of people that were infected but asymptomatic. A key assumption of this framework is that individuals are removed from the immune class and became susceptible again after some period of time in the absence of immunological boosting by new infections. Thus, the rate of acquisition of new infections determines the duration of immunity. Analysis of this model showed that in endemic regions, the suppression of transmission through drug treatment for example, may sometimes increase the prevalence of disease. Although several more recent studies have attempted to incorporate immunity into the Ross-MacDonald framework,[18] few have generated qualitatively different results from Aron and May[16] and until we have a better understanding of the mechanisms underlying the development of immunity, their validity remains un-testable.

Understanding how naturally acquired immunity to the malaria parasite develops remains a major challenge and one that is critical to vaccine design. Figures 2A-C show the age profiles of malaria infection and mild and severe disease in an endemic region (from Snow and Marsh,

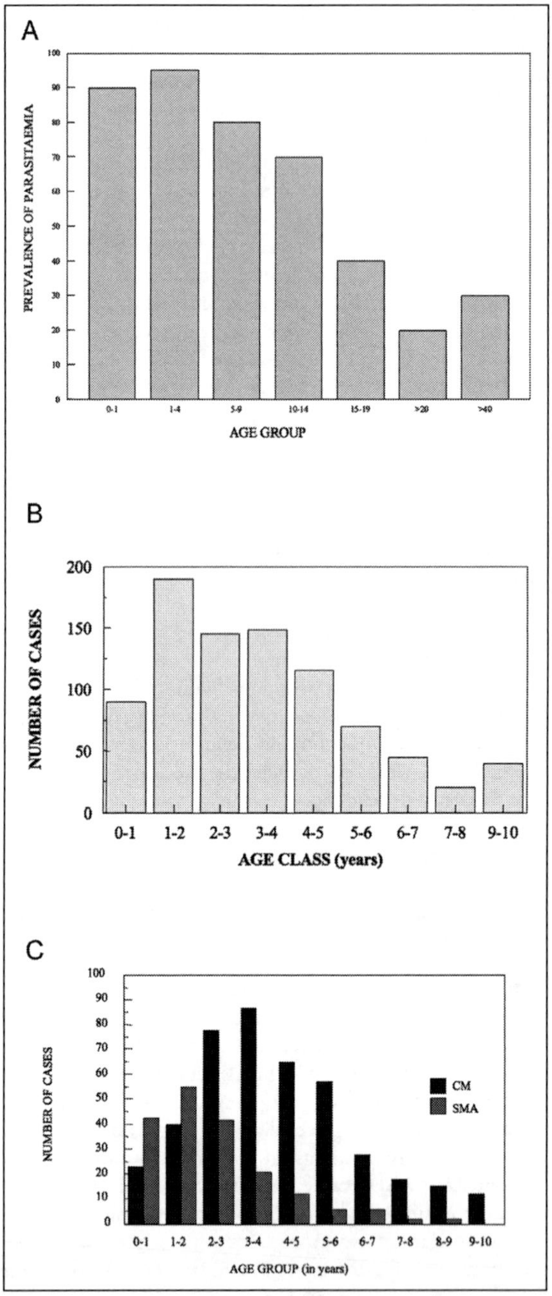

Figure 2. Age profiles of infection and disease in a malaria-endemic region (from Snow and Marsh, 1990, Gupta et al, 1994). A) The prevalence of malaria infection, defined as the percentage of individuals of different age groups harboring blood-stage parasites, measured by microscopy. B) The incidence of mild malaria episodes defined as uncomplicated infections presenting at the out-patient clinic in West Africa. C) The incidence of severe malaria disease in the same region: severe malarial anemia (SMA) and cerebral malaria (CM).

1990). The prevalence of infection (Fig. 2A) is extremely high among infants and young children and drops gradually in adolescence and into adulthood. Clinical disease shows a different pattern (Fig. 2B,C) and can be divided into several distinct syndromes from mild malaria to severe cerebral malaria to severe noncerebral malaria.[19,20] Mild episodes (Fig. 2B) are suppressed among infants under one year, peak at approximately 1-2 years and fall gradually with age into adolescence. Severe noncerebral malaria shows a similar peak among 1-2 year olds, but falls much more rapidly among young children (Fig. 2C). Cerebral malaria also declines rapidly with age, but it shows a delayed peak in incidence (Fig. 2C) compared to severe noncerebral disease. In spite of the number of studies of parasite prevalence and disease, our understanding of naturally acquired immunity remains incomplete and recent mathematical models fitted to data from Tanzania and The Gambia have shown that the generation of immunity is likely to require at least two different mechanisms, both suppression of disease and parasite clearance, occurring on different time scales.[21] In addition, maternal antibodies clearly play an important role in protection against disease in the first few months of life,[22,23] however there may be differences in the period of postnatal protection against cerebral and noncerebral malaria.[24] Gupta et al[25] also identify a "strain-transcending" immunity as being important for protection against severe (noncerebral) disease after the first few infections. More basic research is required to untangle the contribution of different mechanisms of immunity to patterns of infection and disease.

Modelling the Effects of Parasite Population Structure

The observations of the gradual decline in infection and disease and the heterogeneity of disease phenotypes led to the suggestion that immunity may be directed against polymorphic parasite antigens. Under this hypothesis, individuals growing up in endemic regions require exposure to many different parasites in order to build up a repertoire of antibodies to different antigenic types. This idea was supported by a number of serological experiments showing that patients generate agglutinating antibodies specific to their own isolate during infection, but not to isolates from other patients.[26-28] Furthermore, Gupta et al[6] showed that exposure to particular isolates among children

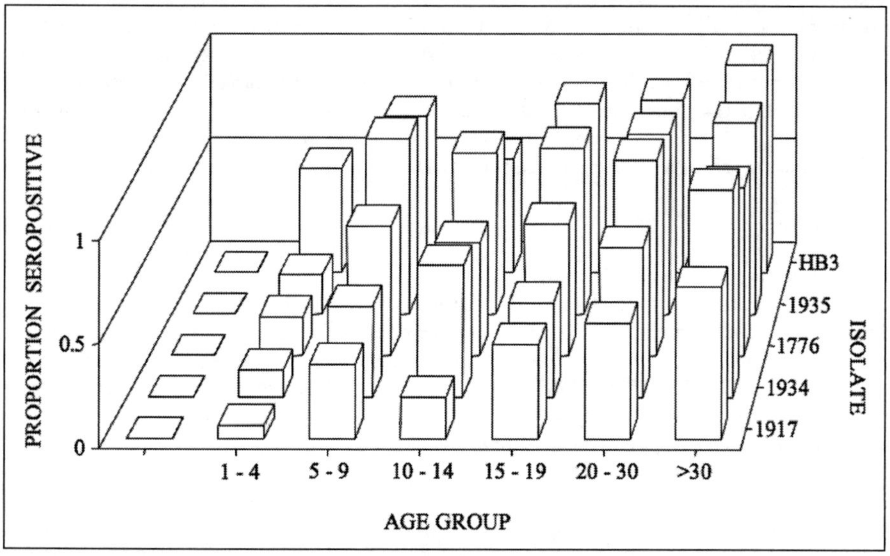

Figure 3. Age profiles of serological responses to different parasite isolates among children from Papua New Guinea (from Gupta et al, 1994). The proportion of people of different age groups with agglutinating antibody responses to each of four wild isolates from PNG and one lab isolate (HB3).

from Papua New Guinea rose gradually with age compared to the rapid rise in exposure to any *P. falciparum* isolate (Fig. 3). In studies of serological responses of Kenyan children, it was shown that the frequency of recognition of particular parasite isolates was associated with both young host age and a limited antibody repertoire,[3,28] suggesting that there may also be a hierarchy of infection with different antigenic types, possibly caused by their different frequencies in the parasite population.

Implications of Antigenic Diversity for Control

An antigenically diverse parasite population has important implications for our understanding of the basic epidemiological parameters of malaria transmission. For example, Gatton and Cheng[29] used a stochastic simulation model to demonstrate that the time taken to develop immunity was directly proportional to parasite antigenic diversity. A model exploring the interaction between immunity to different parasite phenotypes and the rate of evolution of the parasite showed that cross-reactivity in the parental genotypes would delay the appearance of and increase the chances of extinction of, a resulting recombinant form.[30] An extension of this framework to include population effects demonstrated that the presence of increasing numbers of independent parasite genotypes in the parasite population may significantly prolong the persistence of each one in the host population.[31]

Assuming antibodies generated to one antigenic type will not protect a host from infection with a different antigenic type effectively removes competition for hosts between strains. Thus, the parasite population is divided into distinct antigenic systems transmitted essentially independently. Models by Gupta et al.[6,32] explored the implications of this removal of competition between parasite strains. In many theoretical models exploring the relationship between virulence and R_0, competition between parasites with varying virulence and R_0 values will select for a pathogen population characterized by an intermediate level of virulence that maximizes R_0. Under this framework, competitive exclusion of pathogens outside the range of optimum transmission will occur. In a population composed of discrete antigenic types, however, the lack of shared antigenic epitopes between strains reduces direct competition for hosts between different types, facilitating the coexistence of a wider range of R_0 values and levels of virulence.[6] For falciparum malaria, the separation of severe and mild isolates into distinct antigenic systems allows for the coexistence of two different sets of maximized R_0 values. These models predicted that cerebral malaria, with its delayed peak in prevalence, may be caused by a distinct set of circulating strains, whereas severe malarial anemia represents a rare outcome of infection with any number of "mild" strains.

Another important implication of a discretely structured population relates to the calculation of epidemiological parameters from R_0. For simple pathogen systems like measles, the basic reproductive rate of a pathogen is inversely proportional to the average age of first infection:[33]

$$R_0 = \frac{L}{A} \tag{5}$$

where L is the average lifespan of the host and A is the average age of first age of infection. Estimates based on this relationship for malaria yield an extremely high R_0 value, since infants in malaria-endemic regions usually experience their first malaria infection before they are one year-old. In a system composed of discrete antigenic types, however, R_0 for the whole parasite population is calculated as the sum of the R_0 values of individual strains. Thus, rather than assuming that immunity to the malaria parasite is short-lived, Gupta et al[32] showed that long-lived strain-specific immunity can lead to a young average age of first exposure to any of n circulating strains, A_n:

$$A_n = \frac{L}{\sum R_{0_i}} \tag{6}$$

Here, relatively low R_0 values for each antigenic Type i contribute additively to the probability of becoming infected with any antigenic type. In this case, immunity to each strain is assumed to be life-long. Data from Papua New Guinea showing a gradual rise in antibodies to particular parasite isolates compared to antibodies to any isolate support this hypothesis, as discussed in the previous section. The recalculation of R_0 as a cumulative value also has implications for control programs. For most pathogens the proportion of the population that must be vaccinated (p) in order to eradicate a disease, for example, can be related to R_0 by the following equation:

$$p > 1 - \frac{1}{R_0} \tag{7}$$

Hence, to eradicate a disease with a very high reproductive rate, nearly everyone in a population must be vaccinated. Assuming a number of discrete antigenic types each with a relatively low R_0, however, the targeting of particular types in a vaccine, those associated with cerebral malaria for example, becomes a more realistic possibility.

Recent attempts to estimate R_0 in the field have made painstaking efforts to parameterize a modification of the Ross-MacDonald model using data from around the world.[34] Their models predicted the range of R_0 values from over a hundred African countries to vary between less than one and close to ten thousand, several orders of magnitude greater than previous estimates. They assume a system with infinite broods and independent clearance however (see ref. 18 for details) and make no attempt to model the effects of population structure on immunity. Furthermore, they include immunity only as a reduction in infectivity of both humans and mosquitoes (due to transmission-blocking and liver-stage immunity, respectively) that varies as a function of the entomological inoculation rate. These simplifications largely ignore the important aspects of population structure described above and lead to the extreme values of R_0 they calculate.

Emergence and Maintenance of Strain Structure

In order for distinct antigenic strains of the malaria parasite to persist in spite of high rates of recombination, however, some mechanism for preventing the breakdown of associations between antigenic epitopes must exist. Mathematical models have shown how discrete antigenic types may be structured and maintained in the presence of recombination, however, through immune selection.[35,36] These models assume that when infected individuals become immune, they may gain cross-immunity to parasites that share antigenic variants with the infecting isolate. Hosts infected with a particular parasite gain 'exposure' to any other parasites expressing the same epitopes, depending on γ, the strength of cross-immunity. In the simplest hypothetical case, for example, each antigenic type is defined by two loci, each with two possible allelic variants, giving four possible types (see Box 1). This model is defined by a system of ordinary differential equations the changing proportions of the population that are either infected, immune, or exposed to different antigenic variants are modelled for different levels of cross-immunity. When cross-immunity is very weak, infected hosts will not gain protection to different parasites even if they share antigenic epitopes and all antigenic types will coexist at a prevalence determined by their intrinsic transmissibility. This situation could arise due to epitope shedding or weak immune responses, for example. At intermediate levels of cross-immunity, unstable population structure emerges, displaying cyclical or chaotic patterns of dominance of different antigenic types. When cross-immunity is strong, however, immunity to one antigenic type will confer substantial protection against isolates that share antigenic epitopes, leading to the dominance of a set of genotypes with non-overlapping antigenic repertoires (which will not be competing for susceptible hosts). Discrete, non-overlapping antigenic repertoires will persist over time in spite of recombination between parasites, because immune selection will suppress the prevalence of recombinant combinations that share epitopes with dominant antigenic types.

Testing this hypothesis remains difficult while the immuno-dominant parasite epitopes responsible for acquired immunity remain unknown, however. The serological studies described above, showing isolate-specific immune responses in patients from endemic regions, are suggestive of strong antigenic structuring. However, until a system of defining antigenic repertoires based on known epitopes is developed and we understand more about expression patterns within the host, it remains an untested hypothesis.

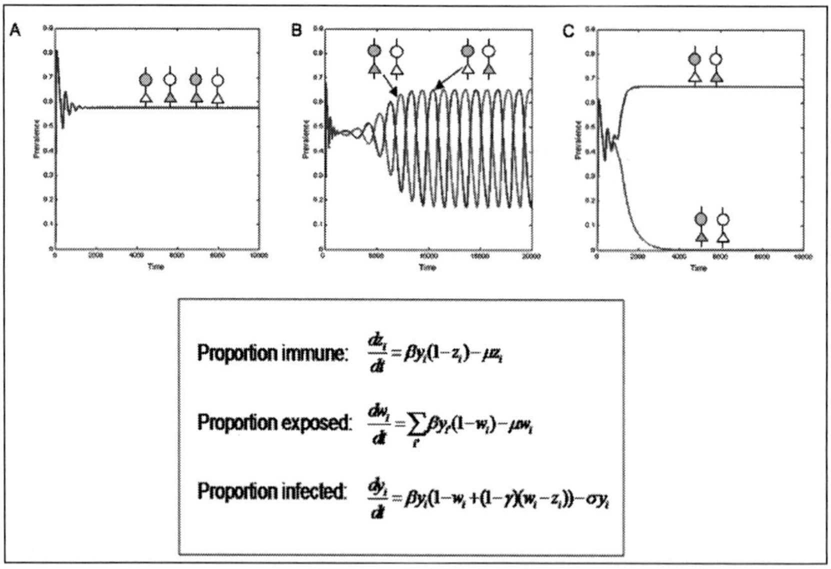

Box 1. The effects of immune selection on antigenically diverse pathogens (Gupta et al[35]). Pathogen strains are defined by two loci (shown as either the circle or the triangle in the figures), each with two alleles (yellow or white, in this case). Once infected with a particular strain, hosts gained partial immunity to any other strains with shared antigenic determinants. The strength of cross-immunity was determined by the parameter gamma (γ) and all strains were assumed to have the same transmissibility, β. For each strain, the host population consisted of three overlapping compartments: the proportion that were infectious to other hosts, x_i, the proportion that were completely immune, z_i and the proportion partially and completely immune, w_i. Immunity was gained immediately upon infection and infectiousness was lost at a rate of σ, such that those who were infected (x) were a subset of those who were completely immune (z), who were in turn a subset of those who were partially and completely immune (w). It was assumed that the duration of infectiousness ($1/\sigma$) was short compared to the average host life-span ($1/\mu$) and that immunity was life-long. The effect of recombination was not explicitly included in the model, however all possible strains were present from the start in order to investigate the competitive interactions between them. The figures show the effects of immune selection assuming (A) weak, (B) medium and (C) strong cross-immunity (γ). At low levels of cross-immunity, all strains coexist at the same prevalence. When cross-immunity is very strong, the pathogen population is dominated by two strains with non-overlapping antigenic repertoires, since these will not be competing for hosts and the remaining strains are suppressed. At intermediate levels of cross-immunity, cyclical behavior is observed and pairs of strains with non-overlapping antigenic repertoires oscillate with respect to each other. Chaotic behavior is also observed for some parameter regions. A color version of this image is available at www.landesbioscience.com/curie.

Sequence-Based Analysis of Population Structure of Malaria Parasite Antigens

Despite these difficulties, now that three full genome sequences of *Plasmodium falciparum* lab strains are available and sequencing technology continues to become cheaper and easier, we have the opportunity to start exploring the genes underlying parasite population structure. One of the best candidate targets of the isolate-specific agglutinating antibodies is the highly diverse, polymorphic family of proteins *Plasmodium falciparum* erythrocyte protein 1 (PfEMP1), expressed on the surface of infected red blood cells.[37] PfEMP1 is implicated as a virulence factor due to its role in the cytoadherence of infected cells to the endothelial lining, mediated by host cell receptors such as CD36 and ICAM1.[38] The *P. falciparum* genome contains approximately 60 *var* genes encoding PfEMP1. These undergo clonal antigenic variation during infection; different variants in the parasite *var* repertoire are expressed sequentially in a mutually exclusive manner,[39] prolonging infection and increasing the parasite's chances of transmission. *Var* genes also exhibit extremely high rates of recombination, not only between parasites during the sexual stages in the mosquito but also in the human host between different genes within the same parasite.[40] This antigenic diversity and complexity of expression during infection complicates our understanding of the structuring of these loci, both within individual parasite genomes and across the parasite population.

Despite the incredible diversity of *var* genes across the parasite population, some constrained structure is apparent. Serological studies have shown that PfEMP1 proteins from parasite isolates infecting young children tend to be recognized by most adults in an endemic region and that by the time children reach the age of about fifteen in areas of high transmission they have gained exposure to most PfEMP1 variants circulating in the local parasite population.[3] Following the sequencing of the *Plasmodium falciparum* genome, analysis of *var* repertoires has also hinted at limits to their diversity. Attempts to classify *var* genes into meaningful groups based on fully sequenced lab strains have led to the identification of approximately 6 different *var* groups, differentiated based on protein domain structure, different upstream promoters, the direction of transcription and chromosomal position.[41-44] This system of classification identifies three major groups (A, B and C) and two minor groups (B/A and B/C). Different groups appear to show marked differences in their cytoadherent properties, for example groups B and C bind to the host receptor CD36, whereas group A does not.[45] They also have different immunological properties, with group A *var* genes being better recognized by antibodies from sera of children growing up in malaria-endemic regions.[46] The diversity of the sequences themselves has made sequencing whole *var* repertoires from clinical isolates extremely difficult, however. A sequence-based approach to *var* gene classification has been developed which uses short sequence fragments.[47] Regions of homology at the beginning and end of these tags allow for the design of primers that can amplify most *var* genes. The sequence tags are grouped according to a series of semi-conserved regions and the number of cysteine residues present between them. A study by Bull et al[48] analyzing these sequence fragments from wild isolates showed that *var* genes belonging to the 6 different groups were represented in the same proportions within different parasite genomes, but their expression patterns varied considerably in different hosts. Figure 4 illustrates this finding, showing the genomic structure of *var* gene repertoires from 12 wild Kenyan isolates and their profiles of *var* expression. The expression of particular groups of *vars* was also found to be associated with different disease phenotypes, such as rosetting and the immune status of the host.[48] The groups identified using these sequence-based techniques are therefore linked with phenotypic characteristics and represent a useful system of classification.

Understanding the evolutionary relationships between *var* genes and *var* gene groups remains a major challenge, however. Phylogenetic analysis of the relationship between *var* sequences is confounded by the high rates of homologous and nonhomologous recombination; most phylogenetic techniques rely on an initial alignment of sequences, which is not possible for most *var* genes. Bockhorst et al[49] have overcome this problem for a small group of conserved *var* genes associated with malaria in pregnant women, by identifying segments of correlated nucleotides within sequences and using whole segments in a phylogenetic analysis and generating "population trees", however this technique cannot be applied to the vast majority of *vars*. Recently, a technique for visualizing the relationships between *var* sequence fragments has been developed by Bull et al[50] which is not reliant on an alignment. Sequence tags are represented as nodes in a network, with edges between nodes representing exact sequence matches at variable regions termed "position

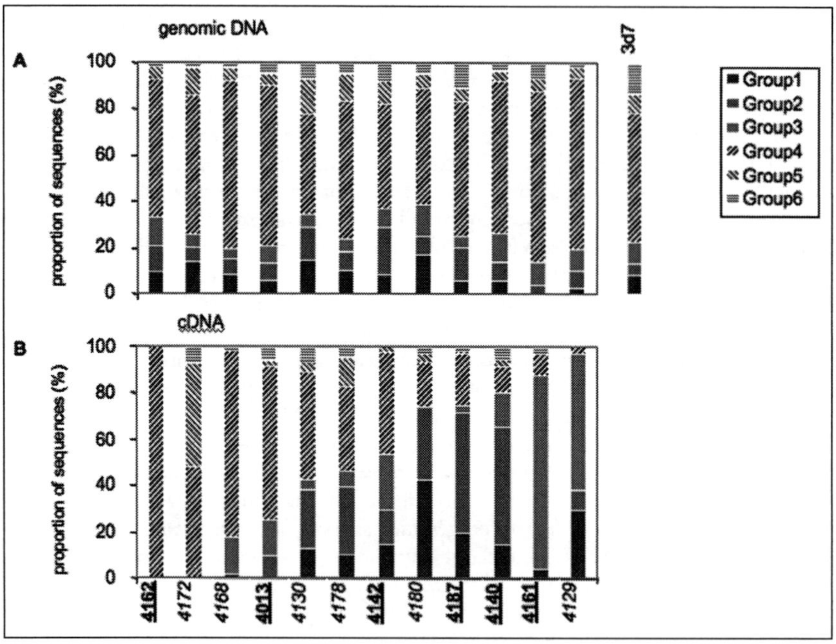

Figure 4. The genomic (above) and expressed (below) *var* profiles of wild Kenyan isolates (from Bull et al[48]). The proportion of *var* sequence tags among 12 wild isolates assigned to six sequence group from genomic DNA (A) and cDNA (B). The genomic DNA profile is shown for the lab strain 3D7 to the right of (A).

specific polymorphic blocks" (PSPBs). These regions show considerable mosaicism, such that otherwise unrelated sequences may share one or more PSPB. Figure 5 illustrates an example of some of these sequence fragments, the network they form and the results of an analysis of over 1000 *var* genes from clinical and lab isolates, as well as several *P. reichenowi* (the primate malaria parasite) homologues. The large network is coloured according to the 6 groups previously identified. The network has two distinct lobes with loose links between them and these lobes correspond well to different groups of *var* genes based on both classification systems described above. All group 1 sequences fall within the smaller lobe, for example, whereas group 4 sequences are found in the larger lobe. Furthermore, group A genes (using the Lavstsen et al[44] system) are all found within the smaller lobe. Thus, the separation of groups seems to represent a recombination hierarchy, with certain types of *var* gene recombining with each other more frequently than others. Interestingly, *P. reichenowi* genes fall within the network, indicating that many of these polymorphic regions are relatively ancient. The diversity of the malaria parasite is therefore generated by recombination, which shuffles variable regions between *var* genes.

A major field of research in the next few years will be the search for antigenic epitopes within *var* gene sequences and the investigation of how the expression of different *var* epitopes relates to parasite phenotype and disease patterns in the host. Without this link between genotype and phenotype, it will be difficult to directly relate the sequencing projects discussed above with the mathematical models of population structure presented in previous sections.

Conclusion

The use of mathematical models in the design of malaria control programmes dates back to the beginning of the 20th century. The Ross-MacDonald model of malaria transmission provided the basis for most modelling efforts since then and gave valuable insights into the importance of

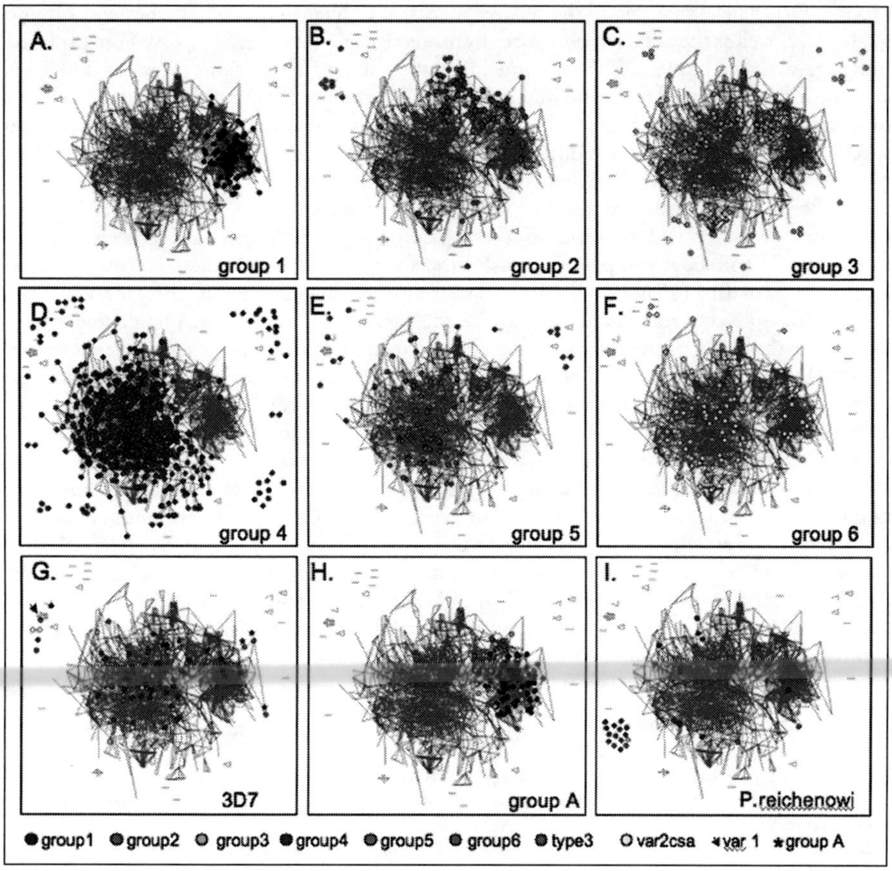

Figure 5. Network showing the relationship between the *var* sequence tags for wild isolates, lab strains and *P. reichenowi* (from Bull et al[50]). Locations of *var* tags from different sequence groups within a network constructed primarily from Kenyan wild isolates. A-F) Locations of sequence tags belonging to groups defined by Bull et al.[47] G) Locations of sequences from the fully-sequenced lab strain 3D7. H) Locations of Group A (defined by Lavstsen[44]) reference sequences. I) Locations of *P. reichenowi* homologues.

epidemiological variables such as the mosquito vector lifespan for determining the spread of the disease. Many variations on this framework have been developed, however we still lack a thorough understanding of the acquisition of immunity to *Plasmodium falciparum* and the relationship between infection, disease and immunity in malaria-endemic populations.

The models discussed in this chapter have aimed to understand mechanisms behind malaria epidemiology and dynamics. Many studies have also explored models focusing on specific interventions, however, comparing the effects of different types of control programs and the practicalities involved in their implementation. For example, Saul[51] developed a simulation model of vaccination in which he examined the effects of different types of vaccines—antigametocyte or antisporozoite/asexual—and the expected changes in malaria prevalence and disease with variable levels of vaccine coverage and efficacy. He showed that antisporozoite or asexual vaccines were more effective at lower levels of coverage and that in combination with other methods of control, vaccination could be a feasible strategy. His recent models (for example ref. 52) take into account the antibody levels needed to maintain herd immunity towards a specific mosquito-stage transmission-blocking vaccine. Other investigations of the relative merits of different types of vaccines and vaccine strategies, as well as the effects of only partially protective or "leaky" vaccines, have been explored using age-structure models.[53-55] These models included one set of equations for vaccinated hosts and one for unvaccinated hosts and highlighted the importance for successful control programs of vaccine efficacy and duration and boosting of immunity by natural infections. Another more specific model reproduced the conditions of a particular country, Sri Lanka and modeled the effect of transmission-blocking vaccines under epidemic conditions.[56]

We have also not dealt with mathematical models of the spread of drug resistance through the malaria parasite population. This aspect of parasite population structure has been modeled extensively,[57-61] and shows complex relationships between the number of genetic loci involved, the level and type of drug use and the level of malaria transmission in the region. These models either focus on the spread of resistant alleles through the parasite population (i.e., population genetic models) or they incorporate infected hosts and use an epidemiological framework (reviewed in ref. 60). Despite the differences in modelling approaches, most of these studies conclude that the key to preventing rapid emergence and spread of resistance is minimizing drug use and using them in combination with transmission-blocking strategies such as bednets.

Even careful use of malaria drugs will eventually result in the emergence of drug resistance however, making the development of a vaccine a priority. We have shown that the population structure of the malaria parasite at immunodominant loci such as the *var* genes has important implications for the basic epidemiology of malaria and for the design of control programs like vaccination. A major goal will be to link *var* gene repertoire structure and expression patterns with serological and disease profiles in endemic regions. This will enable the re-assessment and validation of the mathematical models discussed in this chapter, providing the opportunity to design appropriate control programs in the future.

References

1. Snow RW, Guerra CA, Noor AM et al. The global distribution of clinical episodes of Plasmodium falciparum malaria. Nature 2005; 434(7030):214-7.
2. Marsh K, Sherwood JA, Howard RJ. Parasite-infected-cell-agglutination and indirect immunofluorescence assays for detection of human serum antibodies bound to antigens on Plasmodium falciparum-infected erythrocytes. J Immunol Methods 1986; 91(1):107-15.
3. Bull PC, Lowe BS, Kortok M et al. Antibody recognition of Plasmodium falciparum erythrocyte surface antigens in Kenya: evidence for rare and prevalent variants. Infect Immun 1999; 67(2):733-9.
4. Bull PC, Marsh K. The role of antibodies to Plasmodium falciparum-infected-erythrocyte surface antigens in naturally acquired immunity to malaria. Trends Microbiol 2002; 10(2):55-8.
5. Gupta S, Day KP. A strain theory of malaria transmission. Parasitol Today 1994; 10(12):476-81.
6. Gupta S, Hill AV, Kwiatkowski D et al. Parasite virulence and disease patterns in Plasmodium falciparum malaria. Proc Natl Acad Sci USA 1994; 91(9):3715-9.
7. Burkot TR. Nonrandom host selection by anopheline mosquitoes. Parasitol Today 1988; 4(6):156-62.
8. Lacroix R, Mukabana WR, Gouagna LC et al. Malaria infection increases attractiveness of humans to mosquitoes. PLoS Biol 2005; 3(9):e298.
9. Dye C, Hasibeder G. Population dynamics of mosquito-borne disease: effects of flies which bite some people more frequently than others. Trans R Soc Trop Med Hyg 1986; 80(1):69-77.
10. Woolhouse ME, Dye C, Etard JF et al. Heterogeneities in the transmission of infectious agents: implications for the design of control programs. Proc Natl Acad Sci USA 1997; 94(1):338-42.
11. Smith DL, Dushoff J, McKenzie FE. The risk of a mosquito-borne infection in a heterogeneous environment. PLoS Biol 2004; 2(11):e368.
12. Smith DL, Dushoff J, Snow RW et al. The entomological inoculation rate and Plasmodium falciparum infection in African children. Nature 2005; 438(7067):492-5.
13. Koella JC. On the use of mathematical models of malaria transmission. Acta Trop 1991; 49(1):1-25.
14. Dietz K. Density-dependence in parasite transmission dynamics. Parasitol Today 1988; 4(4):91-7.
15. Aron J. Mathematical modeling of immunity to malaria. Mathematical Biosciences 1988; 90:385-98.
16. Aron JL, May RM. The population dynamics of malaria. In: Anderson RM, editor. Population Dynamics of Infectious Disease. London: Chapman and Hall, 1982:139-79.
17. Aron JL. Dynamics of acquired immunity boosted by exposure to infection. Math Biosci 1983; 64:249-59.

18. Aguas R, White LJ, Snow RW et al. Prospects for malaria eradication in sub-Saharan Africa. PLoS ONE 2008; 3(3):e1767.
19. Mwangi TW, Ross A, Snow RW et al. Case definitions of clinical malaria under different transmission conditions in Kilifi District, Kenya. J Infect Dis 2005; 191(11):1932-9.
20. Brewster DR, Kwiatkowski D, White NJ. Neurological sequelae of cerebral malaria in children. Lancet 1990; 336(8722):1039-43.
21. Filipe JA, Riley EM, Drakeley CJ et al. Determination of the processes driving the acquisition of immunity to malaria using a mathematical transmission model. PLoS Comput Biol 2007; 3(12):e255.
22. Sehgal VM, Siddjiqui WA, Alpers MP. A seroepidemiological study to evaluate the role of passive maternal immunity to malaria in infants. Trans R Soc Trop Med Hyg 1989; 83:105-6.
23. Hogh B, Marbiah NT, Burghaus PA et al. Relationship between maternally derived anti-Plasmodium falciparum antibodies and risk of infection and disease in infants living in an area of Liberia, West Africa, in which malaria is highly endemic. Infect Immun 1995; 63(10):4043-8.
24. Gupta S, Snow RW, Donnelly C et al. Acquired immunity and postnatal clinical protection in childhood cerebral malaria. Proc Biol Sci 1999; 266(1414):33-8.
25. Gupta S, Snow RW, Donnelly CA et al. Immunity to noncerebral severe malaria is acquired after one or two infections. Nat Med 1999; 5(3):340-3.
26. Marsh K, Howard RJ. Antigens induced on erythrocytes by P. falciparum: expression of diverse and conserved determinants. Science 1986; 231(4734):150-3.
27. Giha HA, Staalsoe T, Dodoo D et al. Nine-year longitudinal study of antibodies to variant antigens on the surface of Plasmodium falciparum-infected erythrocytes. Infect Immun 1999; 67(8):4092-8.
28. Bull PC, Kortok M, Kai O et al. Plasmodium falciparum-infected erythrocytes: agglutination by diverse Kenyan plasma is associated with severe disease and young host age. J Infect Dis 2000; 182(1):252-9.
29. Gatton ML, Cheng Q. Modeling the development of acquired clinical immunity to Plasmodium falciparum malaria. Infection and Immunity 2004; 72(11):6538-45.
30. McKenzie FE, Ferreira MU, Baird JK et al. Meiotic recombination, cross-reactivity and persistence in Plasmodium falciparum. Evolution 2001; 55(7):1299-307.
31. McKenzie FE, Bossert WH. An integrated model of Plasmodium falciparum dynamics. J Theor Biol 2005; 232(3):411-26.
32. Gupta S, Trenholme K, Anderson RM et al. Antigenic diversity and the transmission dynamics of Plasmodium falciparum. Science 1994; 263(5149):961-3.
33. Anderson RM, May RM. Infectious Diseases Of Humans: Dynamics And Control. London: Oxford University Press, 1991.
34. Smith DL, McKenzie FE, Snow RW et al. Revisiting the basic reproductive number for malaria and its implications for malaria control. PLoS Biol 2007; 5(3):e42.
35. Gupta S, Maiden MC, Feavers IM et al. The maintenance of strain structure in populations of recombining infectious agents. Nat Med 1996; 2(4):437-42.
36. Gupta S, Ferguson N, Anderson R. Chaos, persistence and evolution of strain structure in antigenically diverse infectious agents. Science 1998; 280(5365):912-5.
37. Bull PC, Lowe BS, Kortok M et al. Parasite antigens on the infected red cell surface are targets for naturally acquired immunity to malaria. Nature Medicine 1998; 4(3):358-60.
38. Baruch DI, Gormely JA, Ma C et al. Plasmodium falciparum erythrocyte membrane protein 1 is a parasitized erythrocyte receptor for adherence to CD36, thrombospondin and intercellular adhesion molecule 1. Proc Natl Acad Sci USA 1996; 93(8):3497-502.
39. Voss TS, Healer J, Marty AJ et al. A var gene promoter controls allelic exclusion of virulence genes in Plasmodium falciparum malaria. Nature 2006; 439(7079):1004-8.
40. Freitas-Junior LH, Bottius E, Pirrit LA et al. Frequent ectopic recombination of virulence factor genes in telomeric chromosome clusters of P. falciparum. Nature 2000; 407(6807):1018-22.
41. Kraemer SM, Kyes SA, Aggarwal G et al. Patterns of gene recombination shape var gene repertoires in Plasmodium falciparum: comparisons of geographically diverse isolates. BMC Genomics 2007; 8:45.
42. Kraemer SM, Smith JD. Evidence for the importance of genetic structuring to the structural and functional specialization of the Plasmodium falciparum var gene family. Mol Microbiol 2003; 50(5):1527-38.
43. Kyes SA, Kraemer SM, Smith JD. Antigenic variation in Plasmodium falciparum: gene organization and regulation of the var multigene family. Eukaryot Cell 2007; 6(9):1511-20.
44. Lavstsen T, Salanti A, Jensen AT et al. Sub-grouping of Plasmodium falciparum 3D7 var genes based on sequence analysis of coding and noncoding regions. Malar J 2003; 2:27.
45. Kraemer SM, Smith JD. A family affair: var genes, PfEMP1 binding and malaria disease. Curr Opin Microbiol 2006; 9(4):374-80.
46. Jensen AT, Magistrado P, Sharp S et al. Plasmodium falciparum associated with severe childhood malaria preferentially expresses PfEMP1 encoded by group A var genes. J Exp Med 2004; 199(9):1179-90.

47. Bull PC, Kyes S, Buckee CO et al. An approach to classifying sequence tags sampled from Plasmodium falciparum var genes. Mol Biochem Parasitol 2007; 154(1):98-102.
48. Bull PC, Berriman M, Kyes S et al. Plasmodium falciparum variant surface antigen expression patterns during malaria. PLoS Pathog 2005; 1(3):e26.
49. Bockhorst J, Lu F, Janes JH et al. Structural polymorphism and diversifying selection on the pregnancy malaria vaccine candidate VAR2CSA. Mol Biochem Parasitol 2007; 155(2):103-12.
50. Bull PC, Buckee CO, Kyes S et al. Plasmodium falciparum antigenic variation. Mapping mosaic var gene sequences onto a network of shared, highly polymorphic sequence blocks. Mol Microbiol 2008; 68(6):1519-34.
51. Saul A. Minimal efficacy requirements for malarial vaccines to significantly lower transmission in epidemic or seasonal malaria. Acta Trop 1993; 52(4):283-96.
52. Saul A. Mosquito stage, transmission blocking vaccines for malaria. Curr Opin Infect Dis 2007; 20(5):476-81.
53. Halloran ME, Struchiner CJ. Modeling transmission dynamics of stage-specific malaria vaccines. Parasitol Today 1992; 8(3):77-85.
54. Halloran ME, Struchiner CJ, Spielman A. Modeling malaria vaccines. II: Population effects of stage-specific malaria vaccines dependent on natural boosting. Math Biosci 1989; 94(1):115-49.
55. Struchiner CJ, Halloran ME, Spielman A. Modeling malaria vaccines. I: New uses for old ideas. Math Biosci 1989; 94(1):87-113.
56. de Zoysa AP, Herath PR, Abhayawardana TA et al. Modulation of human malaria transmission by anti-gamete transmission blocking immunity. Trans R Soc Trop Med Hyg 1988; 82(4):548-53.
57. Curtis CF, Otoo LN. A simple model of the build-up of resistance to mixtures of anti-malarial drugs. Trans R Soc Trop Med Hyg 1986; 80(6):889-92.
58. Hastings IM. A model for the origins and spread of drug-resistant malaria. Parasitology 1997; 115 (Pt 2):133-41.
59. Hastings IM, Watkins WM. Intensity of malaria transmission and the evolution of drug resistance. Acta Trop 2005; 94(3):218-29.
60. Mackinnon MJ. Drug resistance models for malaria. Acta Trop 2005; 94(3):207-17.
61. O'Meara WP, Smith DL, McKenzie FE. Potential impact of intermittent preventive treatment (IPT) on spread of drug-resistant malaria. PLoS Med 2006; 3(5):e141.

CHAPTER 9

Mathematical Modelling of the Epidemiology of Tuberculosis

Peter J. White* and Geoff P. Garnett

Abstract

Despite the infectious agent that causes tuberculosis having been discovered in 1882, many aspects of the natural history and transmission dynamics of TB are still not fully understood. This is reflected in differences in the structures of mathematical models of TB, which in turn produce differences in the predicted impacts of interventions. Gaining a greater understanding of TB transmission dynamics requires further empirical laboratory and field work, mathematical modelling and interaction between them. Modelling can be used to quantify uncertainty due to different gaps in our knowledge to help identify research priorities. Fortunately, the present moment is an exciting time for TB epidemiology, with rapid progress being made in applying new mathematical modelling techniques, new tools for TB diagnosis and genetic analysis and a growing interest in developing more-effective public-health interventions.

Introduction

Despite the availability of effective treatment tuberculosis remains a major global cause of morbidity and mortality, with around one-third of the world's population believed to be infected. It caused an estimated 1.7 million deaths and 8.9 million new cases of infection in 2004.[1] The highest incidence of disease is in sub-Saharan Africa, in part due to interactions with HIV,[2-4] which has fuelled dramatic rises in incidence of the disease in many countries. (Globally, TB is the proximate cause of many HIV-related deaths, particularly in Africa.[3]) Even in many countries where its overall incidence is low TB remains a problem: there has been an outbreak in the New York in the recent past[5,6] and incidence is currently rising in the UK.[7]

TB Natural History

TB's natural history is summarized briefly here; it is discussed in more detail below with respect to its representation in mathematical models.

Tuberculosis is caused by the slowly-replicating bacterium *Mycobacterium tuberculosis* (*M tuberculosis, Mtb*). *Mtb* is in the same genus as *M bovis*, the cause of the disease of cattle bovine TB, which humans can acquire from contaminated milk; it can also infect other species such as badgers (*Meles meles*). Other members of the genus, such as *M avis*, can cause human disease, particularly in the immunocompromised host.

*Corresponding Author: Peter J. White—Modelling and Economics Unit, Health Protection Agency Centre for Infections, 61 Colindale Avenue, London, NW9 5EQ, UK and MRC Centre for Outbreak Analysis and Modelling, Department of Infectious Disease Epidemiology, Faculty of Medicine, Imperial College London, London W2 1PG, UK. Email: peter.white@hpa.org.uk and p.white@imperial.ac.uk

Modelling Parasite Transmission and Control, edited by Edwin Michael and Robert C. Spear.

Person-to-person transmission of *Mtb* is via the respiratory route, which can occur both through close contact between persons and through infectious bacilli being carried throughout buildings by air currents (which makes ventilation an important preventative measure;[8] this can be supplemented by ultraviolet irradiation).

It is important to distinguish between TB *infection* and TB *disease*. Most people infected with TB never develop disease during their lifetime, but remain latently infected—and uninfectious. The progression of an infection to disease is more likely in the years shortly after infection, known as primary disease. Thereafter, there is a much lower rate of progression to disease, called endogenous reactivation, resulting in a lifetime risk of roughly 10% of developing TB disease (in the absence of HIV—see below). Endogenous reactivation can take decades to occur[9,10] and is the main cause of disease in older people in settings where the prevalence of TB disease is very low. However, those with latent TB infection are also subject to exogenous reinfection, where they acquire another TB infection; this can cause rapid progression and can be an important cause of incident disease in settings with a high prevalence of TB disease and consequently a high force of infection.[11]

Active TB disease can affect other parts of the body besides the lungs, e.g., causing Pott's disease in the spine and tuberculosis meningitis in the brain, causing morbidity and mortality. However, only those with *pulmonary* (lung) disease can be infectious. (Most mathematical models of TB transmission only consider pulmonary TB, although nonpulmonary infection causes significant morbidity. Also, since the natural history of TB is different in children,[12] most models consider only adults.) Pulmonary TB can be divided into smear-negative and smear-positive, according to whether a sputum sample gives a negative or positive result in microscopic observation. Smear-negativity is caused by a lower concentration of bacilli in the sputum and is associated with lower, if any, transmissibility.

Without treatment, individuals may recover from active TB disease, but there is disagreement over whether natural immune recovery clears the infection, or leads to (re-establishment of) latent infection. In some individuals, untreated active disease leads to chronic, persistent, infectious active disease.

Treatment requires long-term use of antibiotics (at least 6 months is recommended for short-course therapy), but is generally highly effective, including in those with HIV, provided the patient is adherent.[4] Lack of adherence can result in the bacterium acquiring drug resistance; transmission of drug-resistant strains is a significant problem in many parts of the world: globally an estimated 4% of patients have multi-drug resistant TB,[4] but this proportion can be much higher in particular settings. Treatment of drug-resistant TB takes longer and is less effective.[4] Importantly, most patients—even those with HIV—become uninfectious for TB soon after treatment commencement.[3]

Chest X-ray screening can identify pulmonary lesions that potentially indicate active tuberculosis. Definitive diagnosis requires sputum smear microscopy, or laboratory culture of sputum specimens (which can take several weeks), since a low bacterial load may result in a 'false negative' sputum smear test; nucleic-acid amplification tests also lack sensitivity for smear-negative disease.[4] Diagnosis of latent TB infection previously relied on the tuberculin skin test (TST) where inflammation associated with antigen challenge was used to identify those previously exposed. Unfortunately, this test lacks sensitivity (particularly in the early stages of primary tuberculosis, in immunocompromised persons and in disseminated tuberculosis) and specificity (due to confounding by BCG vaccination or other mycobacterial infections). Fortunately, several tests with far superior sensitivity and specificity have recently become available.[4,13]

The immune suppression associated with HIV infection increases the rate of progression of latent TB infection, thus increasing the incidence of TB disease.[3] However, it is debatable how much this increases TB transmission, since HIV increases the proportion of active disease that is smear-negative and shortens the infectious period due to more-rapid death associated with AIDS.

Currently, there is one vaccine against TB, which uses a live attenuated strain of *M bovis* called Bacillus Calmette-Guérin (BCG). Unfortunately, this offers only partial protection against acquisition of infection and progression to active disease and a more-effective vaccine is badly needed. There is also need for new antibiotics. Fortunately, in recent years there has been renewed research interest in TB,[14] in the areas of diagnostics, drug therapy and vaccines,[15,16] promoted by initiatives such as the Global Alliance for TB Drug Development (GATB), which is supported by the Bill and Melinda Gates Foundation and the Rockefeller Foundation. Greatly-improved diagnostic tests have recently become are available[4,13,17,18] and have already offered new insight.[19]

Mathematical Models of TB Transmission Dynamics

The first model of TB was published by Waaler et al in 1962,[20] followed by others.[21-29] Early models tended to treat infection incidence as a parameter, rather than modelling the transmission of infection per se. Subsequent modelling work was limited until a number of circumstances generated substantial interest in tuberculosis and modelling of the infection. Firstly, there was the realization that infectious diseases had not been defeated by effective antimicrobial treatment in developing countries, where crowded accommodation and poor nutrition facilitated transmission and progression to disease and where diagnosis and treatment were inadequate. Secondly, the emergence of AIDS and associated increases in tuberculosis disease incidence renewed concern over the disease. Thirdly, the emergence of resistance to the limited number of antituberculosis drugs available lead to the need for more-complex and expensive treatments and to concerns that new drugs were not in development.

A new generation of mathematical models have been developed and analyzed exploring many aspects of tuberculosis. Unfortunately, the lack of a clear and cheap tool to diagnose latent infection and uncertainty about the natural history of tuberculosis have led to a range of different assumptions about the infection and disease being adopted, which affects the findings of the models. These assumptions may be adopted for the sake of simplicity or alternatively may reflect the authors' understanding of the infection and disease. It is the assumptions about natural history which distinguish many of the models rather than the mathematical structure employed. Most mathematical models of tuberculosis have been deterministic compartmental models using ordinary or partial differential equations or difference equations; these have tended to be either relatively simple models for algebraic analysis of equilibria and stability conditions, or more-realistic, more-complex models analyzed numerically. In addition, there have been some stochastic, discrete-event simulation models developed[30-33] and this approach is likely to increase in popularity. The following discussion of the natural history of tuberculosis is applicable across the model types.

Modelling the Natural History of TB

Latent Period

Different approaches have been taken to modelling the latent period of TB. Some models have a single latent stage which all newly-infected individuals enter, meaning that all individuals have the same exponentially-distributed period between initial infection and progression to active disease (e.g., see refs. 32, 34-36). Other models have more-realistic heterogeneity in rates of progression: some have a proportion of newly-infected individuals immediately developing active disease with the remainder developing slower-progressing latent infection,[37-52] others divide newly-infected individuals into fast- and slow-progressors,[33,53-60] whilst others have all newly-infected persons passing through an early latent phase which has a relatively high risk of progression to disease before entering a late latent state which has a lower risk of progression.[61-63] The models of Porco et al[64] and West and Thompson[53] have several latent stages of TB infection before the development of active disease.

Exogenous Reinfection

Exogeneous reinfection is where an individual with latent tuberculosis infection is subsequently infected with TB bacilli from another source. Models differ in the extent to which this occurs and its consequences: some models do not incorporate this process[32,34,37,38,41-44,46,56,62,64] and others implement it in different ways—in some models it causes progression to active disease,[33,40,45,47,48,50-52,

[54,59,61] whilst in others it causes individuals to move from slow- to fast-progressing latent infection, from where they move quickly to active disease.[58,60,63,65]

The effect of exogenous reinfection in tuberculosis epidemiology can be profound. Empirical studies have found that it is an important cause of disease in not only high-incidence areas,[11] but also medium-[66] and even low-incidence[67,68] ones. This last counter-intuitive finding is due to the presence of high-incidence foci of transmission even in low-incidence areas (see Contact patterns, below). Gomes et al[47,63,69,70] found that a large range of incidence of disease in different settings is explicable by a relatively smaller range of contact rates. The greater the importance of exogenous re-infection in causing new episodes of disease in a particular setting, the more rapidly infection can be brought under control; conversely, if much disease is occurring due to endogenous reactivation of previously-acquired, latent infection then disease cases will continue to arise for some considerable time—unless latent infection can be effectively detected and treated. However, modelling work has shown that where exogenous reinfection is the dominant cause of disease, a vaccine that is only partially protective against reinfection (as is likely to be the case with vaccines developed in the near future[71]) will be of limited benefit.[47,69,70]

Active Disease

Individuals with active tuberculosis disease are not necessarily infectious. To be infectious, one must have pulmonary disease, which can then be sub-classified classified as smear-negative or smear-positive, according to whether the infection can be detected by sputum-smear microscopy (which is less sensitive than sputum culture). Some believe that smear-negative individuals are uninfectious, others that they are infectious, but much less so than smear-positive individuals. Whilst many models consider only infectious active TB, some also consider non-infectious disease (e.g., see refs. 32, 40, 45, 56, 60); in some of these models non-infectious active TB can be a precursor to infectious disease.[32,45] Typically, models of TB do not consider sputum smear status (e.g., see refs. 33-35, 37, 38, 41-44, 46, 48, 51-53, 55, 57, 62, 63, 65, 69). However, the model of Salomon et al[58] considers both smear-negative and smear-positive individuals, with the former being less infectious, having a lower detection rate and converting to smear-positivity over time.

Recovery from TB Disease

There is uncertainty over whether those who recover from active disease through natural recovery or antibiotic treatment completely clear their infection, or whether they return to a state of latent infection. In some models, there is no natural recovery,[32,38,39,42-48,62,63,64] whilst in other models natural recovery from active disease returns the individual to latent infection,[33,37,40,41,50,51,54-59,61,65] and in others natural recovery clears infection from the host.[35,53,60] Of those models that incorporate exogenous reinfection, some allow it to occur to those who have recovered naturally from infection[35,51,54,57,58,61] whilst others do not.[40,60,65]

Models also vary in their implementation of antibiotic treatment. In Blower et al 1996[38] and Ziv et al 2001[62] treatment is effectively lifelong, whilst in Lietman and Blower[42] and Ziv et al 2004[48] treatment is effectively instantaneous and removes individuals from the model population. However, most models have a 'treatment' compartment that is entered when individuals commence treatment and exited when they cease treatment, due to successful completion, treatment failure, loss-to-follow-up, or mortality. Typically, models assume either that successful treatment clears infection completely,[32,36,47,56,58,60,65] or, more often, that it returns the individual to latent infection.[33,34,40,44,45,51,52,54,57,59,63,72] Some models explicitly consider those who have experienced treatment failure.[40,50,57,59,65] Of those models that incorporate exogenous reinfection, some allow it to occur to those who have been successfully treated[32,45,47,51,52,54,57,58,63,72] whilst others do not.[36,40,60,65]

In some models, natural recovery and successful antibiotic treatment have the same effect—e.g., both clearing infection in Dye and Williams 2008,[60] both returning to latent infection in Rodrigues et al[51]—but in others they do not. In the model of Salomon et al,[58] 'natural cure' does not clear infection, but returns individuals to slow-progressing latent infection, whilst successful therapy does clear infection; exogenous reinfection can subsequently occur in both cases. In Dye and Williams 2000[65] and Dye and Espinal 2001,[56] natural recovery moves the individual into a distinct latent state, from where relapse—but not exogenous reinfection—can occur. In the models of Blower et al 1995[37] and Porco and Blower 1998,[41] individuals who recover from disease enter a 'Recovered' compartment, which is a latent state from which relapse can occur (these models do not consider treatment), whilst the model of Blower et al 1996[38] considers treatment interventions and does not feature 'natural recovery' (none of these models incorporate exogenous reinfection). In two models of Gomes et al[47,63] there is no natural recovery; in the former model[47] treatment clears infection, whilst in the latter it does not[63] (in both models exogenous reinfection can occur to treated individuals). In Dye et al 1998,[40] Cohen and Murray 2004[57] and Cohen et al 2007,[33] both natural recovery and treatment result in distinct, separate latent-infection states (Dye et al 1998[40] also distinguishes between "good" and "bad" treatment).

Vaccination

A key area of uncertainty in tuberculosis epidemiology concerns immunity induced by vaccination. It has long been believed that vaccination with BCG (a live attenuated strain of *M bovis*), did not protect against acquisition of *Mtb* infection but gave partial protection against progression to active disease, thus protecting the vaccinated individual against disease and reducing onward transmission of infection. Consistent with this, Vynnycky and Fine argued, based on their modelling analysis,[61] that prior tuberculosis infection does not reduce susceptibility to subsequent (re) acquisition of *Mtb* but does reduce the probability of progression to active disease. However, recent studies using a new TB diagnostic test that can discriminate between an immune response to BCG and to infection with *Mtb* have shown that BCG does offer partial protection against acquisition of *Mtb* infection as well as against progression.[19] BCG vaccination has been found to have variable efficacy in different settings, for reasons that are unclear.[73-75] One possibility is that there has been differential attenuation in the different laboratories where it is maintained for vaccine production.[76] Another possibility is that in some populations there is a relatively high rate of exposure to environmental mycobacteria, which might elicit a partially-protective immune response to which BCG vaccination adds negligible additional protection.[47,77,78]

The assumed effects of vaccination have varied amongst models. Note that most models considered hypothetical future vaccines rather than BCG. It is likely that any new TB vaccine developed in the near future will be only partially protective against acquisition of infection and/or progression to disease.[71] In the models of Lietman and Blower[42] and Ziv et al 2004,[48] vaccination partially protects uninfected individuals against acquisition of infection and reduces the proportion who progress instantly to active disease in the event of their becoming infected; in those who have already acquired TB and are latently infected, vaccination reduces the rate of progression to active disease. In the models of Garcia et al[36] and Dye et al 1998,[40] vaccination offers complete protection against acquisition of infection by naïve individuals but does not affect the course of infection once acquired, whilst in Dye and Williams 2008[60] vaccination protects, not only against acquisition of infection, but also prevents progression to active disease in latently-infected individuals. In Gomes et al 2004,[47] vaccination is partially protective against acquisition of infection and exogenous reinfection, but does not affect progression, whilst Gomes et al 2007[63] consider the impact of vaccines that promote recovery from latent infection, without affecting susceptibility to (re)infection. Murray and Salomon[54] and Murray 2002[30] consider vaccines that protect against acquisition or progression. Models also differ with regard to whether the effect of vaccination wanes over time[36,40,42,48] or not.[47,60] Clearly, the modeled impact of vaccination depends upon the assumed natural history of tuberculosis (particularly whether exogenous reinfection occurs), the properties of the vaccine—whether it offers complete or only partial protection; whether it protects against initial acquisition, exogenous reinfection, or progression; whether it promotes recovery from latent infection; and whether its effects wane or are life-long—and the assumed vaccine coverage.

Population Age Structure

For simplicity, many models do not incorporate population age-structure. However, there are several reasons why incorporation of age-structure is desirable. Firstly, there is evidence that the risk of active disease is age-dependent (see refs. 40 and 61 for references). Secondly, protection from BCG vaccination wanes with time since vaccination (and effectively with age, therefore). Thirdly, since infection can last for many decades, TB's transmission dynamics have long time-scales and incorporating age structure can be important, particularly when the annual risk of infection changes over time, resulting in a changing age-prevalence profile. Fourthly, patterns of contact between individuals are often age-dependent.[79] Models that do incorporate age-structure include refs. 30, 32, 40, 60, 61 and 80.

Interactions with HIV

Papers modelling the interaction of HIV and TB have recently been reviewed by Bacaër et al[81] Modelling in detail the natural histories of both HIV and TB leads to complex models—e.g., see ref. 64. A more-common approach is to simplify the modeled natural history of HIV[40,45,58,81] and/or to make HIV incidence a parameter, rather than modelling explicitly the dynamics of HIV transmission.[49,58]

Contact Patterns

Vynnycky and Fine developed a detailed, age-structured model of TB natural history and published a series of papers applying the model to England and Wales.[61,82-85] Similar to the earlier models of TB, this model does not incorporate the dynamic transmission process per se, but rather estimates annual rates of tuberculosis acquisition by fitting to case-notification surveillance data. The authors argued that long-term declines in TB disease in the UK were primarily due to declining contact rates between individuals.[84]

Currently, what constitutes a potentially-infectious contact is poorly understood, which limits our understanding of relevant contact patterns. Whilst it is clear that tuberculosis is often clustered in households, schools and workplaces and much transmission occurs in populations living at high density in poorly-ventilated conditions such as prisons, homeless hostels and many hospitals, with transmission further associated with prolonged contact,[86-89] there is also evidence that TB can be transmitted through 'casual' contacts.[90-92]

To date, most models of TB transmission have assumed a homogeneously-mixing population, although a model considering a population divided into households has been developed[93] and then extended to incorporate both household and casual contacts.[94] A recent model by Cohen et al 2007[33] combined a more-sophisticated network structure (using the method of Read and Keeling[95]) with a more-realistic natural history of tuberculosis and found that exogenous reinfection can play an important role even in 'low incidence' settings, because tuberculosis infection tends to be clustered, meaning that those with TB infection typically experience a 'local' prevalence of active TB that is much higher than the overall population prevalence and hence they experience a much higher force of infection that promotes their progression to active disease, which in turn promotes a high local prevalence for others who are infected. The same authors[96] found that heterogeneity in the network of contacts between individuals results in localized outbreaks occurring stochastically, without requiring a particular 'cause'.

In developed countries with a low overall incidence of tuberculosis, the disease has been found to have a relatively high incidence in specific sub-populations such as homeless persons, problem drug users and prisoners (which are often overlapping groups).[88,97-99] It is not clear at present how much transmission occurs from these high-risk populations to the rest of the population. A combination of empirical work and modelling has shown that targeted screening by mobile X-ray unit followed by confirmatory testing can be an effective strategy for tackling TB in these particular groups.[100-102]

The Basic and Effective Reproductive Numbers of TB

Estimation and interpretation of TB's basic reproductive number, R_0, is complicated, for several reasons.[83] Significant changes in the contact rate can occur over the long timescale of a tuberculosis infection, affecting the potential for a new infection to spread. Also, the proportion of infected individuals who progress to infectious disease is also affected by age-specific rates of progression and mortality.

Furthermore, exogenous reinfection complicates the relationship between the basic reproductive number, R_0 and the effective reproductive number, $R(t)$. Without exogenous reinfection, in a homogeneous population at least, the relationship between R_0 and $R(t)$, is simple, with the latter being the product of the former and the proportion of the population that is uninfected. However, this relationship does not hold for TB because exogenous reinfection affects $R(t)$ but plays no part in R_0, the invasion threshold for a totally-susceptible population. Exogenous reinfection increases the proportion of infected individuals who progress to disease and hence become infectious, which tends to increase $R(t)$; however, this increases the prevalence of infection, which reduces the proportion of the population that is susceptible, which in contrast tends to reduce $R(t)$.

It was proposed that there may be a 'backward bifurcation' resulting in exogenous reinfection enabling TB to persist even where $R_0 < 1$,[103] but this phenomenon is not relevant for reasonable parameter values,[104] as it requires those already infected to have increased susceptibility to subsequent reinfection, rather than reduced susceptibility due to immunity.

Modelling Strains of TB

There are three main reasons for modelling distinct strains of TB: to understand the emergence of drug resistance at the population level, to develop methods of analysis of DNA-fingerprinting data to elucidate transmission patterns and to explore the evolution of the bacterium and its relationship with pathogenesis and transmissibility.

Inference of Tuberculosis Transmission Patterns from DNA Fingerprinting Data

Tuberculosis's variable and usually long latent period makes investigating its transmission dynamics challenging. Recent developments in DNA fingerprinting techniques for TB offer an exciting new opportunity to gain new insights into TB transmission patterns by identifying transmission clusters.[105] The inference of transmission patterns from identification of clusters of related isolates is not simple, however and modelling has an important role to play in developing methods. Some isolates may have the same fingerprint without being part of the same local transmission cluster. Conversely, incomplete sampling may mean that isolates from recent transmission or newly-imported strains may appear unlinked;[106] however, independent importations of a strain endemic elsewhere may falsely appear to be a local transmission cluster. Modelling studies have shown that clustering due to recent transmission will tend to be underestimated, due to incomplete sampling.[31,107] (However, a modelling study applied to the Netherlands found that in the case of older individuals—who were usually infected early in life when TB was more prevalent and who experienced reactivation—clustering overestimated the amount of recent transmission because those individuals were more likely to have been sources of infection, rather than recently-infected cases.[108]) Murray's thorough modelling study[31] also found that odds ratios for risk factors associated with clustering are also typically underestimated, particularly if the odds ratios are high, 'true' cluster sizes are small, or the proportion of cases sampled is low. She further concluded that, due to variation in these factors between studies, comparison of studies is difficult.

Drug Resistance

A key concern of tuberculosis control is combating drug resistance, which can arise through poor adherence to treatment and can then be transmitted. Resistance to isoniazid is relatively common; multi-drug resistant tuberculosis (MDR-TB) is resistant to both isoniazid and rifampicin. (It is important to note that these are phenotypic properties—i.e., there are multiple strains of TB that are drug-resistant.) Treatment of drug-resistant infection is much more difficult and expensive than normal treatment[4] and, with the emergence of extensively-drug-resistant (XDR) tuberculosis there is a concern that some infections may become untreatable.[109,110] Understanding transmission patterns is important to the development of effective control strategies.

Unfortunately, the expense of testing for drug resistance has meant that in developing countries it has been common to place tuberculosis patients on first-line therapy unless they have certain risk factors for drug-resistant infection (e.g., likely acquisition from a known drug-resistant case or a past history of drug-resistant infection); only if therapy fails are they tested for drug resistant infection. Clearly, this promotes the evolution of further resistance and the transmission of drug-resistant strains by providing ineffective therapy for a period of months. A new diagnostic test, MODS, offers both faster diagnosis of infection and simultaneously determines its drug-sensitivity profile and is much cheaper than traditional methods.[17] The most cost-effective way to use MODS remains to be determined[18] and modelling has an important role to play in estimating numbers of infections averted using different protocols.

Blower et al 1996[38] presented the first model of the transmission dynamics of drug-sensitive and MDR TB; a later paper[39] discussed its application to control strategies. Clearly the fitness of MDR-TB strains will determine how difficult MDR-TB is to combat and will determine whether the problem may remain localised or become more widespread. Commonly it is assumed that drug-resistant strains must 'pay a price' for their resistance in terms of reduced transmission efficiency. For example, the fitness of MDR-TB relative to drug-sensitive wild-type strains was varied in the range 70-100% in the model of Dye and Williams 2000[65] and in the range 10-70% in the model of Salomon et al,[58] whilst Dye and Espinal 2001[56] assumed that MDR-TB's fitness was 70% fitness of wild-type.

Worryingly, an in vitro study by Gagneux et al[111] found that resistant strains may not necessarily be less fit. Furthermore, even if there is an 'initial' fitness cost to acquisition of drug resistance, compensatory mutations acquired over time may negate it. Cohen and Murray 2004[57] presented a model that incorporates heterogeneity in the fitness of MDR-TB, which can change through mutation; they found that relatively fit MDR-TB strains may emerge over time, meaning that short-term measures of the burden of drug resistance may not be indicative of the long-term scale of the problem. Blower and Chou[46] developed a multi-strain model that allowed for heterogeneity in strain fitness and incorporated the process of 'amplification' of resistance, where strains can sequentially acquire resistance to multiple drugs as a consequence of suboptimal treatment. They recommended that second-line drugs should be rapidly deployed to potential 'hot zones' where amplification of resistance by sequential acquisition of mutations could lead to the emergence of fit MDR strains.

Rodrigues et al[51] report that when exogenous reinfection is relatively common, resulting in individuals being infected with both drug-sensitive and drug-resistant strains, the outcome of the resulting within-host competition which determines which strain dominates can have important consequences for the population-level emergence of resistant strains.

The recent emergence of extensively drug resistant tuberculosis (XDR-TB) has prompted new modelling work by Blower and Supervie,[112] which applied the model of Blower and Chou[46] and a more-detailed analysis by Basu et al[59] of the prospects for reducing the incidence of XDR-TB through measures to combat nosocomial transmission in a rural South African setting. As XDR-TB becomes better-understood it will clearly be an important concern for further modelling work.

Host Genetic Factors and Within-Host Modelling

An area of mathematical modelling that we will not consider in detail here, but which is an area of growing importance, is within-host modelling.[113] This aims to gain a greater understanding of the interaction of the pathogen and the immune system, including how latent infection is established and maintained, how progression to active disease occurs, the process of recovery and the effect of vaccination. It has been applied to the fundamentals of the interaction of *M tuberculosis* and the immune system,[114-118] the impact of antibiotic therapy and the emergence of resistance[119,120] and the effect of HIV co-infection.[121] Many host-genetic factors have been associated with susceptibility or resistance to tuberculosis infection and there is growing interest in the possibility that strains of TB may be adapted to particular host populations and that this may have implications for the efficacy of vaccines and antibiotic therapies in different settings.[4,122-127] However, there has been relatively little modelling work to date on the effect of host genetic factors on the epidemiology of

TB.[43,44] Future modelling work is likely to combine within-host and between-host (transmission) modelling to get a better understanding TB evolution.

TB-Control Strategies

The first attempts to model TB were motivated by a desire to improve control programs[20-29] and this has continued to inspire much work. Models have been used to set targets for (e.g.) the proportion of cases found and successfully treated and how long gaining control of TB is likely to take under different scenarios. Since resources are inevitably limited, models can be used to determine the most efficient allocation of those resources. Of course, it should be remembered that there is uncertainty in the predicted impact of interventions, due not only to uncertainty in parameter estimates but also uncertainty in model structure—i.e., how the natural history of TB is represented in the model. Improvements in our understanding of TB's natural history and transmission dynamics will obviously reduce the uncertainty of model predictions.

Modelling has been used to inform WHO's Stop TB Strategy guidelines for the proportion of cases of active TB that need to be diagnosed and treated successfully in order to control TB[40,60,65,128] and to assess the impact of interventions by predicting numbers of cases averted (e.g., see ref. 128). Salomon et al modelling the impact of potential new, shorter DOTS regimens which may have lower rates of default and make more-efficient use of limited treatment resources.[58] Resch et al[50] found that providing second-line drugs for resistant TB infections in Peru was cost-effective.

Treating latent infection has been recognized to have an important role because the prevalence of latent infection is high relative to that of active disease and there is the risk of endogenous reactivation (which may be prompted by HIV), as well as by exogenous reinfection, leading to active disease. Even if new acquisition of TB infection were to cease immediately then there would continue to be cases of TB disease due to endogenous reactivation for decades to come. Currie et al 2003[43] modeled the interaction between HIV and TB in Uganda, Kenya and South Africa and found that the most effective way to control TB epidemics fuelled by HIV was to focus on TB control through preventive therapy and treatment of active TB disease. Cohen et al 2006[129] found that isoniazid preventive therapy (IPT) for HIV-positive persons with latent TB in a sub-Saharan African setting would reduce TB incidence at the cost of increased drug resistance, requiring policies for the effective detection and treatment of resistant TB, which would be costly. They also found that HAART treatment for HIV would have little short-term impact on TB transmission, although it would avert HIV-associated deaths. It is interesting to note that a modelling paper from the preHAART era also concluded that prophylaxis and treatment of active TB disease could also be highly effective in averting TB deaths in a population with high HIV prevalence.[130]

Ziv et al 2001[62] modeled the impact of targeted treatment for early latent TB infection, which has the greatest hazard of progression to active disease (with these recently-infected individuals identified through contact tracing, although this is not modeled explicitly) and compared it with a strategy of treating longer-term latent infections (which would tend to be identified through screening programs) and found that the former may be more efficient since it targets those at greatest risk of progression. They found that to achieve TB elimination by combining treatment of latent infection with treatment of active disease, the proportion of cases of early latent infection that need to be treated is lower than the proportion of cases of long-term latent infection.

Typically, combining interventions is potentially much more effective than using any one alone due to synergies.[60] For example, even if new acquisition of infection could be halted immediately, endogenous reactivation would cause new cases of active disease to arise in those who are already infected for decades, so effective treatment strategies would still be required and any new vaccine (which is unlikely to be completely protective against acquisition, in any case) will become part of an ensemble of interventions against TB.

The most cost-effective combinations of interventions are likely to vary amongst settings and are likely to change with time, as the transmission dynamics of TB change as a consequence of interventions and other factors. For example, Ziv et al 2004[48] found that postexposure vaccines (protecting latently-infected individuals against progression to active disease) would initially have a greater impact than pre-exposure vaccines (protecting against acquisition of infection) on reducing the number of new cases of disease in a high-prevalence setting, but that over time the impact of the former would decline and the latter increase.

A new vaccine may have a role to play in combating the emergence of drug-resistant strains, by limiting transmission and the opportunity to acquire 'amplified' resistance or compensatory mutations increasing transmissibility. Modelling would be required to determine the necessary vaccine properties (minimum efficacy against acquisition or progression, duration of protection) and coverage.

Conclusion

Recent years have seen intense activity modelling tuberculosis, with some interesting insights about modeled behavior emerging. Unfortunately, much of this work is speculative because of our incomplete understanding of tuberculosis natural history and the slow timescales of tuberculosis epidemics. This slow spread of infection and emergence of disease hinders efforts to compare model predictions with observed patterns. Many important questions remain to be resolved regarding the natural history of TB infection and gaining a greater understanding of TB transmission dynamics requires further empirical laboratory and field work, mathematical modelling and interaction between them. Modelling can be used to quantify uncertainty due to different gaps in our knowledge to help identify research priorities. Fortunately, the present moment is an exciting time for TB epidemiology, with rapid progress being made in applying new mathematical modelling techniques, new tools for TB diagnosis and genetic analysis and a growing interest in developing more-effective public-health interventions.[14] Since future field trials of vaccine candidates will occur against a background of BCG vaccination and TB treatment[71] a key application of mathematical modelling will be in trial design and analysis, as has been advocated for HIV-vaccine trials.[131]

References

1. World Health Organization. Global tuberculosis control: surveillance, planning, financing. Geneva, 2006: WHO/HTM/TB/2006.362.
2. Nunn P, Williams B, Floyd K et al. Tuberculosis control in the era of HIV. Nat Rev Immunol 2005; 5(10): 819-826.
3. Corbett EL, Marston B, Churchyard GJ et al. Tuberculosis in sub-Saharan Africa: opportunities, challenges and change in the era of antiretroviral treatment. Lancet 2006; 367:926-937.
4. Maartens G, Wilkinson RJ. Tuberculosis. Lancet 2007; 370:2030-2043.
5. Brudney K, Dobkin J. Resurgent tuberculosis in New York City: human immunodeficiency virus, homelessness and the decline of tuberculosis control programs. Am Rev Respir Dis 1991; 144:745-9.
6. Frieden TR, Fujiwara PI, Washko RM et al. Tuberculosis in New York City—turning the tide. N Engl J Med 1995; 333:229-33.
7. Health Protection Agency. Tuberculosis in the UK: Annual report on tuberculosis surveillance and control in the UK 2007. London: Health Protection Agency Centre for Infections, 2007.
8. Escombe AR, Oeser CC, Gilman RH et al. Natural ventilation for the prevention of airborne contagion. PLoS Med 2007; 4(2):309-317.
9. Lillebaek T, Dirksen A, Baess I et al. Molecular evidence of endogenous reactivation of Mycobacterium tuberculosis after 33 years of latent infection. J Infect Dis 2002; 185(3):401-404.
10. Lillebaek T, Dirksen A, Vynnycky E et al. Stability of DNA patterns and evidence of Mycobacterium tuberculosis reactivation occurring decades after the initial infection. J Infect Dis 2003; 188(7):1032-1039.
11. Van Rie A, Warren R, Richardson M et al. Exogenous reinfection as a cause of recurrent tuberculosis after curative treatment. N Engl J Med 1999; 341(16):1174-1179.
12. Feja K, Saiman L. Tuberculosis in children. Clin Chest Med 2005; 26(2):295-312.
13. Lalvani, A. Diagnosing tuberculosis infection in the 21st century—New tools to tackle an old enemy. Chest 2007; 131(6):1898-1906.
14. Onyebujoh P, Rodriguez W, Mwaba P. Priorities in tuberculosis research. Lancet 2006; 367:940-942.
15. Andersen P. Tuberculosis vaccines—an update. Nat Rev Microbiol 2007; 5(7):484-487.
16. Ly LH, McMurray DN. Tuberculosis: vaccines in the pipeline. Expert Rev Vaccines 2008; 7(5):635-650.
17. Moore DAJ, Evans CAW, Gilman RH et al. Microscopic-observation drug-susceptibility assay for the diagnosis of TB. N Engl J Med 2006; 355(15):1539-1550.

18. Moore DAJ. Future prospects for the MODS assay in multidrug-resistant tuberculosis diagnosis. Future Microbiol 2007; 2(2):97-101.
19. Soysal A, Millington KA, Bakir M et al. Effect of BCG vaccination on risk of Mycobacterium tuberculosis infection in children with household tuberculosis contact: a prospective community-based study. Lancet 2005; 366(9495):1443-1451.
20. Waaler HT, Gese A, Anderson S. The use of mathematical models in the study of the epidemiology of tuberculosis. Am J Pub Health 1962; 52:1002-1013.
21. Waaler HT, Piot MA. The use of an epidemiological model for estimating the effectiveness of tuberculosis control measures: Sensitivity of the effectiveness of tuberculosis control measures to the coverage of the population. Bull World Health Organ 1969; 41:75-93.
22. Waaler HT, Piot MA. Use of an epidemiological model for estimating the effectiveness of tuberculosis control measures: Sensitivity of the effectiveness of tuberculosis control measures to the social time preference. Bull World Health Organ 1970; 43:1-16.
23. Brogger S. Systems analysis in tuberculosis control: a model. Am Rev Respir Dis 1967; 95:419-434.
24. Ferebee SH. An epidemiological model of tuberculosis in the United States. Bull Natl Tuberc Respir Dis Assoc 1967; 4-7.
25. Ferebee S. Controlled chemoprophylaxis trials in tuberculosis a general review. Adv Tuberc Res 1970; 17:28-106.
26. ReVelle CS. The economics allocation of tuberculosis control activities in developing nations. Ph.D. Thesis, 1967. Cornell University, Ithaca, NY.
27. ReVelle CS, Lynn WR, Feldmann F. Mathematical models for the economic allocation of tuberculosis control activities in developing nations. Am Rev Respir Dis 1967; 96:893-909.
28. Revelle C, Feldmann F, Lynn W. An optimization model of tuberculosis epidemiology. Manage Sci 1969; 16:B190-B211.
29. Azuma Y. Simple simulation-model of tuberculosis epidemiology for use without large-scale computers. Bull World Health Organ 1975; 52(3):313-322.
30. Murray M. Determinants of cluster distribution in the molecular epidemiology of tuberculosis. Proc Natl Acad Sci USA 2002; 99(3):1538-1543.
31. Murray M. Sampling bias in the molecular epidemiology of tuberculosis. Emerg Infect Dis 2002; 8(4):363-369.
32. Hughes GR, Currie CSM, Corbett EL. Modeling tuberculosis in areas of high HIV prevalence. Proceedings of the 2006 Winter Simulation Conference, Vols 1-5. New York, IEEE:459-465.
33. Cohen T, Colijn C, Finklea B et al. Exogenous re-infection and the dynamics of tuberculosis epidemics: local effects in a network model of transmission. J R Soc Interface 2007; 523-531.
34. Castillo-Chavez C, Feng ZL. To treat or not to treat: The case of tuberculosis. J Math Biol 1997; 35(6):629-656.
35. Jia ZW, Tang GY, Jin Z et al. Modeling the impact of immigration on the epidemiology of tuberculosis. Theor Popul Biol 2008; 73(3):437-448.
36. Garcia AJ, Maccario J, Richardson S. Modelling the annual risk of tuberculosis infection. Int J Epidemiol 1997; 26(1):190-203.
37. Blower SM, McLean AR, Porco TC et al. The intrinsic transmission dynamics of tuberculosis epidemics. Nat Med 1995; 1(8):815-821.
38. Blower SM, Small PM, Hopwell PC. Control strategies for tuberculosis epidemics: new models for old problems. Science 1996; 273:497-500.
39. Blower SM, Gerberding JL. Understanding, predicting and controlling the emergence of drug-resistant tuberculosis: a theoretical framework. J Mol Med 1998; 76(9):624-636.
40. Dye C, Garnett GP, Sleeman A et al. Prospects for worldwide tuberculosis control under the WHO DOTS strategy. Lancet 1998; 352(9144):1886-1891.
41. Porco TC, Blower SM. Quantifying the intrinsic transmission dynamics of tuberculosis. Theor Popul Biol 1998; 54(2):117-132.
42. Lietman T, Blower SM. Potential impact of tuberculosis vaccines as epidemic control agents. Clin Infect Dis 2000; 30:S316-S322.
43. Murphy BM, Singer BH, Anderson S et al. Comparing epidemic tuberculosis in demographically distinct heterogeneous populations. Math Biosci 2002; 180:161-185.
44. Murphy BM, Singer BH, Kirschner D. On treatment of tuberculosis in heterogeneous populations. J Theor Biol 2003; 223(4):391-404.
45. Currie CSM, Williams BG, Cheng RCH et al. Tuberculosis epidemics driven by HIV: is prevention better than cure? AIDS 2003; 17(17):2501-2508.
46. Blower SM, Chou T. Modeling the emergence of the 'hot zones': tuberculosis and the amplification dynamics of drug resistance. Nat Med 2004; 10(10):1111-1116.

47. Gomes MGM, Franco AO, Gomes MC et al. The reinfection threshold promotes variability in tuberculosis epidemiology and vaccine efficacy. Proc R Soc Lond Ser B-Biol Sci 2004; 271(1539):617-623.
48. Ziv E, Daley CL, Blower SM. Potential public health impact of new tuberculosis vaccines. Emerg Infect Dis 2004; 10(9):1529-1535.
49. Dowdy DW, Chaisson RE, Moulton LH et al. The potential impact of enhanced diagnostic techniques for tuberculosis driven by HIV: a mathematical model. AIDS 2006; 20(5):751-762.
50. Resch SC, Salomon JA, Murray M et al. Cost-effectiveness of treating multidrug-resistant tuberculosis. PLoS Med 2006; 3(7):1048-1057.
51. Rodrigues P, Gomes MGM, Rebelo C. Drug resistance in tuberculosis—a reinfection model. Theor Popul Biol 2007; 71(2):196-212.
52. Bhunu CP, Garira W, Mukandavire Z et al. Tuberculosis transmission model with chemoprophylaxis and treatment. Bull Math Biol 2008; 70(4):1163-1191.
53. West RW, Thompson JR. Modeling the impact of HIV on the spread of tuberculosis in the United States. Math Biosci 1997; 143(1):35-60.
54. Murray CJL, Salomon JA. Modeling the impact of global tuberculosis control strategies. Proc Natl Acad Sci USA 1998; 95(23):13881-13886.
55. Debanne SM, Bielefeld RA, Cauthen GM et al. Multivariate Markovian modeling of tuberculosis: Forecast for the United States. Emerg Infect Dis 2000; 6(2):148-157.
56. Dye C, Espinal MA. Will tuberculosis become resistant to all antibiotics? Proc R Soc Lond Ser B-Biol Sci 2001; 268(1462):45-52.
57. Cohen T, Murray M. Modeling epidemics of multidrug-resistant M tuberculosis of heterogeneous fitness. Nat Med 2004; 10(10):1117-1121.
58. Salomon JA, Lloyd-Smith JO, Getz WM et al. Prospects for advancing tuberculosis control efforts through novel therapies. PLoS Med 2006; 3(8):1302-1309.
59. Basu S, Andrews JR, Poolman EM et al. Prevention of nosocomial transmission of extensively drug-resistant tuberculosis in rural South African district hospitals: an epidemiological modelling study. Lancet 2007; 370(9597):1500-1507.
60. Dye C, Williams BG. Eliminating human tuberculosis in the twenty-first century. J R Soc Interface 2008; 5(23):653-662.
61. Vynnycky E, Fine PEM. The natural history of tuberculosis: the implications of age-dependent risks of disease and the role of reinfection. Epidemiol Infect 1997; 119(2):183-201.
62. Ziv E, Daley CL, Blower SM. Early therapy for latent tuberculosis infection. Am J Epidemiol 2001; 153(4):381-385.
63. Gomes MGM, Rodrigues P, Hilker FM et al. Implications of partial immunity on the prospects for tuberculosis control by post-exposure interventions. J Theor Biol 2007; 248(4):608-617.
64. Porco TC, Small PM, Blower SM. Amplification dynamics: Predicting the effect of HIV on tuberculosis outbreaks. J Acquir Immune Defic Syndr 2001; 28(5):437-444.
65. Dye C, Williams BG. Criteria for the control of drug-resistant tuberculosis. Proc Natl Acad Sci USA 2000; 97(14):8180-8185.
66. Caminero JA, Pena MJ, Campos-Herrero MI et al. Exogenous reinfection with tuberculosis on a European island with a moderate incidence of disease. Am J Respir Crit Care Med 2001; 163(3):717-720.
67. Bandera A, Gori A, Catozzi L et al. Molecular epidemiology study of exogenous reinfection in an area with a low incidence of tuberculosis. J Clin Microbiol 2001; 39:2213-2218.
68. de Boer AS, Borgdorff MW, Vynnycky E et al. Exogenous re-infection as a cause of recurrent tuberculosis in a low-incidence area. Int J Tuberc Lung Dis 2003; 7(2):145-152.
69. Gomes MGM, White LJ, Medley GF. Infection, reinfection and vaccination under suboptimal immune protection: epidemiological perspectives. J Theor Biol 2004; 228(4):539-549.
70. Gomes MGM, White LJ, Medley GF. The reinfection threshold. J Theor Biol 2005; 236(1):111-113.
71. Gupta UD, Katoch VM, McMurray DN. Current status of TB vaccines. Vaccine 2007; 25(19):3742-3751.
72. Currie CSM, Floyd K, Williams BG et al. Cost, affordability and cost-effectiveness of strategies to control tuberculosis in countries with high HIV prevalence. BMC Public Health 2005; 5:14.
73. Fine PEM, Rodrigues LC. Modern vaccines: mycobacterial diseases. Lancet 1990; 335:1016-1020.
74. Bloom BR, Fine PE. The BCG experience: implications for future vaccines against TB. In: Bloom BR, ed. Tuberculosis Pathogenesis, Protection And Control. Washington, DC: American Society for Microbiology 1994:531-552.
75. Colditz GA, Brewer TF, Berkey CS et al. Efficacy of BCG vaccine in the prevention of tuberculosis. JAMA 1994; 271:698-702.
76. Behr MA, Wilson MA, Gill WP et al. Comparative genomics of BCG vaccines by whole-genome DNA microarray. Science 1999; 284:1520-1523.

77. Palmer CE, Long MW. Effects of infection with atypical mycobacteria on BCG vaccination and tuberculosis. Am Rev Respir Dis 1966; 94:553-568.
78. Fine PEM. Variation in protection by BCG: implications of and for heterologous immunity. Lancet 1995; 346:1339-1345.
79. Mossong J, Hens N, Jit M et al. Social contacts and mixing patterns relevant to the spread of infectious diseases. PLoS Med 2008; 5(3):e74. doi:10.1371/journal.pmed.0050074.
80. Schulzer M, Radhamani MP, Grzybowski S et al. A Mathematical Model for the Prediction of the Impact of HIV Infection on Tuberculosis. Int J Epidemiol 1994; 23(2):400-407.
81. Bacaër N, Ouifki R, Pretorius C et al. Modeling the joint epidemics of TB and HIV in a South African township. J Math Biol 2008; 57:557-593.
82. Vynnycky E, Fine PEM. The annual risk of infection with Mycobacterium tuberculosis in England and Wales since 1901. Int J Tuberc Lung Dis 1997; 1(5):389-396.
83. Vynnycky E, Fine PEM. The long-term dynamics of tuberculosis and other diseases with long serial intervals: implications of and for changing reproduction numbers. Epidemiol Infect 1998; 121(2):309-324.
84. Vynnycky E, Fine PEM. Interpreting the decline in tuberculosis: the role of secular trends in effective contact. Int J Epidemiol 1999; 28(2):327-334.
85. Vynnycky E, Fine PEM. Lifetime risks, incubation period and serial interval of tuberculosis. Am J Epidemiol 2000; 152(3):247-263.
86. Kenyon TA, Valway SE, Ihle WW et al. Transmission of multidrug-resistant Mycobacterium tuberculosis during a long airplane flight. N Engl J Med 1996; 334:933-938.
87. Raffalli J, Sepkowitz KA, Armstrong D. Community-based outbreaks of tuberculosis. Arch Intern Med 1996; 156:1053-1060.
88. Barnes PF, Yang ZH, Pogoda JM et al. Foci of tuberculosis transmission in central Los Angeles. Am J Respir Crit Care Med 1999; 159(4):1081-1086.
89. Becerra MC, Pachao-Torreblanca IF, Bayona J et al. Expanding tuberculosis case detection by screening household contacts. Public Health Rep 2005; 120(3):271-277.
90. Verver S, Warren RM, Munch Z et al. Transmission of tuberculosis in a high incidence urban community in South Africa. Int J Epidemiol 2004; 33(2):351-357.
91. Horna-Campos OJ, Sanchez-Perez HJ, Sanchez I et al. Public transportation and pulmonary tuberculosis, Lima, Peru. Emerg Infect Dis 2007; 13(10):1491-1493.
92. Ohkado A, Nagamine M, Murase Y et al. Molecular epidemiology of Mycobacterium tuberculosis in an urban area in Japan, 2002-2006. Int J Tuberc Lung Dis 2008; 12(5):548-554.
93. Aparicio JP, Capurro AF, Castillo-Chavez C. Transmission and dynamics of tuberculosis on generalized households. J Theor Biol 2000; 206(3):327-341.
94. Song BJ, Castillo-Chavez C, Aparicio JP. Tuberculosis models with fast and slow dynamics: the role of close and casual contacts. Math Biosci 2002; 180:187-205.
95. Read JM, Keeling MJ. Disease evolution on networks: the role of contact structure. Proc R Soc Lond Ser B-Biol Sci 2003; 270:699-708.
96. Colijn C, Cohen T, Murray M. Emergent heterogeneity in declining tuberculosis epidemics. J Theor Biol 2007; 247(4):765-774.
97. Malakmadze N, Gonzalez IM, Oemig T et al. Unsuspected recent transmission of tuberculosis among high-risk groups: Implications of universal tuberculosis genotyping in its detection. Clin Infect Dis 2005; 40(3):366-373.
98. Story A, van Hest R, Hayward A. Tuberculosis and social exclusion—Developed countries need new strategies for controlling tuberculosis. Br Med J 2006; 333(7558):57-58.
99. Story A, Murad S, Roberts W et al. Tuberculosis in London: The importance of homelessness, problem drug use and prison. Thorax 2007; 62(8):667-671.
100. de Vries G, van Hest RA. From contact investigation to tuberculosis screening of drug addicts and homeless persons in Rotterdam. Eur J Public Health 2006; 16(2):133-136.
101. de Vries G, van Hest RAH, Richardus AH. Impact of mobile radiographic screening on tuberculosis among drug users and homeless persons. Am J Respir Crit Care Med 2007; 176(2):201-207.
102. White PJ, Abubakar I, Garnett GP et al. Averting TB transmission in London by providing screening to prisoners, homeless people and problem drug users. (in preparation).
103. Feng ZL, Castillo-Chavez C, Capurro AF. A model for tuberculosis with exogenous reinfection. Theor Popul Biol 2000; 57(3):235-247.
104. Lipsitch M, Murray MB. Multiple equilibria: Tuberculosis transmission require unrealistic assumptions. Theor Popul Biol 2003; 63(2):169-170.
105. Murray M, Nardell E. Molecular epidemiology of tuberculosis: achievements and challenges to current knowledge. Bull World Health Organ 2002; 80(6):477-482.

106. Glynn JR, Bauer J, de Boer AS et al. Interpreting DNA fingerprint clusters of Mycobacterium tuberculosis. Int J Tuberc Lung Dis 1999; 3(12):1055-1060.
107. Glynn JR, Vynnycky E, Fine PEM. Influence of sampling on estimates of clustering and recent transmission of Mycobacterium tuberculosis derived from DNA fingerprinting techniques. Am J Epidemiol 1999; 149(4):366-371.
108. Vynnycky E, Nagelkerke N, Borgdorff MW et al. The effect of age and study duration on the relationship between 'clustering' of DNA fingerprint patterns and the proportion of tuberculosis disease attributable to recent transmission. Epidemiol Infect 2001; 126(1):43-62.
109. Centers for Disease Control and Prevention. Emergence of Mycobacterium tuberculosis with extensive resistance to secondline drugs—worldwide, 2000-2004. Morb Mortal Wkly Rep 2006; 55:301-305.
110. Raviglione M. XDR-TB: entering the post-antibiotic era? Int J Tuberc Lung Dis 2006; 10(11):1185-1187.
111. Gagneux S, Long CD, Small PM et al. The competitive cost of antibiotic resistance in Mycobacterium tuberculosis. Science 2006; 312:1944-1946.
112. Blower S, Supervie V. Predicting the future of XDR tuberculosis. Lancet Infect Dis 2007; 7(7):443-443.
113. Andrew SM, Baker CTH, Bocharov GA. Rival approaches to mathematical modelling in immunology. J Comput Appl Math 2007; 205(2):669-686.
114. Ganguli S, Gammack D, Kirschner DE. A metapopulation model of granuloma formation in the lung during infection with Mycobacterium tuberculosis. Math Biosci Eng 2005; 2(3):535-560.
115. Kirschner D, Marino S. Mycobacterium tuberculosis as viewed through a computer. Trends Microbiol 2005; 13(5):206-211.
116. Lin PL, Kirschner D, Flynn JL. Modeling pathogen and host: in vitro, in vivo and in silico models of latent Mycobacterium tuberculosis infection. Drug Discov Today: Disease Models 2005; 2(2):149-154.
117. Sud D, Bigbee C, Flynn JL et al. Contribution of CD8(+) T-cells to control of Mycobacterium tuberculosis infection. J Immunol 2006; 176(7):4296-4314.
118. Young D, Stark J, Kirschner D. Systems biology of persistent infection: tuberculosis as a case study. Nat Rev Microbiol 2008; 6(7):520-528.
119. Magombedze G, Garira W, Mwenje E. Mathematical modeling of chemotherapy of human TB infection. J Biol Syst 2006; 14(4):509-553.
120. Alavez-Ramirez J, Castellanos JRA, Esteva L et al. Within-host population dynamics of antibiotic-resistant M. tuberculosis. Math Med Biol 2007; 24(1):35-56.
121. Kirschner D. Dynamics of co-infection with M tuberculosis and HIV-1. Theor Popul Biol 1999; 55(1):94-109.
122. Bellamy R, Hill AV. Genetic susceptibility to mycobacteria and other infectious pathogens in humans. Curr Opin Immunol 1998; 10:483-487.
123. Hill AVS. The immunogenetics of human infectious diseases. Annu Rev Immunol 1998; 16:593-617.
124. Meyer CG, May J, Stark K. Human leukocyte antigens in tuberculosis and leprosy. Trends Microbiol 1998; 6(4):148-154.
125. Kramnik I, Dietrich WF, Demant P et al. Genetic control of resistance to experimental infection with virulent Mycobacterium tuberculosis. Proc Natl Acad Sci USA 2000; 97(15):8560-8565.
126. Gagneux S, DeRiemer K, Van T et al. Variable host-pathogen compatibility in Mycobacterium tuberculosis. Proc Natl Acad Sci USA 2006; 103(8):2869-2873.
127. Gagneux S, Small PM. Global phylogeography of Mycobacterium tuberculosis and implications for tuberculosis product development. Lancet Inf Dis 2007; 7(5):328-337.
128. Dye C, Zhao FZ, Scheele S et al. Evaluating the impact of tuberculosis control: number of deaths prevented by short-course chemotherapy in China. Int J Epidemiol 2000; 29(3):558-564.
129. Cohen T, Lipsitch M, Walensky RP et al. Beneficial and perverse effects of isoniazid preventive therapy for latent tuberculosis infection in HIV-tuberculosis coinfected populations. Proc Natl Acad Sci USA 2006; 103(18):7042-7047.
130. Heymann SJ. Modelling the efficacy of prophylactic and curative therapies for preventing the spread of tuberculosis in africa. Trans Roy Soc Trop Med Hyg 1993; 87(4):406-411.
131. Boily MC, Abu-Raddad L, Desai K et al. Measuring the public-health impact of candidate HIV vaccines as part of the licensing process. Lancet Inf Dis 2008; 8(3):200-207.

Chapter 10

Modelling Trachoma for Control Programmes

Manoj Gambhir,* María-Gloria Basáñez, Isobel M. Blake and Nicholas C. Grassly

Abstract

Trachoma is a major cause of blindness in the developing world and 63 million people are currently infected. Large-scale control programmes are being implemented to clear ocular *Chlamydia trachomatis* infection—the causative agent of trachoma—and improve environmental conditions to reduce transmission. Chemotherapeutic intervention involves antibiotic administration and the effectiveness of this treatment is currently under investigation. A mathematical model has been developed to allow the impact of control programmes on infection and blinding disease sequelae to be predicted. The model has a structure that allows an important aspect of trachoma pathogenesis to be taken into account, namely the effect of repeated cycles of infection and recovery leading to scarring and the damaging disease sequelae. This novel model structure reproduces many age- and time-dependent epidemiological patterns observed in endemic settings and allows the dynamic effect of treatment on infection and disease sequelae to be gauged.

Introduction

Trachoma is the leading cause of infectious blindness in the world[1] and in 1996 the World Health Organization (WHO) adopted a resolution for the Global Elimination of blinding Trachoma by the year 2020 (GET 2020). In addition, trachoma has been included among several 'neglected tropical diseases' affecting the developing world[2] which are currently being targeted for very large integrated control programmes. Mass drug administration (MDA) with the antibiotic azithromycin is the cornerstone of the WHO 'SAFE'[a,1] strategy for trachoma control and there have been several recent studies whose aim has been to investigate the effect of one or several rounds of mass treatment for populations with varying baseline levels of disease endemicity.[3-5] One of the main findings of these studies has been that communities respond heterogeneously to mass treatment and in some communities a persistent reduction in infection prevalence is achieved whereas others return very quickly to pretreatment levels. The likelihood of eliminating the infection appears to be strongly related to the baseline prevalence and, therefore, the transmission environment but this relationship remains to be properly quantified. Mathematical models that include the salient features of disease pathogenesis and the mechanisms of contact and infection transmission can, when fitted to baseline data, allow control interventions to be simulated; the long term consequences of treatment can therefore be investigated.

[a] S—Surgery for trichiasis; A—Antibiotic for infection; F—Facial cleanliness; E—Environmental improvement.

*Corresponding Author: Manoj Gambhir—Department of Infectious Disease Epidemiology, Faculty of Medicine, St. Mary's campus, Imperial College London, Norfolk Place, London W2 1PG, UK. Email: m.gambhir@imperial.ac.uk

Modelling Parasite Transmission and Control, edited by Edwin Michael and Robert C. Spear.

This chapter begins with a brief review of the mechanisms behind trachoma disease pathogenesis and we explain that these mechanisms can lead to distributions of bacterial infection and disease that are observed in endemic populations. The facts pertaining to disease pathogenesis are then included in the construction of a simple mathematical model of the transmission of infection in a population and the progression of individuals to disease sequelae. The structure of the mathematical model is of interest because it combines aspects of the simple SIS approach[6]—most often used for viral and bacterial microparasitic infections—with the infection burden approach more commonly used to model macroparasitic infections such as helminths. The simple model of disease pathogenesis is shown to reproduce the population patterns of infection, active disease, bacterial load and trichiasis (the final disease sequela before blindness) by age. Once these static patterns have been reproduced, a treatment scenario can be applied to the model, which is then examined dynamically as it rebounds from a state of low infection back to the original endemic state. Mass-antibiotic treatments are modelled by assuming that the transmission context remains essentially the same but infected individuals are returned to their susceptible states. We examine the effect of antibiotic treatment on infection and disease and its implications on programmes to control blinding trachoma and we discuss the usefulness of modelling for control programme impact projections.

Disease Pathogenesis and Epidemiology

The causative organism of trachoma is the infectious bacterial agent *Chlamydia trachomatis* and ocular infection with this agent can eventually lead to a chronic conjunctivitis the most severe sequelae of which are corneal opacity (CO) and blindness. A first infection with the bacteria leads to some inflammation and usually a complete clearance; however, individuals are left hypersensitive to subsequent exposure and repeated infections cause chronic inflammation and, eventually, scarring that leads to the severe disease sequelae.[7-10] The WHO simplified grading scheme—shown in Table 1 —outlines the five progressive disease stages for use by nonspecialists working in a field setting (omitting the final stage, blindness). These WHO disease grades will be used throughout this chapter.

Until recently, with the advent of sensitive Polymerase Chain Reaction (PCR)-based tools (compared with e.g., Giemsa staining and culture-based methods) to detect infection, the connection between infection with *C. trachomatis* and progress to the various stages of disease has not been clear.[11-15] However, recent qualitative and quantitative PCR studies in which it has been possible to detect the presence of infection with *C. trachomatis* and estimate the bacterial load in conjunctival swabs,[13,14] have made it possible to examine the distribution of the infection load over the community and compare with disease diagnoses. These new data can, in turn, be used to construct and calibrate mathematical models of the pathogenesis and epidemiology of trachoma,

Table 1. WHO simplified clinical grading of trachoma disease and disease sequelae, adapted from Taylor[56]

Criterion Definition (and WHO Acronym)	Criterion Description
Trachomatous Inflammation-Follicular (TF)	At least five follicles (of at least 0.5 mm in diameter) on upper tarsal conjunctiva
Trachomatous Inflammation-Intense (TI)	Papillary hypertrophy and inflammation covering more than half of deep tarsal vessels
Trachomatous Scarring (TS)	Clear trachomatous scarring of upper tarsal conjunctiva
Trachomatous Trichiasis (TT)	One, or more than one, eyelash rubbing on the eyeball or signs of the eyelash's removal
Corneal Opacity (CO)	Corneal opacity caused by trachoma, some of which obscures the pupil

as will be shown in the following. These studies have also highlighted discrepancies between cases of disease and infection and these are probably brought about by the kinetics of infection in which, following clearance, disease may persist.[15-17] Infection without disease, detected by PCR, may be a result of contamination with infections that are not replicating, an idea recently put forward in a new quantitative PCR study for 16s RNA expression.[18]

The prevalence of active disease (TF, TI in the WHO grading, Table 1) peaks for children aged 1-5 years[19,20] and drops to much lower levels in adults. Often, the larger number of adult cases of TF and TI are female and this is thought to be because of their more frequent contact with infectious children compared with males.[5] The prevalence of scarring (TS), trichiasis (TT), corneal opacity (CO) and blindness in the youngest age groups is often found to be very low—though some recent studies have shown high levels in very compromised populations[21,22]—with a rising trend with age. Only a small fraction of the trachomatous disease burden for older adults is generally found to be corneal opacity and blindness, but the prevalences of scarring can be high.[23,24] When the prevalence of active disease is high, the prevalence of the disease sequelae is also high, which lends support to the idea that repeated episodes of inflammation lead to disease progression.[25-27]

Infection and clinical disease tend to have a shorter duration in older adults from trachoma endemic communities and this may be due to an acquired immunity[7] that involves cellular and humoral components.[28,29] It is also possible that a decreased exposure to infection among adults may play a role in their observed shorter durations of infection. The acquired immunity tends to accelerate the clearing of infection and may have only a very small protective effect against re-infection,[30,31] though some genital *C. trachomatis* animal models show significant levels of protection.[32]

Studies investigating infection loads have found strong age-dependences and in some cases close to half of the community bacterial load has been found in children under 1 year.[14] The infection loads of adults are often lower than those of children which is thought to be due to the acquisition of clearing immunity with age.[13] This imbalance between adults and children is particularly marked in communities in which active disease is hyperendemic (here considered to be where the prevalence is greater than 20%) and the infection tends to be more evenly distributed in lower prevalence populations.[14] Acquired immunity is less stimulated in lower prevalence communities due to lower exposure levels and so the immunological reaction to infection is more similar in children and adults.

Another proposed mechanism for chronic inflammation and subsequent scarring disease progression involves persistent chlamydial infection and antigen release,[33-38] though studies have found this idea very difficult to distinguish from repeated infection that is clustered in households.[39]

Antibiotic-Based Control Programmes

The main weapon in the arsenal against trachoma is azithromycin and this antibiotic is currently being deployed across trachoma endemic countries.[40] Both theoretical and empirical studies have sought to answer questions regarding, for example, the frequency with which the antibiotic should be administered, the routes of reinfection in a community and the rate of re-emergence of infection and disease and what these rates tell us about the transmission environment.

The mathematical model of Lietman et al[41-44] has investigated the frequency of antibiotic treatment needed to eliminate *C. trachomatis*. In this study, the rate of re-emergence of infection leads to an estimate of the initial exponential rate of increase of the infection prevalence in the community. This rate then forms the basis of an analysis to determine the necessary treatment frequency to achieve a steady prevalence decline[41] and the value thus obtained has been compared with data from a field trial (carried out by the same authors) in the Gurage zone of Ethiopia,[45] where annual treatment was found to be sufficient for elimination at an 80% or higher coverage. This study has informed the WHO policy for community antibiotic treatment frequency and duration, but it does not include details pertaining to the natural history of infection that allow the number of infections experienced to be tracked. The gradual experience of infection and associated inflammation, leads to scarring and consequently the disease sequelae and these phenomena are explicitly included in the models presented in this chapter.

Field studies include those by Burton et al[4] in the Gambia and by West et al[3] and Solomon et al[5] in Tanzania, who all look at either one-off or multiple single dose azithromycin treatments. Burton et al[4] found that most treated villages cleared their infection 12 months after treatment for low initial infection levels, but that, in some, infection was reintroduced externally. Solomon et al[5,46] have reported that, at a low endemic level, a single dose at very high coverage caused infection to almost completely disappear and a second treatment 24 months later has eliminated infection, though here azithromycin coverage was supplemented by the use of tetracycline. West et al[3] found that, after an initially large drop in infection, for a hyperendemic community, infection was reemerging 18 months after treatment.

Methods

A First Mathematical Model of Trachoma

Here, we outline two mathematical models; the first is a basic model accounting for repeat infection and the accumulation of scarring and the disease sequelae TS and TT (Fig. 1) and the second includes an active disease compartment that corresponds to the WHO grading TF/TI but does not model sequelae (Fig. 2). The conceptual mathematical model is based on the following tenets (1) trachoma is a disease for which progress to the severe sequelae is dependent upon multiple infections; (2) while the accumulation of scars on the way to a TS or TT diagnosis may occur gradually, with each successive infection, it is useful to think of there being a threshold number of infections, beyond which an average individual might be diagnosed as TS or TT; (3) the duration of infection and disease tends to decrease with age, which is likely due to the acquisition of immunity that more rapidly clears infection following the experience of a greater number of infections; (4) bacterial infection load also tends to decrease with age, which is also likely to be due to a stronger infection-clearing acquired immunity in those who have experienced a larger number of prior infections. These tenets, centred on the response of an individual to re-infection

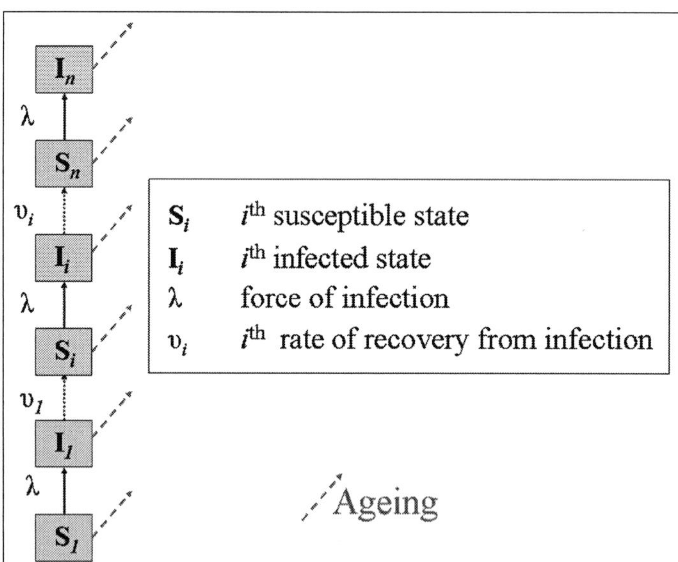

Figure 1. A compartmental diagram illustrating the model described in the text. Each susceptible and infected compartment is connected to the compartment above so that the population passes up a 'ladder' of infection. The subscript *i* corresponds to the number of prior infections experienced. Adapted from Gambhir et al.[57]

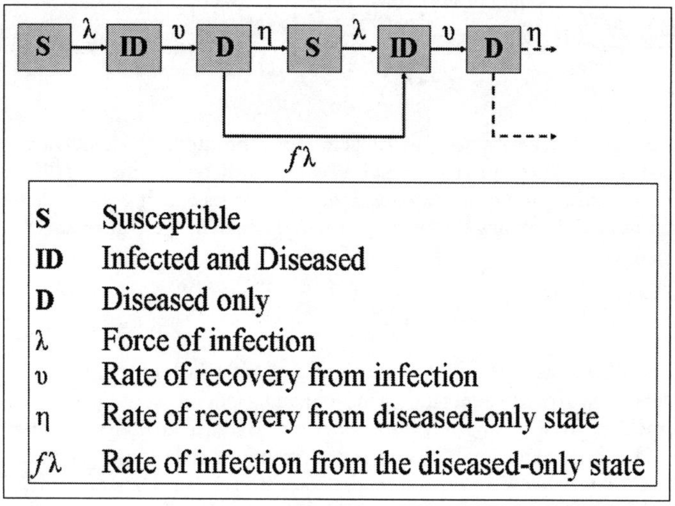

Figure 2. Diagram showing a second model that includes a diseased-only class, as well as an infected and diseased class. This model allows the population to clear the infection yet still show signs of disease, a phenomenon commonly observed in field-studies.

with the bacteria, can be incorporated into a simple mathematical model. The model proposed here is of the Susceptible-Infected (SI) framework,[6] with two basic compartments; however, here each of the compartments is indexed by the number of infections that an individual has experienced prior to the current one i (see Fig. 1).

The model illustrated in Figure 1 is constructed as successive susceptible and infected stages, indexed by suffix i. Individuals pass up a 'ladder' of infection through susceptible (S_i) and infected (I_i) compartments, each of which is followed by the next compartment above it. On progression up the ladder, individuals' acquired immune response is assumed to result in more rapid clearance of infection, reduced bacterial load and more severe disease, in accordance with the relevant aspects of *Chlamydia* immunobiology outlined above. The rate of clearance of infection is assumed to be a function of i, with parameters estimated by fitting to available data on the duration of infection with age.[7,47] The probability of transmission of infection from an infected individual by flies, fomites, or direct contact is assumed to be proportional to the bacterial load of that individual.

The following equations (adapted from Gambhir et al[57]) describe the rate of change in the numbers of susceptible and infected individuals of age a and at time t, experiencing their ith infection. The equations presented here are continuous partial differential equations in age and time and these were discretised for numerical solution using the Euler integration scheme in Matlab®. The three terms on the right hand side of each equation represent: (1) transmission of infection through contact between susceptible and infected individuals; (2) mortality of hosts; (3) recovery from infection.

$$\frac{\partial S_i(a,t)}{\partial t}+\frac{\partial S_i(a,t)}{\partial a}=-\beta S_i(a,t)\int_{a'}\frac{w(a,a')\sum_j \rho_j I_j(a',t)}{\sum_j I_j(a',t)+S_j(a',t)}da'-\mu(a)S_i(a,t)+v_{i-1}I_{i-1}(a,t \qquad (1)$$

$$\frac{\partial I_i(a,t)}{\partial t}+\frac{\partial I_i(a,t)}{\partial a}=\beta S_i(a,t)\int_{a'}\frac{w(a,a')\sum_j \rho_j I_j(a',t)}{\sum_j I_j(a',t)+S_j(a',t)}da'-\mu(a)I_i(a,t)-v_i I(a,t) \tag{2}$$

where β is the transmission parameter, which determines the rate at which individuals move from susceptible to infected states when they come into contact with one another and this parameter was estimated from fitting the model to the available data; $w(a, a')$ is a mixing matrix (in the discretised version of the model) describing the rate at which individuals of age a and a' mix,[6]

$$w(a,a')=\varepsilon\delta_{a,a'}+(1-\varepsilon)\frac{N_{a'}}{\sum_{a'}N_{a'}}, \tag{3}$$

where $\delta_{a,a'}$ is the Kronecker Delta;[48] $N_{a'}$ is the number of individuals of age a' and ε is a mixing parameter ranging from 0 (to represent random mixing among age groups) to 1 (to represent assortative mixing, i.e., each age group mixing only with itself, which is set here to 0.5) for an intermediate level of random and assortative mixing;[49] ρ_j is the infectivity of infected individuals in compartment I_j; $\mu(a)$ is the death-rate at age a obtained from the 2001 WHO Tanzanian life table;[50] and υ_j is the recovery rate of individuals in compartment I_j. Aside from the 'ladder' structure of the model, the other important features of the disease pathogenesis are the changing recovery rate and the infectivity with number of prior infections. These are outlined next.

Recovery Rate

The recovery rate per individual υ_i from infection i (measured as the rate per year, so that the reciprocal of this value gives the average duration of infection), is assumed to follow a monotonically increasing function of i with lowest value υ_1 (corresponding to the recovery rate from the first infection) with saturating maximum value of υ_∞ (corresponding to the recovery rate following a large number of prior infections),

$$\upsilon_i=(\upsilon_1-\upsilon_\infty)\exp\left[-\gamma\,(i-1)\right]+\upsilon_\infty \tag{4}$$

Parameters υ_1, υ_∞ and γ were estimated by fitting the model to the mean duration of infection for different age groups using data already reported[7] but re-analyzed by Grassly et al[47] with results shown in Table 2 (γ represents the per infection rate of change of the recovery rate).

Table 2. Parameter definitions and values for the model obtained through maximum likelihood using a function combining infection prevalence, bacterial load and recovery rate data from a hyperendemic setting adapted from Gambhir et al[57]

Parameter	Parameter Definition	Value
$1/\upsilon_1$	Mean duration of first infection	13.6 months
$1/\upsilon_\infty$	Mean duration of infection after multiple prior infections	2.8 months
γ	Rate of drop of duration of infection per prior infection	0.75 infection^{-1}
l_1	Infection load per person at first infection	1.03×10^5 copies omp1 per swab
φ	Rate of drop of infection load per prior infection	0.05 infection^{-1}
β	Transmission coefficient: the rate of transmission (per year) of infection between individuals	29.6 year^{-1}

Infectivity

The infectivity of an individual ρ_i is assumed here to be linearly proportional to the bacterial load l_i carried by that individual, such that the maximum value of the infectivity is 1 and it decays with subsequent infections experienced by an individual and we assume that the load is a function of the number of prior infections experienced by that individual, in accordance with the decline in load with age that has been observed in trachoma endemic communities.[4,5,14,51] We use a simple exponential decay, which describes a decline in the load from its initial value l_1 according to the expression shown below,

$$l_i = l_1 \exp\left[-\varphi\,(i-1)\right] \tag{5}$$

The infectivity ρ_i is then calculated by dividing this expression by l_1 such that the maximum value of ρ_i is 1 and it decays exponentially with the decay constant ϕ. The decay rate of this exponential is estimated from age-dependent data on bacterial load

Parameter Estimation

The model parameters (β, υ_1, υ_∞, γ, l_1 and ϕ) were estimated by simultaneously fitting the model-generated age-structured infection prevalence, rate of recovery from infection and infection load. A fit over six parameters represents a substantial search problem, so several starting points were used for the minimisation algorithm to increase the chance that the global minimum was found.

Data

The model was fitted to the published data of West et al[51] (age-dependent prevalence levels and infection load) and Grassly et al[47] (age-dependent durations of infection), both of which were collected prior to intervention from study sites that were hyperendemic in active disease prevalence.

Prevalence of Disease Sequelae with Age

A simple mechanism for the accumulation of trachomatous scarring (TS) is incorporated into the basic model by assuming that scarring increases with the number of prior infections. With worse scarring comes the more severe disease sequela, trachomatous trichiasis (TT). Corneal opacity (CO), the final disease sequela prior to blindness, is not modelled here because its aetiology differs from that of the other sequelae. While TS and TT are caused by and worsen with repeated infection, CO represents something of a point of no return for individuals who have reached the TT stage and whose eyelids will continue to inflict mechanical corneal damage without surgical intervention. Since the population dynamical model here is a first attempt at capturing the relevant contact-related factors for infection and disease, this separate mechanical process is not included; however, the central importance of CO for the GET2020 goal to eliminate blinding trachoma means that progress to CO should be considered for inclusion in future models.

The simplest possible scheme is used to determine the prevalence of each of the sequelae: thresholds exist, along the ladder of infection, beyond which each of the sequelae are assumed to be present. Beyond the threshold corresponding to a specific sequela, that sequela is assumed to be detectable in the population that has reached this stage e.g., when the threshold for TT has been passed, both TS and TT are present. For simplicity, it is assumed here that the two disease sequelae can coexist with one another i.e., a person may be graded with scarring or scarring and trichiasis.

Disease Sequelae and Prior Infection Number

Once the model has been fitted to the hyperendemic infection data in the endemic setting, the threshold infection levels for the disease sequelae TS and TT are determined. These can be found by comparing the model-generated disease sequelae prevalence curves—given specific threshold infection numbers—with the data from the hyperendemic setting.[51] The data set used for this hyperendemic setting, while from the same district in Tanzania as the baseline set of West et al,[51] was collected several years prior to the infection data, but it is assumed here that since there had been no treatment interventions prior to the MDA study, infection and disease in the district was in a steady state.

A Second Model Including an Active-Disease Class

The assessment of the level of trachoma in an area—a community, district, or country—is usually conducted on the basis of clinical examination. Since the primary method for the treatment of trachoma is antibiotic for the bacterial infection, the relation between infection and disease is important. In addition to the model already presented, which takes into consideration only the infected state, a second model—entirely separate from the first, but very similar in structure—has been constructed to account for active disease (TF/TI). Infection and disease are not perfectly coincident in trachoma and the age-structured model outlined implicitly assumes that they are, an assumption that allows the model to explore, with great simplicity, the epidemiological patterns of trachoma. However, the inclusion of a stage in which individuals are diseased allows the exploration of: (1) the consequences of the existence of a proportion of the population that is diseased without being infected; (2) the time-lag between the onset of infection and the manifestation of disease (as well as the lag between the cessation of infection and disease), which is of particular importance when infection in a community has been treated by antibiotic and, while the infection load may drop substantially, the signs of disease remain among many members of the population.

Due to the added complexity of investigating the effect of a diseased stage, this second model is not age-structured. The addition of the new diseased class allows the model population to include members who are diseased without at the same time being infected. However, for simplicity, individuals cannot be infected without showing clinical signs of disease. The aim here is to include a diseased class, since it is clinical disease (TF/TI) that is measured—more frequently than the infection prevalence—to determine the level of trachoma in a community.

Figure 2 illustrates the model; the main features distinguishing it from the age-structured model are: (1) the diseased-only state (with recovery rate from this state of η) and (2) the new path that individuals can take to advance along the ladder of infection from the diseased-only state, bypassing a recovered state and straight into the next infected-and-diseased state above. The addition of a new way in which individuals can progress along the infection ladder is intended to allow for the possibility that, even after antibiotic distribution has reduced the level of infection in the community, disease remains and there may be individuals who continue to experience disease for a considerable time, possibly even progressing in the severity of the disease they experience. This addition is intended to also capture the lag between the clearance of infection and the eventual clearance of disease from a community. Figure 2 shows that this additional path along the ladder is dependent upon the prevailing force of infection in the population, but that it is attenuated by a factor f. The compartments shown in Figure 2 represent a particular level of the ladder of infection and should have been labelled with subscripts i as for the earlier model (Fig. 1) but, for simplicity, these have been suppressed here.

Fitting the Model to Data

The second model was developed in Berkeley Madonna™ and fitted to data available from three published studies from, Upper Saloum district, Gambia;[4] Rombo district, Tanzania[16] and Kongwa district, Tanzania.[3,52] Pre- and posttreatment data—with varying numbers of follow-up surveys following the baseline assessment—were used to determine the parameter values of the model using least squares fitting of the model-generated time-dependent infection and disease prevalence to the data.

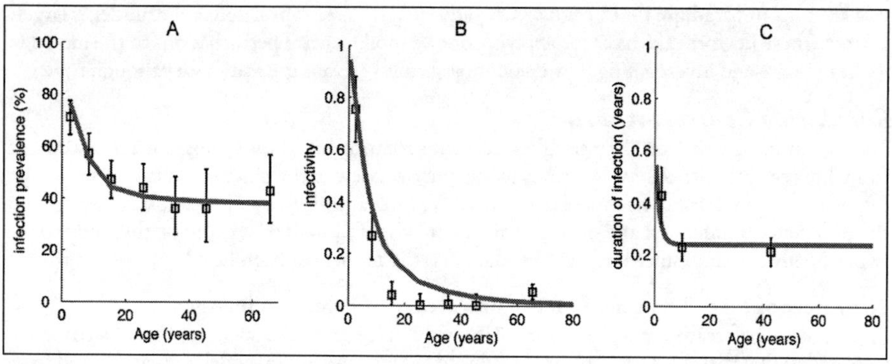

Figure 3. Age-profiles generated by the best-fitting parameters to the hyperendemic data set of West et al[51] and—for the recovery rate—Bailey et al,[7] reanalysed by Grassly et al[47] (data displayed as square data points with 95% confidence intervals and model fits shown as solid lines): (A) the prevalence of chlamydial infection measured by PCR; B) the infectivity, which is proportional to the bacterial load, measured by quantitative PCR; C) duration of infection. The model was fitted simultaneously to all three data sets. Adapted from Gambhir et al.[57]

Results

Fitting the First Model to Static Infection and Disease Sequelae Data

Once the basic model has been simultaneously fitted to the prevalence, infection load and recovery rate data from the hyperendemic setting—with parameter values found to be as in Table 2—the epidemiological model outputs can be generated as shown in Figure 3 and Figure 4. These figures show that the steady state hyperendemic model results in a decreasing prevalence with age that drops from an initially very high value (Fig. 3A), a decreasing age profile of the infectivity (proportional to the bacterial load) (Fig. 3B), a decreasing profile for the duration of infection (giving a monotonic rise of the recovery rate) beginning at a high value and asymptoting to a maximum value (Fig. 3C), a rising profile with a flattening at older ages of the TS scarring sequela (Fig. 4A) and an almost linearly rising profile of the TT sequela starting at a critical age of around 20 years (Fig. 4B). The threshold number of infections for the detection of TS and TT were found to be 88 and 130 respectively, though these numbers would be lower if, for example, age-related changes in contact rate were included in the model. This set of close fits to the epidemiological data suggests

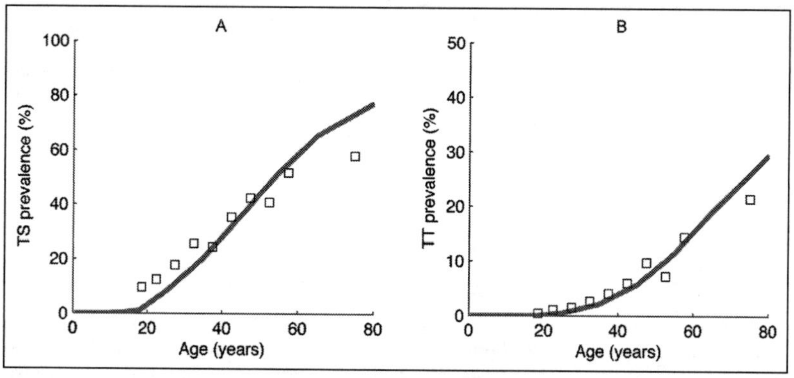

Figure 4. Age-profiles of disease sequelae generated by the best-fitting parameters to the hyperendemic data and using the disease sequelae threshold infection numbers given in the text. Adapted from Gambhir et al.[57]

that the current paradigm for *C. trachomatis* pathogenesis, as encapsulated in the model, is largely correct. These fits form the basis for applying one or more control perturbations to the model in its steady state and investigating how epidemiologically relevant quantities vary though time.

Simulating Control Scenarios

The aim of the models is to simulate control scenarios based on the parameters estimated from different prevalence settings. Here, we give an overview of the kinds of control scenario that can be investigated using the model developed. We look at the effects of antibiotic treatment on the prevalence of infection and on active disease in what follows and we confine the study to the hyperendemic setting, since it is for these data that the model has been fitted.

The Effect of Mass Treatment on the Prevalence of Infection by Age

At the time of treatment, a proportion of individuals (given by the fraction: efficacy × coverage) is treated so that their bacterial carriage is reduced to zero and, in the model, they are moved from their infected state (I_i) back into the susceptible state (S_i); their memory of the prior number of infections is retained. In Figure 5, the plot at time $t = 0$ shows that the prevalence of infection in all age groups has dropped to a much lower level than before treatment (shown by the dotted line). Figure 5 also shows the time development of the prevalence of infection from an initial time point at which mass antibiotic administration (MDA) has occurred. The efficacy, for an individual case, of a single dose of azithromycin is taken to be 95%[53,54] and the community coverage level achieved is assumed to be 86% as given by West et al.[51] The pattern of the spread of infection throughout the population following mass treatment illustrates the following main points: (1) because the prevalence of infection in children is higher than in the rest of the population, the same percentage reduction in the infected individuals in each age-group will leave a larger proportion of children still infected posttreatment than the rest of the population; (2) the rate at which infection returns to children is greater than for the rest of the population. This can be seen in Figure 6 where, in addition to the posttreatment prevalence of infection being greater in children than adults, the return to pretreatment levels of infection is particularly rapid in those under 10 years-old. There are two reasons for this: the rate of recovery from infection in children is lower than for adults and, therefore, children retain their infections for greater lengths of time causing greater prevalence levels; the second reason is dependent on the comparatively greater infectivity of children than adults in the model. Since children are assumed to interact more frequently with those their own age ($\varepsilon = 0.5$ i.e., there is some assortative mixing), this greater infectivity leads to a greater probability of becoming infected and, therefore, a greater prevalence. The greater infectivity among

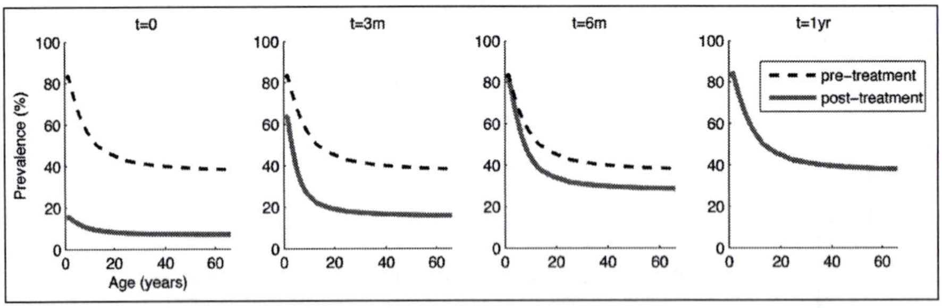

Figure 5. The time development of age-profiles of the prevalence of infection, fitted to hyperendemic data. The pretreatment infection profile is shown at each time point as a dotted line. Mass treatment is administered at t = 0 and so the profile at time t = 0 corresponds with the moment immediately following an antibiotic mass-treatment and the infection has been reduced by the factor coverage x efficacy. Subsequent time point plots show the re-emergence of infection following the treatment.

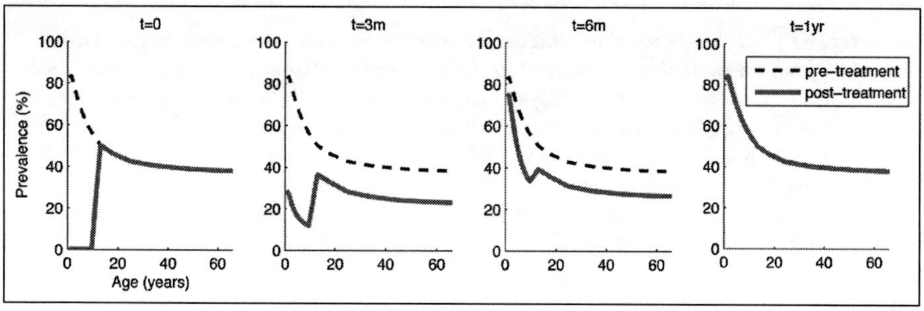

Figure 6. The age-dependent prevalence of infection following a single treatment of the 0-10 year-old children in a hyperendemic setting. The coverage and efficacy have been set to 100% here to illustrate more clearly the effect of treating the 0-10 year group. Each time point following t = 0, immediately following the treatment administration, follows the re-emergence and redistribution of infection over the population. The dotted line corresponds to the infection profile pretreatment.

children is very similar to the effect of an increased frequency of exposure to infection in children compared with adults.

Age-Targeted Treatment

As well as the mass treatment of as large a fraction of the entire community as possible, there are treatment options that target those sections of the community most likely to harbour high infectious bacterial loads. The modelling of protocols in which household members of those who show signs of active trachoma are treated requires a model more fine-grained than the one examined here (as outlined by Blake et al[58]). However, it is instructive to look at the selective treatment of children and the impact of this treatment on the prevalence of infection in the whole community. Figure 5 has illustrated the effect over time of the treatment of the whole age-range at coverage level 86% and 1 year following treatment the hyperendemic community has already returned to its pretreatment prevalence of infection. While it is likely that the treatment of 0-10 year-old children would be a less effective strategy than community-wide treatment, because much of the community infection load is concentrated among children, it is not obvious that the treatment of children alone will not show a similar impact.

The results shown in Figure 6 illustrate the impact over time of a single treatment round of the population between 0-10 years. For the purpose of this illustration, the efficacy and coverage of the round of treatment have both been set to 100%, which, while unrealistic, clarifies the dynamical picture. At the time of treatment, all individuals up to and including the age of 10 years are treated so that their bacterial carriage is reduced to zero and, in the model, they are moved from their infected state back into the susceptible state. The plot at time t = 0 shows that the prevalence of infection in all people under 10 years is zero while the prevalence of infection for those over this age remains as it was prior to treatment (dotted line). As time progresses, the prevalence of infection in the 0-10 year olds rises as these children come into contact with older infected people. Because the youngest children had, in general, experienced fewer infections than older children at the time that they were treated, they have lower levels of acquired immunity. They therefore clear infection more slowly and have higher bacterial loads when infected, resulting in a more rapid increase in the prevalence of infection among these children.

The dotted curves in Figure 6 represent the age-dependent infection prevalence prior to treatment. By comparison with the solid curve, it can be seen that a treatment round administered to children under 10 years not only reduces prevalence among these children (direct impact) but also decreases the prevalence of infection among those of all ages (indirect impact). Moreover, the size of this impact can be significant. By removing infectious children from the population, adults benefit because they are less likely to come into contact with an infectious source. Because

Table 3. The parameters used for the active disease model. These parameters were obtained by fitting the model to the pre and posttreatment data of the three endemic settings analysed in this chapter (a) Upper Saloum district, Gambia; baseline infection prevalence 3.2%, (b) Rombo district, Tanzania; baseline infection prevalence 9.5% and (c) Kongwa district, Tanzania; baseline infection prevalence 57%. Note that the rate of recovery from the diseased-only state, denoted by η, represents the total rate at which individuals recover from this state

Parameter	Parameter Definition	Value
$1/\upsilon_1$	Mean duration of first infection	13.6 months
$1/\upsilon_\infty$	Mean duration of infection after multiple prior infections	2.8 months
$1/\eta_l$	Mean duration of first episode of disease	31 months
$1/\eta_\infty$	Mean duration of disease after multiple prior episodes	6 months
f	Attenuation factor of the force of infection for those in the diseased-only state	a) very low (<0.01) b) 0.35 c) 0.99
β	Transmission rate: the rate of transmission (per year) of infection between individuals	a) 2.5 year^{-1} b) 6.4 year^{-1} c) 20.8 year^{-1}

the younger children tend to harbour greater bacterial loads, a large portion of the community infection load has been eliminated by treating these children.

The Effect of Treatment on Active Disease

The values of the parameters pertaining to the second, active disease, model, obtained through fitting to the pre and posttreatment data, are listed in Table 3. The parameter values for the fraction of the population advancing along the ladder of infection without recovery show that there is no consistent value obtained from fitting to the data. In the low-endemic setting the fraction is found to be very small, whereas for the high-endemic setting the value is very high. The model-generated

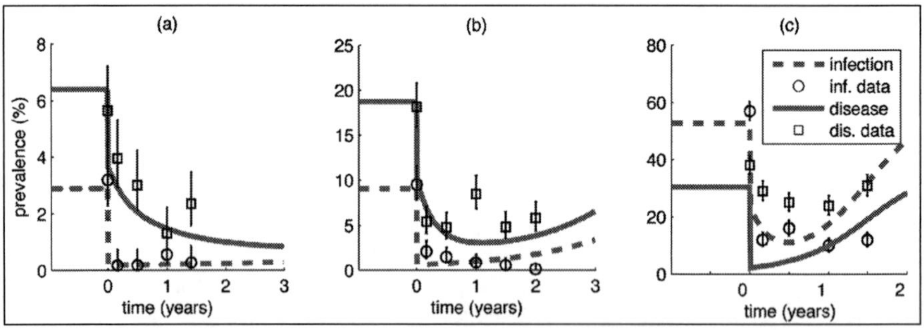

Figure 7. The development through time of the prevalence of infection and active disease for the model and for recent baseline and posttreatment data[4,16,51] in each of three endemic settings (a) Upper Saloum district, Gambia; baseline infection prevalence 3.2%, (b) Rombo district, Tanzania; baseline infection prevalence 9.5%) and (c) Kongwa district, Tanzania; baseline infection prevalence 57%). Treatment causes a sharp drop in the infection and disease prevalence followed by a rebound at differing rates.

pattern obtained with the extreme values of 0 and 1 (close to the values actually obtained through fitting) give very similar outcomes.

Using these parameter values, the model generates outcomes for the trajectory of the prevalence of active disease (TF/TI) and infection pre and posttreatment. Figure 7 shows the model-generated prevalence levels of active disease and infection and the data to which the model was fitted, for the three data sets. The model curves show the two key features of the kinetics of trachoma infection and disease as it affects trachoma epidemiology: (1) the prevalence of infection is different from the prevalence of disease; in this model, the prevalence of disease is always assumed to be higher than that of infection; and (2) once antibiotic treatment has been administered, the prevalence of infection drops to a low level immediately, while the prevalence of active disease lags behind infection at a higher level.

Conclusion

This chapter has detailed the development of a model for ocular infection with *C. trachomatis* in a community setting. New data from qualitative and quantitative PCR (as well as 16s RNA-based PCR), collected at the baseline and follow-up points during trachoma control programmes, are providing new high resolution data on ocular infection with *C. trachomatis* at individual and community levels. Mathematical models of trachoma transmission and control, of a kind able to incorporate this new data, are currently being developed in order to inform control policy for the achievement of the GET 2020 goals. The model outlined here is based upon the idea of a 'ladder' of infection in which individuals progress to greater numbers of infection when they come into contact with infected members of the population. With increasing numbers of infections, individuals in the model become progressively more scarred and develop the damaging disease sequelae. The model, therefore, combines aspects of the SI contact framework—often used for modelling microparasitic infections—and the infection burden framework where, in this case, the number of prior infections is equivalent to the infection 'burden'. This novel approach to modelling disease pathogenesis when multiple infections are important may prove to be useful for modelling other infections, such as malaria.[6] Previous models of trachoma infection have not taken into account the progress of an individual up the ladder of infection and have had no simple way of incorporating disease sequelae into the framework.

The model is shown to represent adequately the pretreatment epidemiological patterns of trachoma in a hyperendemic setting, namely: the prevalence of infection and disease sequelae at all ages and the age-profiles of the bacterial load and the recovery rate.

When the model is fitted to baseline infection prevalence data from a recent antibiotic intervention trial in a hyperendemic setting, the resulting age-dependent infection prevalence curve shows magnitudes over all ages that are very close to the observed data and the maximum point of the prevalence of infection is found for young children.

When a single mass drug treatment is administered to the whole age range, the infection prevalence drops instantaneously. Subsequently, the prevalence rises back up to its pretreatment profile, with the most rapid rise occurring in children due to their partially assortative mixing pattern and their high infectivity. In a hyperendemic setting, this infection rebound is shown to happen over the course of 1 year, after which the profile has returned to its pretreatment level. When only a subset age-range of the population is treated, the impact of the treatment is felt over all ages due to the decreased force of infection. However, the single treatment round is found to result in very little impact after a year. In a hyperendemic setting, due to the rapidity of the infection rebound, the WHO-recommended treatment schedule of three annual antibiotic treatments would therefore probably result in no lasting effect on infection or the disease sequelae. Very few cases of trichiasis and corneal opacity would therefore also be averted by these treatment programmes and the GET 2020 goals in this setting would seem to be very difficult to achieve. It is therefore necessary to implement the full SAFE strategy, which will reduce the transmission level of the population, thereby reducing the endemic infection prevalence and the rate of rebound of infection following treatment allowing the impact of repeated rounds of mass drug administration to accumulate over time. The most recent follow-up surveys from the studies used here to fit the model have found that, in areas with very low initial endemicities one or two high coverage MDAs may be sufficient to achieve elimination,[46] whereas in a high-endemic setting two treatments are not sufficient to result in lasting drops in transmission.[52]

Following a single treatment, a model that includes an active disease stage generates trajectories in which disease prevalence is higher than infection and is found to lag behind it, reproducing these two key observations of field studies. This model also finds that the infection and disease rebound following treatment is very rapid for a setting in which initial endemicity is very high and considerably slower in lower endemic communities. The relation between infection and active disease in this basic model can be used to inform surveys carried out by treatment programmes in which clinical status is measured but infection is treated: following treatment the relation between these two states is not simple. Easy-to-use field tests for determining the level of infection are certainly more reliable for infection measurement[55] but while these tests are not widely available the conversion of active disease prevalence to an estimate of infection prevalence would be useful.

The models discussed here have focused on reproducing the progression of individuals through the infection ladder that appears to be so important for trachoma epidemiology and the progression to ocular disease. However, the models are very simple and the conclusions are subject to their assumptions, including: the mixing and exposure patterns with age, sensitivities to parameter assumptions and estimates and the deterministic nature of the model (which can also be adapted to a stochastic framework, taking, for example, elimination of infection into account). The age-structured model should also be fitted to lower endemic settings and posttreatment data to predict properly the frequency and duration of antibiotic and nonchemotherapeutic treatment interventions (e.g. see Gambhir et al[57]). Further refinements to the model should also focus on important heterogeneities such as risk groups which cannot easily clear infection, those who harbour particularly large infection loads or respond poorly to antibiotic treatment, as well as the household structure of transmission. These refinements will allow better modelling of targeted treatments and the prevention of infection re-emergence. As more data emerge from intervention trials and control programmes, the models presented in this chapter are capable of being fitted to these new data to allow projections to be made of the impact of treatment in trachoma-endemic communities. This may help guide programmes in achieving the elimination of blindness due to trachoma.

Acknowledgements

This chapter benefits from two two-day discussion meetings on trachoma epidemiology and pathogenesis held in London and Geneva, sponsored by the International Trachoma Initiative (ITI). We would like to thank both ITI and participants at this meeting for their contributions. MG would like to thank ITI for funding, MGB acknowledges the Medical Research Council for a Career Establishment Grant, NCG thanks the Royal Society for a University Research Fellowship and IMB thanks the Medical Research Council for a PhD studentship.

References

1. Resnikoff S, Pascolini D, Etya'ale D et al. Global data on visual impairment in the year 2002. Bull World Health Organ 2004; 82(11):844-851.
2. Molyneux DH, Hotez PJ, Fenwick A. "Rapid-Impact Interventions": How a policy of integrated control for Africa's neglected tropical diseases could benefit the poor. PLoS Med 2005; 2(11):e336.
3. West SK, Munoz B, Mkocha H et al. Infection with Chlamydia trachomatis after mass treatment of a trachoma hyperendemic community in Tanzania: a longitudinal study. Lancet 2005; 366(9493):1296-1300.
4. Burton MJ, Holland MJ, Makalo P et al. Re-emergence of Chlamydia trachomatis infection after mass antibiotic treatment of a trachoma-endemic Gambian community: a longitudinal study. Lancet 2005; 365(9467):1321-1328.
5. Solomon AW, Holland MJ, Alexander ND et al. Mass treatment with single-dose azithromycin for trachoma. N Engl J Med 2004; 351(19):1962-1971.
6. Anderson R, May R. Infectious Diseases of Humans: Dynamics and Control: Oxford University Press; 1992.

7. Bailey R, Duong T, Carpenter R et al. The duration of human ocular Chlamydia trachomatis infection is age dependent. Epidemiol Infect 1999; 123(3):479-486.
8. Taylor HR, Johnson SL, Prendergast RA et al. An animal model of trachoma II. The importance of repeated reinfection. Invest Ophthalmol Vis Sci 1982; 23(4):507-515.
9. Grayston JT, Wang SP, Yeh LJ et al. Importance of reinfection in the pathogenesis of trachoma. Rev Infect Dis 1985; 7(6):717-725.
10. Detels R, Alexander ER, Dhir SP. Trachoma in Punjabi Indians in British Columbia: a prevalence study with comparisons to India. Am J Epidemiol 1966; 84(1):81-91.
11. Baral K, Osaki S, Shreshta B et al. Reliability of clinical diagnosis in identifying infectious trachoma in a low-prevalence area of Nepal. Bull World Health Organ 1999; 77(6):461-466.
12. Hayes LJ, Bailey RL, Mabey DC et al. Genotyping of Chlamydia trachomatis from a trachoma-endemic village in the Gambia by a nested polymerase chain reaction: identification of strain variants. J Infect Dis 1992; 166(5):1173-1177.
13. Burton MJ, Holland MJ, Faal N et al. Which members of a community need antibiotics to control trachoma? Conjunctival Chlamydia trachomatis infection load in Gambian villages. Invest Ophthalmol Vis Sci 2003; 44(10):4215-4222.
14. Solomon AW, Holland MJ, Burton MJ et al. Strategies for control of trachoma: observational study with quantitative PCR. Lancet 2003; 362(9379):198-204.
15. Wright HR, Taylor HR. Clinical examination and laboratory tests for estimation of trachoma prevalence in a remote setting: what are they really telling us? Lancet Infect Dis 2005; 5(5):313-320.
16. Solomon AW, Peeling RW, Foster A et al. Diagnosis and assessment of trachoma. Clin Microbiol Rev 2004; 17(4):982-1011, table of contents.
17. Miller K, Schmidt G, Melese M et al. How reliable is the clinical exam in detecting ocular chlamydial infection? Ophthalmic Epidemiol 2004; 11(3):255-262.
18. Burton MJ, Holland MJ, Jeffries D et al. Conjunctival chlamydial 16S ribosomal RNA expression in trachoma: is chlamydial metabolic activity required for disease to develop? Clin Infect Dis 2006; 42(4):463-470.
19. Lansingh VC, Weih LM, Keeffe JE et al. Assessment of trachoma prevalence in a mobile population in Central Australia. Ophthalmic Epidemiol 2001; 8(2-3):97-108.
20. Schemann JF, Sacko D, Banou A et al. [Cartography of trachoma in Mali: results of a national survey]. Bull World Health Organ 1998; 76(6):599-606.
21. Ngondi J, Ole-Sempele F, Onsarigo A et al. Blinding trachoma in postconflict southern Sudan. PLoS Med 2006; 3(12):e478.
22. Ngondi J, Reacher M, Matthews F et al. The epidemiology of low vision and blindness associated with trichiasis in southern Sudan. BMC Ophthalmol 2007; 7:12.
23. Courtright P. Contribution of sex-linked biology and gender roles to disparities with trachoma. Emerg Infect Dis 2004; 10(11):2012-2016.
24. Cumberland P, Hailu G, Todd J. Active trachoma in children aged three to nine years in rural communities in Ethiopia: prevalence, indicators and risk factors. Trans R Soc Trop Med Hyg 2005; 99(2):120-127.
25. Werner GT, Sareen DK. Trachoma in Punjab: a study of the prevalence and of mass treatment. Trop Geogr Med 1977; 29(2):135-140.
26. Regassa K, Teshome T. Trachoma among adults in Damot Gale District, South Ethiopia. Ophthalmic Epidemiol 2004; 11(1):9-16.
27. Zerihun N. Trachoma in Jimma zone, south western Ethiopia. Trop Med Int Health 1997; 2(12):1115-1121.
28. Ramsey KH, Poulsen CE, Motiu PP. The in vitro antimicrobial capacity of human colostrum against Chlamydia trachomatis. J Reprod Immunol 1998; 38(2):155-167.
29. Brunham RC, Rey-Ladino J. Immunology of Chlamydia infection: implications for a Chlamydia trachomatis vaccine. Nat Rev Immunol 2005; 5(2):149-161.
30. Bowman RJ, Jatta B, Cham B et al. Natural history of trachomatous scarring in The Gambia: results of a 12-year longitudinal follow-up. Ophthalmology 2001; 108(12):2219-2224.
31. West SK, Munoz B, Lynch M et al. Risk factors for constant, severe trachoma among preschool children in Kongwa, Tanzania. Am J Epidemiol 1996; 143(1):73-78.
32. Morrison RP. Chlamydial immunology and vaccinology. Proceedings of the Chlamydia Vaccine Development Colloquium 2003; 19-23.
33. Stephens RS. The cellular paradigm of chlamydial pathogenesis. Trends Microbiol 2003; 11(1):44-51.
34. Ward M, Bailey R, Lesley A et al. Persisting inapparent chlamydial infection in a trachoma endemic community in The Gambia. Scand J Infect Dis Suppl 1990; 69:137-148.
35. West SK, Munoz B, Mkocha H et al. Progression of active trachoma to scarring in a cohort of Tanzanian children. Ophthalmic Epidemiol 2001; 8(2-3):137-144.

36. Morrison RP. New insights into a persistent problem—chlamydial infections. J Clin Invest 2003; 111(11):1647-1649.
37. Caldwell HD, Wood H, Crane D et al. Polymorphisms in Chlamydia trachomatis tryptophan synthase genes differentiate between genital and ocular isolates. J Clin Invest 2003; 111(11):1757-1769.
38. Morrison RP, Caldwell HD. Immunity to murine chlamydial genital infection. Infect Immun 2002; 70(6):2741-2751.
39. Mabey DCW, Bailey RL, Hutin YJF. The epidemiology and pathogenesis of trachoma. Review of medical microbiology 1992; 3:112-119.
40. Kumaresan J. Can blinding trachoma be eliminated by 20/20? Eye 2005; 19(10):1067-1073.
41. Lietman T, Porco T, Dawson C et al. Global elimination of trachoma: how frequently should we administer mass chemotherapy? Nat Med 1999; 5(5):572-576.
42. Gill DA, Lakew T, Alemayehu W et al. Complete elimination is a difficult goal for trachoma programs in severely affected communities. Clin Infect Dis 2008; 46(4):564-566.
43. Melese M, Alemayehu W, Lakew T et al. Comparison of annual and biannual mass antibiotic administration for elimination of infectious trachoma. JAMA 2008; 299(7):778-784.
44. Ray KJ, Porco TC, Hong KC et al. A rationale for continuing mass antibiotic distributions for trachoma. BMC Infect Dis 2007; 7:91.
45. Melese M, Chidambaram JD, Alemayehu W et al. Feasibility of eliminating ocular Chlamydia trachomatis with repeat mass antibiotic treatments. JAMA 2004; 292(6):721-725.
46. Solomon AW, Harding-Esch E, Alexander ND et al. Two doses of azithromycin to eliminate trachoma in a Tanzanian community. N Engl J Med 2008; 358(17):1870-1871.
47. Grassly N, Ward ME, Ferris S et al. The natural history of trachoma infection and disease in a Gambian cohort with frequent follow-up. PLoS Neglected Tropical Diseases 2008; 2(12): e341.
48. Boas ML. Mathematical Methods In The Physical Sciences. 2nd ed. New York: John Wiley & Sons, 1983:
49. Gambhir M, Basanez MG, Turner F et al. Trachoma: transmission, infection and control. Lancet Infect Dis 2007; 7(6):420-427.
50. WHO. WHO Statistical Information System (WHOSIS); http://www3.who.int/whosis/menu.cfm?path=whosis,life.
51. West ES, Munoz B, Mkocha H et al. Mass treatment and the effect on the load of Chlamydia trachomatis infection in a trachoma-hyperendemic community. Invest Ophthalmol Vis Sci 2005; 46(1):83-87.
52. West SK, Munoz B, Mkocha H et al. Trachoma and ocular Chlamydia trachomatis were not eliminated three years after two rounds of mass treatment in a trachoma hyperendemic village. Invest Ophthalmol Vis Sci 2007; 48(4):1492-1497.
53. Schachter J, West SK, Mabey D et al. Azithromycin in control of trachoma. Lancet 1999; 354(9179):630-635.
54. Bailey RL, Arullendran P, Whittle HC et al. Randomised controlled trial of single-dose azithromycin in treatment of trachoma. Lancet 1993; 342(8869):453-456.
55. Michel CE, Solomon AW, Magbanua JP et al. Field evaluation of a rapid point-of-care assay for targeting antibiotic treatment for trachoma control: a comparative study. Lancet 2006; 367(9522):1585-1590.
56. Taylor HR. Trachoma grading: a new grading scheme. Rev Int Trach Pathol Ocul Trop Subtrop Sante Publique 1987(64):175-181.
57. Gambhir M, Basánez M-G, Burton MJ et al. The development of an age-structured model for trachoma transmission dynamics, pathogenesis and control. PLoS Neg Trop Dis, doi:10:1371/journal, pntd.0000462, 2009.
58. Blake IM, Burton MJ, Bailey RL et al. Estimating household and community transmission of ocular Chlamydia trachomatis. PLoS Neglected Tropical Diseases 2009, 3(3): e401.

Chapter 11

Transmission Models and Management of Lymphatic Filariasis Elimination

Edwin Michael* and Manoj Gambhir

Abstract

The planning and evaluation of parasitic control programmes are complicated by the many interacting population dynamic and programmatic factors that determine infection trends under different control options. A key need is quantification about the status of the parasite system state at any one given timepoint and the dynamic change brought upon that state as an intervention program proceeds. Here, we focus on the control and elimination of the vector-borne disease, lymphatic filariasis, to show how mathematical models of parasite transmission can provide a quantitative framework for aiding the design of parasite elimination and monitoring programs by their ability to support (1) conducting rational analysis and definition of endpoints for different programmatic aims or objectives, including transmission endpoints for disease elimination, (2) undertaking strategic analysis to aid the optimal design of intervention programs to meet set endpoints under different endemic settings and (3) providing support for performing informed evaluations of ongoing programs, including aiding the formation of timely adaptive management strategies to correct for any observed deficiencies in program effectiveness. The results also highlight how the use of a model-based framework will be critical to addressing the impacts of ecological complexities, heterogeneities and uncertainties on effective parasite management and thereby guiding the development of strategies to resolve and overcome such real-world complexities. In particular, we underscore how this approach can provide a link between ecological science and policy by revealing novel tools and measures to appraise and enhance the biological controllability or eradicability of parasitic diseases. We conclude by emphasizing an urgent need to develop and apply flexible adaptive management frameworks informed by mathematical models that are based on learning and reducing uncertainty using monitoring data, apply phased or sequential decision-making to address extant uncertainty and focus on developing ecologically resilient management strategies, in ongoing efforts to control or eliminate filariasis and other parasitic diseases in resource-poor communities.

Introduction

Parasite transmission modelling is becoming increasingly recognized as a vital tool in guiding the management of parasitic disease control or elimination and for predicting the impacts and benefits of competing management options.[1-3] This recognition arises largely from the difficulty of reliably predicting the long-term impact of repeated chemotherapy on the rate of transmission of parasites and hence reinfection within a community, which has resulted in inducing uncertainty surrounding operational issues, such as the required duration of treatment and determination of

*Corresponding Author: Edwin Michael—Department of Infectious Disease Epidemiology, Faculty of Medicine (St. Mary's campus), Imperial College London, Norfolk Place, London W2 1PG, UK. Email: e.michael@imperial.ac.uk

Modelling Parasite Transmission and Control, edited by Edwin Michael and Robert C. Spear.

infection thresholds below which transmission is controlled or ceases, for most of the current interventions against the major parasitic diseases.[4-6] It also stems from increasing recognition that empirical field studies alone are unlikely to provide the required information due to constraints on resources and time and the effects of variations in local transmission conditions which make generalizations from individual studies difficult. This issue is particularly problematic for parasites with complex life cycles like the lymphatic filarial worms, which exhibit both complicated transmission dynamics—and therefore likely to demonstrate complex outcomes in response to chemotherapeutic interventions—as well as high levels of aleatory and epistemic uncertainties in their transmission processes, parameters and values.[4,7-13]

Here, we focus on recent work in modelling lymphatic filariasis transmission to highlight how mathematical models of parasite transmission dynamics can play useful roles in resolving each of these questions and therefore provide the management decision support framework required for defining optimal intervention strategies and for monitoring and evaluating community-based interventions for controlling or eliminating parasitic diseases. A key focus is on how understanding the nature of transmission dynamics, particularly with regard to complex system transitions, heterogeneities and uncertainties, will be crucial to informing the design of effective parasite management. A secondary objective is also to indicate how the application of such tools can serve to reveal gaps in existing knowledge on the biological controllability or eradicability of parasite transmission by management action and hence help guide critically required research for developing a more robust theory of parasite eradication.

Transmission Models and Decisions in Parasite Management

Informed decision-making in any parasite control programme consists of four key elements: (1) the setting of objectives, (2) specification of management options or actions, (3) an understanding of the structure and dynamics of the parasite system to be managed, i.e., how the parasite population responds dynamically to perturbations, and (4) periodic monitoring of the results of management in order to inform and possibly adjust subsequent management decisions.[5,6,14,15] Figure 1 summarizes how parasite transmission models can be used to address these elements and hence provide a framework for supporting the design of parasite control monitoring and evaluation plans. According to this schema, the key advantage of using dynamic transmission models (over subjective or even empirical assessment by an expert or managers) is that such models will be key to resolving the following four major components of a successful parasite control or elimination programme: (1) conducting rational analysis and definition of *endpoints* for different programmatic aims or objectives, including transmission endpoints for disease elimination, (2) undertaking strategic analysis to aid the *optimal design* of intervention programs to meet set endpoints under different endemic settings, (3) providing support for performing informed *evaluations* of ongoing programs, including aiding the formation of timely adaptive management strategies to correct for any observed deficiencies in program effectiveness, and (4) addressing the impacts of uncertainty and complexity in developing a coherent approach to parasite management.[5,6,15]

Models and Quantifying Intervention Endpoint Targets

Quantifying Parasite Breakpoint Thresholds in the Human Population

Recently, we and other workers have shown that the derivation of parasite infection thresholds or endpoints to serve as parasite elimination targets fundamentally requires an analysis of the dynamic properties of the parasite transmission system, as these variables essentially represent stability or persistence components of (such) dynamical systems.[5,8,16-18] In particular, as pointed out in the chapter by Michael and Gambhir (Chapter 2), the two extinction thresholds (one occurring in the vector-to-host transmission process referred to as the "threshold vector biting rate" and the other in the host-to-vector transmission process known as the "breakpoint worm burden") that exist in the case of vector-borne macroparasitic diseases, such as lymphatic filariasis, arise principally as the result of the operation of positive density dependences acting to regulate infection, which gives rise

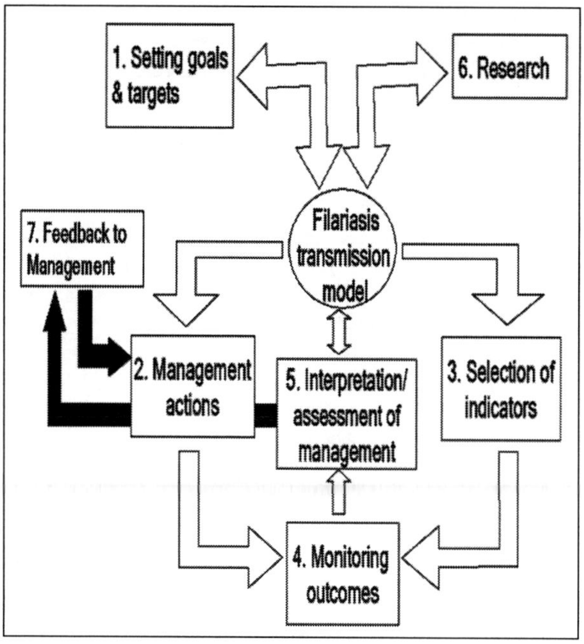

Figure 1. The potential role of parasite transmission models in integrating management actions and monitoring data for providing policy- and management-relevant support to the design and monitoring of parasite control programmes. Models can help define endpoint targets for intervention goals, predict the effects of management actions, aide the selection of outcomes and indicators for monitoring and guide interpretation of monitoring data. Feedback from model-based evaluation of monitoring outcomes will be vital to adapting and refining both management actions and the model via revealing topics for research.[1] Reproduced with permission from Michael E et al. Adv Parasitol 2007; 65:191-237;[6] ©2007 Elsevier.

to the occurrence of multiple stable parasite states and crucially the formation of transmission system boundaries the crossing of which can cause a parasite population to either persist in an endemic state or go to an extinct state. Although our initial work has provided some estimates of the likely numerical values of these thresholds,[8] the finding that elimination thresholds for filariasis are very sensitive to the local level of endemicity, vector species and specific population processes occurring in the local host and vector populations means that ultimately estimating reliable endpoints for a locality will require the fitting of the appropriate vector-species specific transmission model to site-specific infection data. Such data-driven model-based estimation is also necessitated by the often large uncertainties associated with model structure, parameterization (especially when such models are characterized by a relatively large number of parameters, as is typical with dynamic parasite transmission models) and prediction.[3,19-21]

Fitting complex ecological models to data is not a trivial task,[22] especially when there is corresponding uncertainty and lack of detail in the site-specific infection data. However, recently we have extended a methodological approach based on fitting dynamic parasite transmission models to data using computer simulation techniques in conjunction with a Bayesian information melding (BM) method,[3,19-21] to show how such methods may allow the joint use of data and dynamic models to quantify and test locally relevant parasite elimination endpoint values. The BM method (see a more detailed description given by Spear and Hubbard; Chaper 7) takes all available prior information on model inputs and outputs and, where available, likelihood functions for data and generates posterior distributions of model inputs and outputs through statistical comparisons of predictions with data. The essence of the method is to initially assign to each parameter of a model a distribution function reflecting current uncertainty of its value and to refine these estimates from new information as embodied by the data. The form implemented in our analysis used: (1) uniform or vague prior distributions for each of the model input parameters and (2) likelihood functions for the available data, which in the present case are age-prevalence of infection and therefore, are assumed to be binomially distributed. The multidimensional space defined by the set of prior distributions for each input parameter is then evenly sampled 100,000 times. For each instance of a sampled parameter vector, the model is run and likelihoods of predictions compared to observed data are calculated for each of the age prevalence curves generated. We then used the Sampling Importance Resampling (SIR) algorithm[20] to resample 500 parameter sets from the original set of 100,000 parameter generates with a probability given by their output likelihood values. These can then be used to generate distributions of the model outputs, i.e., worm breakpoints, threshold biting rates (TBRs) and the basic reproductive ratios (R_0) estimates for any given study community. Figure 2 shows the 500 resampled age-dependent equilibrium curves obtained by the SIR algorithm plotted against actual age-prevalence data from communities in East Africa (where the anopheline LF

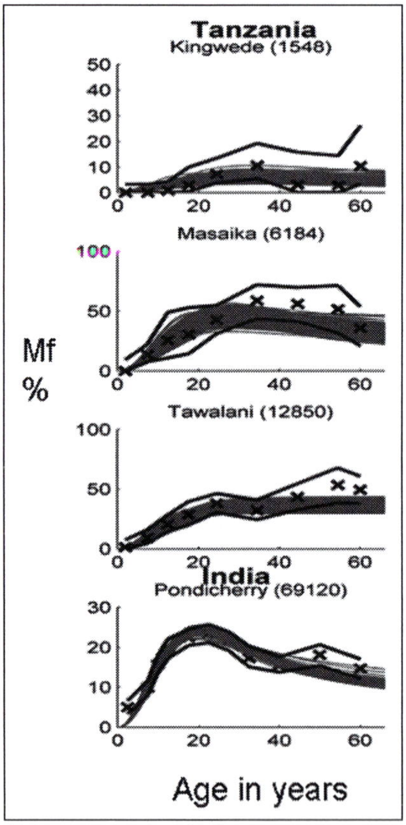

Figure 2. The 500 curves (blue solid lines) generated by importance resampling of the input parameter sets according to their likelihood for each of the datasets investigated via the application of the Bayesian Melding model fitting approach described in the text (black crosses with solid black lines showing upper and lower 95%CIs of the data). The Annual Biting rate is given in parentheses above each of the plots. The appropriate model corresponding to Anopheles or Culex vector transmission species was used as required.[8] A color version of this image is available at www.landesbioscience.com/curie.

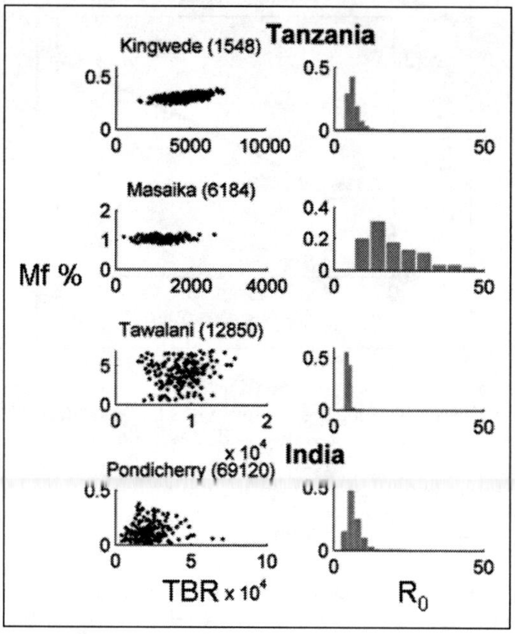

Figure 3. Worm breakpoints vs. TBRs (black scatter plots) and the estimated R_0s from each of the best fitting parameter sets obtained from the Bayesian Melding based model fitting to each of the data sets. R_0 values were obtained essentially by assuming a quasi-equilibrium state for fast-changing variables, setting positive density-dependences to 1 and by finding the largest eigenvalue of the Jacobian matrix.

transmission model was fitted) and India (where the culicine LF transmission model was fitted). Each of the plotted curves is generated by a different model parameter set and the range of curves produced represents the uncertainty remaining in the parameters following the BM updating procedure. Figure 3 portrays the maximum values of worm breakpoints, the corresponding TBR's and a histogram of the R_0 values obtained for each of the data sets. The results highlight how this approach can be used to fit, refine and estimate extant uncertainty in complex dynamic parasite transmission models and how it may be used to estimate locally relevant elimination thresholds. They also show that for both the endpoints (both TBRs and mf prevalence breakpoints), not only can there exist a wide distribution of values consistent with the data in each study community when the model is fitted to mf age prevalence, but also that these thresholds can vary significantly between the study communities.

Vector Infection Thresholds: Theory and Empirical Data

To assess local filarial transmission elimination, it is essential also to not only determine if larval infection thresholds exist in the vector population that could signify parasite transmission elimination but also, as in the case of worm breakpoints above, to quantify what the numerical values of such thresholds might be. There are two principal reasons for this interest in identifying and quantifying these thresholds. First, clearly larval infection thresholds represent the parasite elimination target for drug intervention programmes in vector-borne infections, such as filariasis.[6,23] Second, information regarding the numerical value of such vector infection thresholds is required if xenomonitoring tools, such as parasite DNA-based PCR poolscreen methods,[24-26] can be deemed to be feasibly applicable, on the basis of mosquito sampling sizes, in filariasis monitoring programmes compared to methods based on measurement of infection in humans.[6,23]

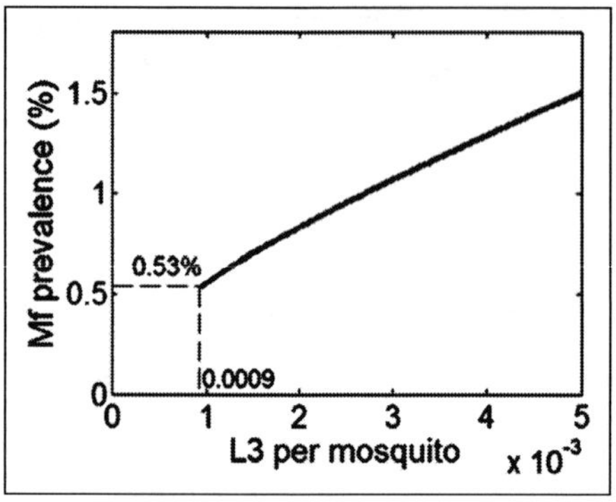

Figure 4. Results from a numerical stability analysis of the coupled deterministic transmission model of lymphatic filariasis indicating the existence of a L3 larval density transmission threshold (value of 0.0009 per mosquito given by the dashed line crossing the X-axis) in culicine mosquitoes corresponding to the existence of a mf prevalence threshold in the human host (value of 0.53% given by the dashed line crossing the Y-axis). Details of model and analysis methods are as given in Gambhir and Michael.[8] Reproduced with permission from Pedersen EM et al. Trends Parasitol 2009; 25:319-27;[23] ©2009 Elsevier.

Recently, we have shown how a joint stability analyses of the deterministic filariasis transmission model described in the chapter by Michael and Gambhir (Chapter 2) and analyses of vector infection data will be critical to address these issues. Figure 4 shows the results from a numerical stability analysis carried out using the deterministic model of lymphatic filariasis transmission . It depicts the existence of a mf prevalence breakpoint threshold of 0.53% prevalence in the human population (shown for culicine transmitted filariasis for given averaged values of each model parameter) that corresponds to a density threshold of 0.0009 L3 per mosquito in the vector population. However, since no L3-specific probes to monitor vector population transmission now exist (these would need to be based on a stage specific RNA transcript, protein or other nonDNA biomarker), control programs are left with estimating the comparable "all-stage infection" (mf, L2 and L3) prevalence figures to gainfully use the theoretical larval breakpoint results depicted in Figure 4. This requires investigating how L3 density per mosquito is related to L3 prevalence and how in turn L3 prevalence is related to prevalence of all-stage larval infection prevalence. Figures 5A,B show how we can use published data on mosquito dissections to derive both these prevalences based on estimating (1) the relationship between average L3 density recorded among dissected mosquitoes versus the prevalence of L3 infection observed in these mosquitoes in order to obtain the threshold L3 infection prevalence (Fig. 5A) and then (2) to use this threshold value in conjunction with estimation of the relationship between vector L3 infection prevalence and overall infection prevalence again observed in published data to derive the vector overall or all-stages infection threshold (Fig. 5B). These results are clearly crude (e.g., the relationships are estimated for both culicine and anopheline mosquitoes combined and the data are derived from a variety of mosquito collection methods, including a mix of resting and/or biting catches);[23] nonetheless they provide a first estimate of the magnitudes of the infective L3 and overall larval infection threshold prevalences that may be used as preliminary transmission endpoint targets in the global LF elimination programme. More specifically, the results in Figures 5A,B suggest that in this regard if L3 infection thresholds obtained through mosquito dissections are to be used as endpoint targets then the relevant endpoint value could be preliminarily set as 0.085%, while for overall larval infection, the corresponding value would be higher at 0.65%. As noted by Michael et al,[6] the importance of these actual threshold values is that they will play a major role in the choice of the infection indicator for monitoring. For example, if dissections are to be used for estimating infection levels in the vector population, the higher value of the overall larval

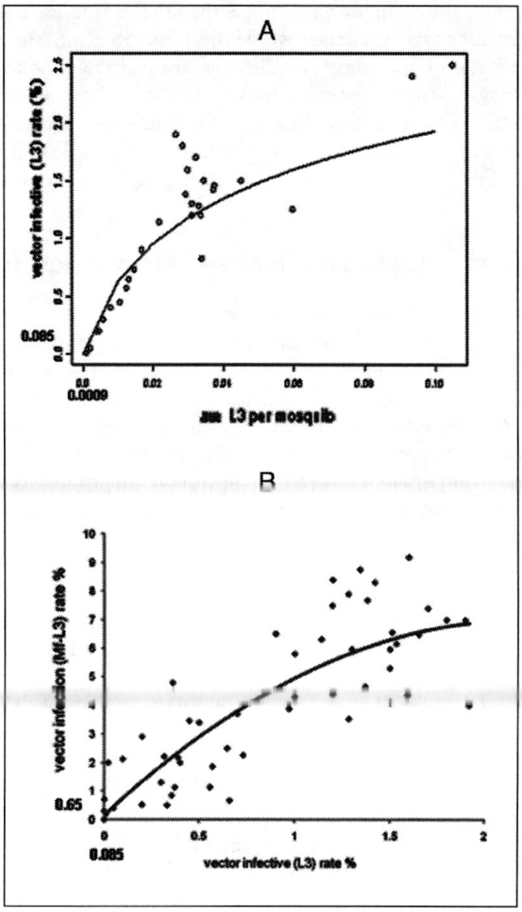

Figure 5. A) The relationship between average L3 per mosquito and prevalence of L3 infection obtained in wild-caught mosquitoes from filarial endemic communities. Each circle in the Figure denotes paired data on these variables obtained from mosquito samples within each study community (n = 29, refs. given in Pedersen et al[23]). The solid curve shows the fit of the negative binomial model relating infection intensity to prevalence given by: $P(M,k)=1-(1+M/k)^{-k}$ where P is the L3 infection prevalence, M is the L3 density and k is the aggregation parameter of the negative binomial distribution. The model was fit to the data by maximum likelihood using binomial errors giving an estimation of k of 0.007. Application of the model to an M value of 0.0009 (the L3 transmission threshold density) gives a corresponding value of 0.085% as a potential L3 threshold prevalence for determining filariasis elimination (shown in bold on the graph). B) The relationship between overall all stages vector infection prevalence and prevalence of L3 infection obtained in wild-caught mosquitoes from filarial endemic communities. Each solid circle in the Figure denotes paired data on these variables obtained from mosquito samples within each study community (n = 63, refs. in Pedersen et al[23]). The solid curve shows the best fit quadratic logistic regression model of the form: $\log[\frac{p}{1-p}] = a + b_1X + b_2X^2$, where p denotes the prevalence of overall all stages infection, X is the prevalence of L3 infection, a represents the intercept and b_1 and b_2 represent the linear and quadratic terms respectively, with estimated values of a = 0.239, b_1 = 0.210 and b_2 = −0.008. Application of this model to the L3 threshold prevalence value of 0.085% gives an empirical value of 0.65% as the corresponding all-stages larval infection prevalence threshold (Figures in bold on the graph). Reproduced with permission from Pedersen EM et al. Trends Parasitol 2009; 25:319-27;[23] ©2009 Elsevier.

infection prevalence indicator, which would reduce the required sample size plus ease the effort of dissections for estimating this prevalence, would mean that this indicator would be the choice for transmission monitoring when using mosquito dissection as a method to monitor the impact of an intervention.[23] Again, as noted for worm breakpoints, model uncertainties as well as heterogeneities in initial conditions between localities, such as in infection aggregation, host immunity and type and magnitudes of positive density dependences, will result in both a distribution of larval infection endpoints within communities as well as significant site-to-site variability in the actual values of this threshold.

Models and Design of Optimal Filariasis Intervention Strategies

Modelling for Choosing Optimal Strategies

A major utility of mathematical models of infectious disease transmission in disease control decision making is that by facilitating the integration of information on pathogen transmission dynamics with programmatic factors such as drug and other intervention efficacies, they may provide a powerful strategic tool for designing and planning community-based control programs against infectious diseases. In particular, if control targets are well defined and model structure, parameter estimates and outputs are validated,[27-31] the predictions of such models can be used for finding the optimal choice among different strategies for meeting a set control-program objective.[1,32] Here, we illustrate and discuss the important role that filariasis transmission models can play in guiding the design of optimal intervention strategies to eliminate culicine-mediated filariasis under different endemic and programmatic conditions, focussing in particular on the optimal design of the two currently proposed mass drug administration (MDA) regimens for filariasis—i.e., combined DEC/ALB and IVM/ALB therapies—and the added impact of including vector control to these regimens.

Figure 6A portrays the results of model predictions of the impacts of these regimens in terms of the number of years required by each option to reach a target threshold of 0.5% mf prevalence (assumed here as the threshold for parasite elimination (but see above)) for a range of precontrol community prevalence or endemicity levels. Three key features of these results illustrate how such model simulations can support the planning and optimal design of parasite elimination programs. First, the dotted droplines in the figure show the feasibility boundaries, i.e., the maximum endemicity level at which it would be feasible to achieve the target threshold of 0.5% mf prevalence within the prescribed 6 years of control, for each intervention and highlight the endemicity ranges over which a particular MDA (at 80% population drug coverage) or MDA plus vector control (incorporating a 90% reduction in vector biting) intervention strategy can successfully meet the above control criterion. The foremost finding here is that annual mass treatments using the combined IVM/ALB regimen may successfully achieve the above elimination criterion only in communities with up to a relatively low precontrol mf prevalence of 3.75% (at the scale of 1 ml blood sampling volume), while in the case of annual mass combined DEC/ALB treatments, the corresponding controllable precontrol community endemicity levels may be only moderately higher at around some 10% mf prevalence (Fig. 6A). These results clearly imply that achieving elimination of filariasis transmission by mass chemotherapy alone, whether by the application of the IVM/ALB or the DEC/ALB regimens, may not be feasible within the prescribed 6 years timeframe in areas of high endemicity, although clearly the DEC/ALB option (largely as a result of its greater macrofilarical activity[1])would achieve parasite elimination for a higher range of precontrol endemicity levels compared to IVM/ALB treatments.

This is highlighted by the second major feature of the results, viz that for both drug regimens, the addition of vector control will extend the range of endemicity levels over which it would still be feasible to achieve the target threshold in the prescribed 6 years compared with that afforded by drug treatments alone (Fig. 6A). Apart from extending the range of controllable endemicity levels, combining vector control with drug administration will also enable the achievement of the set target infection level faster and earlier compared with using each

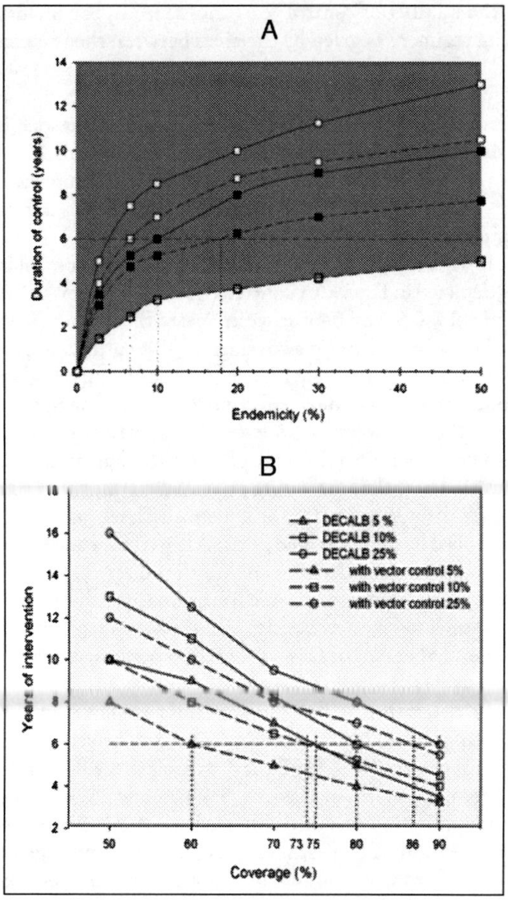

Figure 6. A) Deterministic filariasis transmission model predictions of the number of years of intervention required by various annual MDA and combined annual MDA plus vector control options to achieve the target elimination threshold of 0.5% mf prevalence (44), for a range of precontrol community endemicity (Mf %) levels. Solid curves give predictions for IVM/ALB (open squares) and DEC/ALB (closed squares) annual MDA regimens respectively, while dashed curves portray the corresponding results for each of these regimens combined with vector control (i.e., open squares-dashed line for IVM/ALB plus vector control and closed squares-dashed line for DEC/ALB plus vector control). Drug efficacy values: DEC/ALB—55% worm kill, 95% mf cured and 6 months mf suppression; IVM/ALB—35% worm kill, 99% mf cured and 9 months mf suppression; while vector control is assumed to be 90% effective in reducing vector biting. The lowest dashed curve with transparent squares denote the effects of increasing the frequency of mass treatment to the DEC/ALB plus vector control regimen. The droplines in the Figure serve to indicate the maximum endemicity level at which it would be feasible to achieve the set target threshold of 0.5% mf prevalence within the prescribed 6 years of control by each intervention. All results at 1 ml blood sampling volume. B) Model simulations of the number of years of intervention required by the mass annual DEC/ALB regimen either administered alone (solid lines) or with vector control (dashed lines) to achieve the 0.5% mf prevalence threshold for a mix of precontrol community mf prevalences (5%, 10% and 25%) and drug coverage values. 90% Vector control efficacy assumed. Vertical droplines show the optimal drug coverage required at each endemicity level by each of these options to meet the control criterion of achieving the 0.5% mf prevalence threshold in 6 years. Reproduced with permission from Michael E et al. Trends Parasitol 2006; 22:226-33;[5] ©2006 Elsever.

drug regimen alone. The number of control years thus saved by the inclusion of vector control to each of the two drug regimens is given by the areas between the respective solid and dashed lines in Figure 6A and indicates that on average between ½ to 1 year at low (<10%) and 1½ to 2 years at moderate/high (>20%) precontrol community infection levels may be saved by adding vector control to each of the DEC/ALB and IVM/ALB regimens respectively (Fig. 6A). These findings underscore the critical role that including vector control to MDA programs can play in filariasis elimination programs, especially in areas of high endemicity, where the benefits of this combined strategy would be most apparent and in areas, such as Africa, where MDA using IVM/ALB is to be implemented.

A third important insight of significance to the global filariasis control initiative gained from the model simulations depicted in Figure 6A concerns the finding that when combined with the set control criterion, i.e., reaching 0.5% mf prevalence in 6 years, the addition of vector control (at 90% efficacy) to DEC/ALB mass treatment (at a coverage of 80%) produces a rather moderate upper limit to the endemicity level (approximately 18% at the scale of 1 ml blood sampling volume) that can be feasibly controlled even by this most effective combined strategy. This result implies that we need to either extend the duration or frequency of mass treatment in areas of higher precontrol prevalence or examine other more effective options in these areas. Indeed, model simulations (lower curve in Fig. 6A) indicate that increasing treatment frequency to once in every 6 months may substantially increase the upper limit to the endemicity level (to almost 70% mf prevalence) that can be feasibly controlled by the combined DEC/ALB-vector control strategy, suggesting that this option may well need to be considered in areas with the very highest prevalences if filariasis elimination is to be achieved within a reasonable timeframe in such regions.

Predictions from transmission models can also play a key role in quantifying another important element regarding the rational design of mass treatment programs, viz the optimal drug coverage needed to meet a set control criterion for various endemic conditions.[1,33-35] Figure 6B shows the predictions made by the deterministic filariasis transmission model of the number of years of intervention required using annual MDA with DEC/ALB alone versus mass annual DEC/ALB administration plus vector control (at 90% efficacy) to achieve the 0.5% mf prevalence threshold for a mix of precontrol community mf prevalences (5%, 10% and 25%) and drug coverage patterns and exemplifies how the application of models can facilitate the derivation of optimal drug coverages for drug-based intervention strategies. Several important points regarding optimal program design in this regard emerge from the simulations shown in the figure. A key finding is that for both the MDA alone and the MDA plus vector control strategies, optimal drug coverages will increase with precontrol infection endemicities and for both strategies this impact of variations in drug coverages, particularly in the case of sub-optimal coverage, in achieving a set infection control will be greater at higher precontrol endemicity levels compared to low precontrol infection prevalence levels (Fig. 6B).

The results in Figure 6B also highlight a final key point regarding the inclusion of vector control to MDA programmes, viz that for all endemic situations including vector control will lower the optimal MDA coverage required to meet a set control criterion compared to MDA alone. Given that typical coverages achieved in large-scale community treatment programmes are normally only around 65%,[36] this result again clearly affirms the value and importance of considering the inclusion of vector-management options in current filariasis mass treatment programmes,[1,37-39] especially in areas with high precontrol endemicities.

Table 1 shows how these results can be summarized as a type of decision table to aide the identification of the optimal coverage needed to achieve a programme target for various precontrol community mf endemicity levels. The shaded diagonal region in the table represents the present decision criterion (reaching 0.5% mf threshold in 6 years) for various combinations of endemicity levels and coverage rates. The optimal coverage required to meet the control threshold of 0.5% in 6 years for a particular precontrol prevalence are then found simply by tracing the corresponding paired figures off this diagonal. These results not only contribute to providing a decision support framework for determining the optimal coverage for the successful execution of the mass treatment programme

Table 1. Developing decision support for the design of filariasis elimination programs. Model predictions of the number of years required to reach a 0.5% microfilarial prevalence threshold using combination DEC-ALB mass drug treatment in relation to endemicity and coverage.[1]

	Endemicity (Mf %)			
Coverage	**2.5%**	**5.0%**	**10.0%**	**15.0%**
60%	7	9	10	12
70%	6	7	8	9
80%	5	6	7	8
90%	4	5	6	7
95%	3	4	5	6

under each local endemic setting—such decision tables can also provide an indication to planners of the added number of years of treatment that may be required in each locality if coverage falls off into the sub-optimal zone represented by the area above the shaded entries in Table 1.

Choosing an Optimal Strategy Under Uncertainty

While the above results are pertinent in identifying optimal strategies under conditions of low uncertainty regarding both program objectives as well as biology or ecology of the parasite system, choosing the best option and management decision-making under uncertainty in either these areas presents a more complex problem in parasite management.[14,40,41] Recently, we have shown how uncertainty in transmission endpoints behooves us to take a more flexible sequential approach to achieving filariasis elimination based initially on attaining disease control and then proceeding to the target of parasite elimination when information regarding endpoints becomes better known.[11] This work, which combined analysis and quantification of the disease control threshold, mathematical modelling of the effects of MDA with IVM/ALB and economic analysis, was applied to the control of filariasis eradication in Tanzania, is summarized in Table 2 in terms of expected present costs (EPCs) and option values of strategies with and without a flexible sequential decision option based on exercising disease control first. The results in Table 2 depict that the EPCs of strategies that contain the flexibility of switching between disease control and eradication are lower than that of an eradication strategy in which no flexibility to either switch to disease control if eradication is not feasible or to embark upon eradication after disease control when eradication becomes favourable exists. Since the EPCs of the flexible strategies are lower than that of the inflexible eradication strategy, it is clear that such strategies should be preferable highlighting their optimality when uncertainty regarding eradication occurs. Indeed, the results show that switching to long-term disease control from year 5 (MDAs given every 10 years following an initial 5 annual MDAs to achieve disease control: Strategy 2) can be the cheapest intervention option for LF if eradication can never be achieved (Table 2), while the lower EPCs of Strategies 3a and b compared to Strategy 1 highlight the costs which can be saved by waiting for better information regarding the feasibility of eradication before switching to eradication from disease control. The value of including such flexibilities in the filariasis eradication program for Tanzania is given by option value figures depicted in Table 2 and indicates that including flexible decision-making based initially on disease control can yield savings of between US$ 3.3 to 5.0 million depending on the feasibility of eradication (from never (Strategy 2) to waiting to switch to eradication at different times when knowledge and technology improves (Strategies 3a and 3b)). These results also clearly highlight that when substantial biological and programmatic uncertainty exists,[11] there is value in designing and implementing adaptive management options that will facilitate learning about the parasite system and ultimately the best long-term strategies for achieving overall management goals.

Table 2. Present value costs of LF eradication strategies for the Republic of Tanzania with and without sequential-decision making flexibility based on exercising chronic disease control options[11]

Strategy[1]	Design	EPC[2]	Option Value[3]
S1	*Inflexible*: 10 year annual MDA	157,165,771	
S2	*Flexible*: Switch to long-term disease control after 5 years of annual MDA when it becomes evident eradication is not possible	152,110,572	5,055,199
$S3_a$	*Flexible*: Switch to eradication with 3 extra annual MDAs from year 10 following disease control with 5 initial annual MDAs	153,822,682	3,343,089
$S3_b$	*Flexible*: Switch to eradication with 3 extra annual MDAs from year 15 following disease control with 5 initial annual MDAs	152,968,795	4,196,976

[1]Denotes the three different scenarios described in detail in the text. [2]Expected present value cost of each strategy in US$. [3]The difference between the Expected Present Costs (EPCs) of flexible stategies 2 to 3 over strategy 1. Represents the value (costs saved here) of retaining the option to switch between eradication and disease control depending on the state of eradication feasibility.
From Michael E et al. PLoS One 2008; 3(8):e2936.[11]

However, the finding of the existence of hysteresis loops in the mf prevalence/vector biting rate plane in filariasis transmission dynamics (see Fig. 5 in Chapter 2 by Michael and Gambhir; this volume) suggests that a second tactic to promote filariasis elimination might be to avoid focusing solely on meeting the objective of uncertain elimination, but to exploit the ability of even a relatively poor model to give fairly good guidance to promote good parasite system transitions (e.g., parasite elimination) and prevent bad transitions (e.g., infection re-emergence following control).[41] Given that including vector control with MDA can, by increasing the worm breakpoint threshold value reduce the resilience of the endemic state and by raising the re-emergence infection threshold promote the resilience of the parasite-free state,[8] it is clear that a combined MDA/vector control strategy can play this resilience-enhancing role in facilitating and sustaining filariasis elimination.[8,42]

Models and Design of Monitoring Programs

Recently, it has been shown how by conceptualizing critical components and processes of parasite transmission dynamics, mathematical models of parasite transmission can provide a rational framework for undertaking monitoring and evaluation activities via their ability to unveil information on the following four specific areas: (1) determination of intervention endpoint targets to meet program objectives;[6] (2) predictions regarding the expected magnitude of changes in the states of the parasite system under various interventions to determine if these are on track to meet set endpoint targets, including estimation of optimal frequency of monitoring and suggestion of remedial measures to get the intervention back on track;[1,15] (3) determination and exploration of the roles of key indicators for monitoring effectiveness; and finally (4) guiding interpretation of monitoring results to ascertain if successful control has been achieved.[43]

A difficulty with the above role and use of models in monitoring programs is that it is premised on the notion that it is possible to predict and anticipate the consequences of management decisions, i.e., that once all the necessary information is gathered, a right "scientific" decision can be made. For filariasis (and for most other parasitic systems), given that major uncertainties exist in almost every area of our understanding of transmission dynamics and control, including (1) model and process uncertainty which gives rise to approximate thresholds and predictions of the effects of intervention, (2) observation uncertainty related to diagnostic tool accuracy, (3) the role of spatial and population ecologies on the sampling process, (4) intervention uncertainty in which there is little reliable information available regarding the effectiveness of the various proposed interventions to reduce or interrupt parasite transmission,[5,15] and (5) that the transmission dynamics is likely to be furthermore ecologically complex,[8] it is clear that taking such a traditional anticipatory approach to management is problematic. This highlights that a major role of monitoring programs should be to provide data that can be used to develop our knowledge of parasite system dynamics and responses to management. In particular, in the face of complex infection dynamics, there is a need to plan and use monitoring data to identify the alternate attractors accessible to the parasite infection system, the feedbacks which maintain the system at the attractors, the external ecological influences (environmental, vector species, vector and human migration) which define the context for a specific attractor and conditions, including interventions, which allow or promote "good" flips between attractors.[41]

The above considerations suggest that there is an imperative need to modify existing command and control management models for filariasis control based on globally-set targets in order to account for the complexity and uncertainty in parasite system dynamics and resulting intervention thresholds that are likely to vary from site to site.[8] Specifically, we identify here a need to develop an adaptive management plan that (1) acknowledges uncertainty and system complexity, (2) actively contemplates using monitoring system responses to interventions to learn about and reduce uncertainty in system models and (3) anticipates that future management interventions will be modified as we gather information and learn more about system behaviour. Such an approach must also link monitoring data with robust methods, such as Bayesian analysis, that allow ready and effective updating of new information to refine model structure or reduce parameter uncertainty. This approach will also require a change in attitudes to implementing parasite interventions for at its core it argues for the use of interventions as natural experiments to compare selected policies or practices in order to evaluate alternate hypotheses about the parasite system being managed.[6] This is in sharp contrast to the current rigid top-down anticipatory management strategies used for parasite control, which sees such deviations from set protocols as "errors" to be avoided.

Conclusion

Informed parasite management requires information about both the parasite system state at a given time as well as the impact that potential management actions will bring about on that state. We have shown here how mathematical models of parasite transmission can play a vital role in informing and guiding the development of filariasis elimination programs by their ability to provide quantitative information for decision-making in all key areas of optimal program design and execution, ranging from (1) initial definitions of thresholds to meet various intervention endpoints, (2) optimising program parameters for different endemic conditions and (3) creation and application of rational monitoring and evaluation plans for assessing management success. The results, however, have also highlighted the continuing gaps in understanding and the urgent need to resolve such gaps in both knowledge of parasite population dynamics and hence model formulation and parameterization, and the impact of stochasticity in key infection processes as well as intervention effectiveness—for example, currently there is little reliable information available regarding both drug efficacy rates against the underlying adult worms[1] as well as the rates of efficacy of various vector control measures in reducing vector population size—if modelling results are to be reliably integrated and applied within control programs. We show how a critical advance will also be how best to address the impact of ecological complexity in parasite transmission if we are to develop a more reliable approach to parasite control management.

Whereas new model fitting approaches based on Bayesian updating methods can be used to both reduce model uncertainty and gain an understanding of localized transmission complexities and sequential decision frameworks could be used to guide management policies in the face of uncertainty, perhaps the most important conclusion from the work described here is that this need to resolve uncertainties and address complexities means that control programs should also be sufficiently flexible to adapt to local transmission dynamics and even accommodate change in intervention methods as the program proceeds. Complexity in transmission dynamics also indicate that several intervention methods, for example adding vector control to MDA in filariasis inventions, may need to be applied simultaneously or sequentially during a program to enhance the resilience and sustainability of "good" system transistions.[8,41,42]

Overall, these results thus imply that rather than using a simple "one size fits all" command and control approach, as currently promulgated in parasite control programs, a more nuanced approach focusing on using intervention and monitoring programs to aid greater learning and understanding of parasite system states and the forces that govern critical system transitions and how best to manage such transitions by interventions to achieve desired objectives, will be required if we are to successfully attain the goal of controlling or elimination human parasitic infections. Transmission models will clearly play a major role in the development and application of this adaptive management framework. We indicate that if science, including mathematical models, is to effectively support the management of large-scale control or elimination of parasitic diseases, including lymphatic filariasis and other major human helminths, from endemic communities, such management methods closely linked to and aided by models should now be urgently examined and deployed at all programmatic levels.

Acknowledgements

The authors gratefully acknowledge the financial support of NIH grant no. RO1 AI069387-01A1 in facilitating this research.

References

1. Michael E, Malecela-Lazaro MN, Simonsen PE et al. Mathematical modelling and the control of lymphatic filariasis. Lancet Infect Dis 2004; 4:223-34.
2. Liang S, Seto EY, Remais JV et al. Environmental effects on parasitic disease transmission exemplified by schistosomiasis in western China. PNAS 2007; 104:7110-5.
3. Spear RC, Hubbard A, Liang S et al. Disease transmission models for public health decision making: toward an approach for designing intervention strategies for Schistosomiasis japonica. Environ Health Perspect 2002; 110:907-15.
4. Michael E. The epidemiology of filariasis control. In: Klei TR, Rajan TV, eds. The Filaria. Boston: Kluwer Academic Publishers, 2002;60-74.
5. Michael E, Malecela-Lazaro MN, Kabali C et al. Mathematical models and lymphatic filariasis control: endpoints and optimal interventions. Trends Parasitol 2006; 22:226-33.
6. Michael E, Malecela-Lazaro MN, Kazura JW. Epidemiological modelling for monitoring and evaluation of lymphatic filariasis control. Adv Parasitol 2007; 65:191-237.
7. Chan MS, Srividya A, Norman RA et al. Epifil: a dynamic model of infection and disease in lymphatic filariasis. Am J Trop Med Hyg 1998; 59:606-14.
8. Gambhir M, Michael E. Complex ecological dynamics and eradicability of the vector borne macroparasitic disease, lymphatic filariasis. PLoS One 2008; 3(8):e2874.
9. Norman RA, Chan MS, Srividya A et al. EPIFIL: the development of an age-structured model for describing the transmission dynamics and control of lymphatic filariasis. Epidemiol Infect 2000; 124:529-41.
10. Plaisier AP, Subramanian S, Das PK et al. The LYMFASIM simulation program for modeling lymphatic filariasis and its control. Methods Inf Med 1998; 37:97-108.
11. Michael E, Malecela MN, Zervos M et al. Global eradication of lymphatic filariasis: the value of chronic disease control in parasite elimination programmes. PLoS One 2008; 3(8):e2936.
12. Subramanian S, Stolk WA, Ramaiah KD et al. The dynamics of Wuchereria bancrofti infection: a model-based analysis of longitudinal data from Pondicherry, India. Parasitology 2004; 128:467-82.
13. Stolk WA, Swaminathan S, van Oortmarssen GJ et al. Prospects for elimination of bancroftian filariasis by mass drug treatment in Pondicherry, India: a simulation study. J Infect Dis 2003; 188:1371-81.
14. Kendall WL. Using models to facilitate complex decisions. In: Shenk TM, Franklin AB, eds. Modeling in Natural Resource Management. Washington: Island Press, 2001;147-70.

15. Michael E, Malecela-Lazaro MN, Maegga BTA et al. Mathematical models and lymphatic filariasis control: monitoring and evaluating interventions. Trends Parasitol 2006; 22:529-35.
16. Anderson R, May R. Infectious Diseases of Humans: Dynamics and Control. Oxford: Oxford University Press, 1992.
17. Pugliese A, Tonetto L. Thresholds for macroparasite infections. Math Biol 2004; 49:83-110.
18. Duerr HP, Dietz K, Eichner M. Determinants of the eradicability of filarial infections: a conceptual approach. Trends Parasitol 2005; 21:88-96.
19. Codeco CT, Luz PM, Coelho F et al. Vaccinating in disease-free regions: a vaccine model with application to yellow fever. J Roy Soc Interface 2007; 4:1119-25.
20. Poole D, Raftery AE. Inference for deterministic simulation models: the Bayesian melding approach. J Amer Stat Assoc 2000; 95:1244-55.
21. Radtke PJ, Burk TE, Bolstad PV. Bayesian melding of a forest ecosystem model with correlated inputs. Forest Sci 2002; 48:701-11.
22. Oreskes N. The role of quantitative models in science. In: Canham CD, Cole JJ, Luauenroth WK, eds. Models in Ecosystem Science Princeton: Princeton University Press, 2003;13-31.
23. Pedersen EM, Stolk WA, Laney SJ et al. The role of monitoring mosquito infection in the Global Programme to Eliminate Lymphatic Filariasis. Trends Parasitol 2009; 25:319-27.
24. Williams SA, Laney SJ, Bierwert LA et al. Development and standardization of a rapid, PCR-based method for the detection of Wuchereria bancrofti in mosquitoes, for xenomonitoring the human prevalence of bancroftian filariasis. Ann Trop Med Parasitol 2002; 96:S41-6.
25. Fischer P, Wibowo H, Pischke S et al. PCR-based detection and identification of the filarial parasite Brugia timori from Alor Island, Indonesia. Ann Trop Med Parasitol 2002; 96:809-21.
26. Rao RU, Atkinson LJ, Ramzy RM et al. A real-time PCR-based assay for detection of Wuchereria bancrofti DNA in blood and mosquitoes. Am J Trop Med Hyg 2006; 74:826-32.
27. Medley GF. Predicting the unpredictable (Perspective). Science 2001; 294:1663-64.
28. Green LE, Medley GF. Mathematical modelling of the foot and mouth disease epidemic of 2001: strenghts and weaknesses. Res Vet Sci 2002; 73:201-05.
29. Taylor N. Review of the use of models in informing disease control policy development and adjustment. A report for DEFRA. 2003; Available from: http://www.defra.gov.uk/science/publications/2003/useofmodelsindiseasecontrolpolicy.pdf.
30. Pfeiffer D. Science, epidemiological models and decision making. Vet J 2004; 167:123-24.
31. Kitching RP. Predictive models and FMD: the emporer's ew clothes? Vet J 2004; 167:127-28.
32. Clark CW. Bioeconomic Modelling and Fisheries Management. London: Wiley-Interscience, 1985.
33. Anderson RM, May RM. Helminth infections of humans: mathematical models, population dynamics and control. Adv Parasitol 1985; 24:1-101.
34. Woolhouse ME. On the application of mathematical models of schistosome transmission dynamics. II. Control. Acta Trop 1992; 50:189-204.
35. Winnen M, Plaisier AP, Alley ES et al. Can ivermectin mass treatments eliminate onchocerciasis in Africa? Bull World Health Organ 2002; 80:384-91.
36. Plaisier AP, Stolk WA, van Oortmarssen GJ et al. Effectiveness of annual ivermectin treatment for Wuchereria bancrofti infection. Parasitol Today 2000; 16:298-302.
37. Maxwell CA, Mohammed K, Kisumku U et al. Can vector control play a useful supplementary role against bancroftian filariasis? Bull World Health Organ 1999; 77:138-43.
38. Curtis CF, Malecela-Lazaro M, Reuben R et al. Use of floating layers of polystyrene beads to control populations of the filaria vector Culex quinquefasciatus. Ann Trop Med Parasitol 2002; 96:S97-104.
39. Burkot T, Ichimori K. The PacELF programme: will mass drug administration be enough? Trends Parasitol 2002; 18:109-15.
40. Norton JP, Reckhow KH. Modelling and monitoring environmental outcomes in adaptive management. In: Jakeman AJ, Voinov AA, Rizzoli AE, Chen SH, eds. Environmental Modelling: Software and Decision Support. Amsterdam: Elsevier, 2008;181-204.
41. Scheffer M. Critical Transitions in Nature and Society. Princeton: Princeton University Press, 2009.
42. Bockarie MJ, Pedersen EM, White GB et al. Role of vector control in the global program to eliminate lymphatic filariasis. Ann Rev Entomol 2009; 54:469-87.
43. Maddox DE, Polani K, Unnasch R. Evaluating management success: using ecological models to ask the right monitoring questions. In: Johnson NC, Malk AJ, Szaro RC, Sexton WT, eds. Ecological Stewardship A Common Reference for Ecosystem Management. Oxford: Elsevier Science Ltd, 1999;563-84.

Chapter 12

Disease Transmission Models for Public Health Decision-Making:

Designing Intervention Strategies for *Schistosoma japonicum*

Edmund Y.W. Seto* and Elizabeth J. Carlton

Abstract

The purpose of infectious disease transmission modelling is often to understand the factors that are responsible for the persistence of transmission, the dynamics of the infection process and how best to control transmission. As such, there should be great potential to use mathematical models to routinely plan and evaluate disease control programs. In reality, there are many challenges that have precluded the practical use of disease models in this regard. One challenge relates to the mathematical complexity of the models, which has made it difficult for field workers and health officials to understand and use them. Another challenge is that, despite their mathematical complexity, models typically do not have sufficient structural complexity to consider many of the site-specific epidemiologic and disease control details that the practicing health official routinely considers. Moreover, most modelling studies have not been sufficiently explicit or exemplary in explaining how field data may be incorporated into the models to impact public health decision-making. In this chapter, we start with a classic model of schistosomiasis transmission and relate its key properties to the more detailed model of *Schistosoma japonicum* model presented in chapter by Remais and chapter by Spear and Hubbard. We then discuss how various controls (e.g., chemotherapy, snail control and sanitation) may be evaluated via the detailed model. We then demonstrate in a practical manner, using *S. japonicum* data from China, how field data may be incorporated to inform the practice of disease control. Finally, we present a new model structure that considers how heterogeneous populations are interconnected, which has particular relevance to understanding disease control and emergence in today's highly mobile world.

Introduction

Nearly two decades ago, Woolhouse reviewed the modelling literature for schistosomiasis and how various disease controls may be evaluated.[1,2] We remark upon his rather depressing introduction, which stated that since Macdonald developed the first model in 1965 and subsequent reviews of the modelling literature in the 1977 and 1982 by Cohen and Barbour, respectively, the models had very limited impact on actual schistosomiasis control.[3-5] One of Woolhouse's reasons for this, which we tend to agree with, is that there exists a disconnect between disease modelers who focus their work on theoretical mathematics and field workers who make real-world public health decisions related to disease control. Another reason is that, while mathematically complex,

*Corresponding Author: Edmund Y.W. Seto—School of Public Health, University of California, Berkeley, CA 94720, USA. Email: seto@berkeley.edu

Modelling Parasite Transmission and Control, edited by Edwin Michael and Robert C. Spear.

the models still lacked the realism to capture local variations and logistical constraints that affect the implementation and efficacy of controls. Woolhouse argues that this does not need to be the case. Instead, models may be one of the tools that health planners use, in addition to field experience and epidemiology in designing control programs. With this as background, we begin where Woolhouse left off, using the same simple schistosomiasis modelling framework to consider disease control. We discuss how the *S. japonicum* model presented in Chapters and 5 and 10 relates to the basic modelling framework and allows us to consider various disease controls, including chemotherapy, snail control, sanitation and health education. We then demonstrate in a practical manner, using *S. japonicum* data from China, how field data may be incorporated to inform the practice of disease control.

We also present a new development in schistosomiasis modelling that explicitly considers how heterogeneous populations are interconnected. These connections may occur via natural physical processes (e.g., hydrological connections via watersheds) and/or via social-economic processes (e.g., migrant labor or trading of buffalo). Such a model offers exciting new possibilities to consider how disease control may be best practiced in highly connected environments. As many areas are becoming increasingly connected due to economic development, it also provides opportunities to evaluate the impact of changes in connectivity due improvements to road and irrigation networks, for example, on disease transmission and (re)introduction.

Model Framework

Schistosomiasis infection persists in over 70 countries and remains a cause of great morbidity despite several opportunities in the lifecycle of the parasite to disrupt transmission.[6] Figure 1 illustrates the lifecycle for *S. japoncium* in China and points of possible intervention, which include the treatment of human hosts via chemotherapy, snail control through molluscides or more permanent habitat removal, improvements in sanitation and reductions in exposure (e.g., via health education). We describe each of these controls in more detail below.

Woolhouse's basic model framework is based on the simplification of the worm-snail-cercariae-miracidia 4-state system to a 2-state system of just mean number of schistosome worms per person m and the prevalence (as a proportion) of patent infections in snails y. This is reasonably assumed as the lifespans of the larval stages are considerably shorter than worms and snails (hours versus months and years). Hence, it is assumed that larvae are at equilibrium and we can simply model the remaining states via the following coupled differential equations (see Table 1 for a description of the model parameters):

$$
\begin{aligned}
\frac{dm}{dt} &= \alpha N y - \gamma m \\
\frac{dy}{dt} &= \beta H m(1-y) - \mu y
\end{aligned}
\tag{0.1}
$$

where each equation consists of a gain and loss term. For the worm burden equation, the gain term consists of the product of the number of snails N, prevalence of patent snail infection y and the rate of infection per person per patent infected snail α. The loss term consists of the product of per capita worm mortality γ and the mean worm burden per person m. For the snail infection prevalence equation, the gain term consists of the product of the number of humans H, mean worm burden per person m, the proportion of uninfected snails $(1-y)$ and the per capita rate of infection of snails per schistosome β. The loss term consists of the product of per capita infected snail mortality μ and the patent snail infection prevalence y.

The simple model above has a number of assumptions that are inherent in the underlying Macdonald model, upon which (0.1) is based. First, it is assumed that the helminthes are bisexual and that the mating probability of these worms is incorporated in the β term. Moreover it is generally assumed that there is equal probability of male and female larval infectivity and worms are monogamously paired. More importantly, the basic model ignores possible density dependent effects that may arise through acquired immunity. Furthermore, it is assumed that the loss of

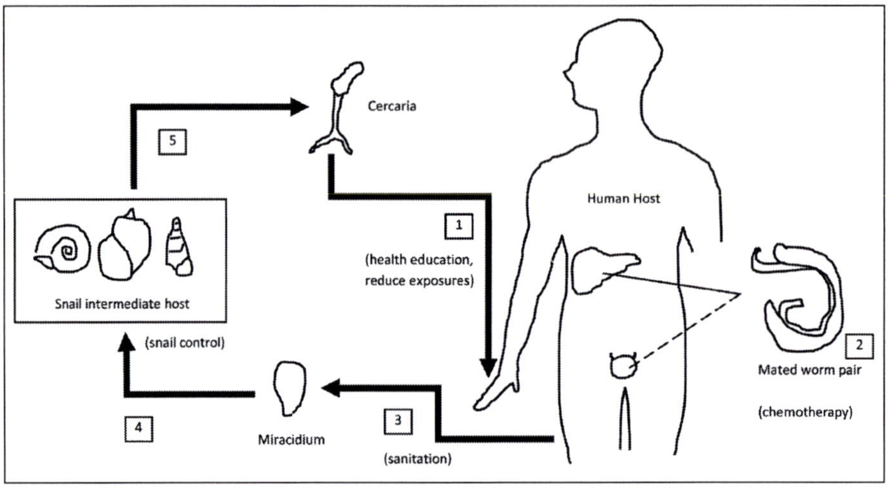

Figure 1. Schistosome lifecycle with main opportunities for control shown in parentheses. 1) Humans become infected by the free-swimming cercarial form of the parasite from skin contact with contaminated surface water. 2) Once infected, adult schistosome worms will develop in the portal veins of the liver or bladder depending on the species of schistosome. Male and female worms will mate and produce eggs, which are released in either the feces or urine. 3) Inadequate sanitation results in parasitic eggs being deposited into an aquatic environment (e.g., ponds, irrigation ditches), where they hatch into a free-swimming miracidial form of the parasite. 4) Miracidia infect the snail intermediate host, which lives in these aquatic environments. 5) After a prepatency period, infected snails will begin to release cercariae, which infect humans to complete their lifecycle.

miracidia from the environment due to snail infection is negligible. Snail numbers are assumed to be held constant, which does not account for seasonal variations or increased morbidity imposed by larval infection. Multiple infections of snails are ignored, as is the issue of infection latency due to a prepatency period. Additionally, human population numbers are assumed to be held constant, with homogeneous exposure to larvae occurring for the population. It also assumes humans are the single definite host. We comment on various extensions of the basic Macdonald model that relax some of these assumptions later.

Table 1. Variables in the transmission models

m	Mean number of schistosome worms per person
y	Prevalence of patent (shedding) infections in snails
α	Rate of infection per person per patent infected snail
N	Number of snails
γ	Worm mortality
β	Rate of infection of snails per schistosome
H	Number of humans
μ	Snail mortality
k	Aggregation parameter for schistosomes in the human host
ϕ	Mating probability of schistosomes
T_{SH}	Snail to human transmission rate
T_{HS}	Human to snail transmission rate
P	Prevalence of human infection

Classically, the approach to studying the Macdonald system is first to evaluate the equilibrium solutions of the model. Equilibrium for each state denoted as m^* and y^* are reached when both the mean worm and patent snail infection prevalence reach a stable unchanging level (i.e., where $dm/dt = 0$ and $dy/dt = 0$). In discussing this equilibrium it is helpful to define the following terms:

$$T_{SH} = \frac{\alpha N}{\gamma}$$
$$T_{HS} = \frac{\beta H}{\mu} \tag{0.2}$$

where T_{SH} considers those parameters that relate to snail-to-human infection and T_{MS} considers those that relate to human-to-snail infection.

The equilibia then can be described according to a quantity known as the basic reproduction number, R_0:

$$R_0 = T_{SH} T_{HS} \tag{0.3}$$

When $R_0 < 1$ then the system does not sustain transmission and the equilibrium values are:

$$m^* = 0$$
$$y^* = 0 \tag{0.4}$$

However, when $R_0 > 1$ there exists a positive equilibrium at:

$$m^* = T_{SH} - \frac{1}{T_{HS}}$$
$$y^* = 1 - \frac{1}{T_{SH} T_{HS}} = 1 - \frac{1}{R_0} \tag{0.5}$$

Hence, an important property of the basic schistosomiasis model is that endemic infection will exist for a population when the Basic Reproduction Number, $R_0 > 1$.

We note that R_0 is explicitly defined by the snail-to-human and human-to-snail infection parameters. However, these terms can be factored into more specific terms, as we saw in the Macdonald-like *S. japonicum* model of Chapter 7. Indeed, we note even in Macdonald's early work, there was a distinction between biologic and site-specific variables (synonymous with the *Ps* and *Pb* variables, respectively mentioned in Chapter 7). For instance, the probability of a cercaria finding and infecting a suitable host, (Macdonald's so-called "exposure factor") may depend upon a biologic constant representing the efficiency of cercarial penetration and a site-specific variable that relates to the number and characteristics of water contacts made by hosts.

In Macdonald's work, considerable detail is spent in describing the effect of adding more detailed mating probability function ϕ as one of the factors that makes up the β term. It is assumed that male and female schistosomes infect with equal probability, but mate at random, with at least two worms, one of each sex required for mating. The results of adding the mating probably function is the establishment of break point in the system, which concerns the resulting equilibrium when the parasite is introduced into a community. As described by Woolhouse,[1] a constant mating probability results in any parasite introduction establishing endemic transmission (positive m^* and y^*). However, when a more realistic assumption that the mating function depends upon the aggregation of worms in hosts (typically regarded as negative binomial distributed), an unstable break point occurs in the system, such that parasite introductions below a threshold do not establish transmission (i.e., the system will fall to $m^* = 0$ and $y^* = 0$ equilibrium). Conversely, when there is a sufficient introduction of parasites, then endemic transmission will be established (i.e., the system will settle on the $m^* > 0$ and $y^* > 0$ equilibrium).

From a control perspective there are thus two theoretical goals in eliminating transmission. One goal may be to reduce the longer-term potential for transmission such that $R_0 < 1$. Alternatively, another goal may be to perturb the system so much as to bring the system below the break point, whereby it will fall to the 0, 0 equilibrium. We note, however, that accomplishing either says nothing of the dynamics (how long it will take) for transmission to be eliminated. We deal with this issue of system dynamics later, after we first explore how the simple Macdonald model performs for real-world data.

A Crude Estimate of R_0 for S. japonicum *Based on Macdonald's Model*

Based on the simple Macdonald model's equilibria equations (0.5) for endemic transmission, we note a rather surprising finding: the only field data required to estimate R_0 is snail infection data. Note that y^*, the equilibrium patent snail infection prevalence (expressed as a proportion) is the sole estimator for R_0:

$$y^* = 1 - \frac{1}{R_0}$$

In China, as in many other places the proportion of snail surveyed that are patent infected typically does not exceed 0.05. In our study of 20 villages in Xichang county in Sichuan Province,[7] we found the highest snail prevalence to be 0.03. Substituting 0.03 for y^*, results in a solution for R_0 of 1.03. Because $R_0 > 1$, the condition for endemic transmission is satisfied.

Control Needed to Terminate S. japonicum *Transmission Based on Macdonald's Model*

The above estimate of $R_0 \cong 1.03$ is surprising because it seems so small. One way to sustainably eliminate transmission is by reducing the potential for transmission by driving $R_0 < 1$. We must effectively transition from a R_0 of 1.03 to an effective reproduction number R (the common term for R_0 after control has been implemented) of <1. We note that because of the following definition of R_0, there are many possible control options to bring $R < 1$:

$$R_0 = T_{SH} T_{HS} = \frac{\alpha N \beta H}{\gamma \mu} \tag{0.6}$$

In other words we need the following solution:

$$R_0 = T_{SH} T_{HS} = \frac{\alpha N \beta H}{\gamma \mu} = 1.03 \Rightarrow R = R_0 L = \frac{\alpha N \beta H L}{\gamma \mu} < 1 \tag{0.7}$$

where L represents a factor that reduces R_0 sufficiently to bring $R < 1$. In this case, to terminate transmission, L must be <0.97. It is quite shocking that only a 3% reduction in R_0 would terminate transmission. It is even more shocking when one considers that factor L could come from any of the possible controls and even combinations of controls.

Routine chemotherapy can be implemented as an increase in γ, the per capita mortality of schistosomes:[8]

$$\gamma' = \gamma - \ln(1 - g'h) \tag{0.8}$$

where g' is the coverage rate and h the efficacy, both as proportions. The increased mortality essentially results in a decrease to T_{SH}, thereby creating $R < R_0$. And our goal is to obtain:

$$R_0 = T_{SH} T_{HS} = \frac{\alpha N \beta H}{\gamma \mu} = 1.03 \Rightarrow R = R_0 \left(\frac{\gamma}{\gamma'}\right) = \frac{\alpha N \beta H L}{\gamma \mu}\left(\frac{\gamma}{\gamma'}\right) < 1$$

Indeed, if we assume a fairly non-aggressive control strategy that consists of a coverage rate of single treatment every 2 years and 70% efficacy when the natural lifespan of worms is 3 years, working through the calculation suggests this strategy would be more than sufficient to eliminate transmission (at least eventually, at equilibrium)! Similar arguments could be made for relatively easy elimination through snail control by modifying $N \Rightarrow N'$, improved sanitation by modifying $\beta \Rightarrow \beta'$ and reducing exposure by modifying $\alpha \Rightarrow \alpha'$ (see[2] for examples of how these additional controls may be implemented).

Problems Inherent in the Macdonald Model Assumptions

The above analyses suggest that with such low R_0 (i.e., slightly above 1), it should be easy to terminate transmission. If only it were so easy in the real world! Indeed, because the Macdonald model is unrealistic in its assumptions, the equilibria conditions are thus unrealistic.

Barbour was quite vocal about the limitations of the basic Macdonald model, attempting to improve the model (or "rescue it" as he states) by accounting for various heterogeneities that may scale R_0 upwards.[9,10] Specifically, Barbour accounts for snail latency to patency by the addition of a full snail population submodel based on *S. japonicum* data,[11] spatial heterogeneity in water contacts and age-dependent immunity.[9] Despite various attempts to rescue the model, the values of R_0 from variations on Macdonald's model may still be unrealistically low.[10]

We note that in Chapter 7, for our calibrated Macdonald-like model for *S. japonicum*, we arrived at an approximate R_0 (or R' representing the prefiltering calibration criterion) of approximately 4. Moreover, as this is a prefilter criterion, it is an upper limit on R_0! Again, using chemotherapy as an example, two treatments per year with 70% efficacy should eliminate transmission for this region of China—something that is already being conducted in many endemic areas of China. However, it is important to note that we still have not considered how long it would take to reach elimination, which we discuss later in terms of the dynamics of the system. We only note here that there may be benefits to combining chemotherapy with other more sustainable disease control strategies.

Still, there remain other heterogeneities that have not been widely considered. For one, more work is needed to consider age-acquired immunity, for which there is evidence for various *Schistosome* species, although the role of age-acquired immunity in *S. japonicum*, in particular, remains unclear.[12] In addition, the presence of animal reservoirs may need to be considered for some environments particularly for *S. japonicum*, as over 40 different wild and domestic mammals have been shown to play a role in transmission.[13] Stochastic effects and the transport of parasites between connected environments may also create heterogeneities that modify R_0, the latter discussed in greater detail below. Space and time-varying effects, such as attenuations that may occur through the synchrony of exposure may also modify R_0. In these respects, recent studies by various modelling groups may be considered (e.g., for *S. japonicum*, age-specific or population group-specific models in,[14,15] models of parasite persistence in connected populations,[16] time-varying factors (Chapter 6) and inclusion of animal hosts[17]).

Barbour has suggested an alternative to the Macdonald model (the so-called Ross model) based on human infection prevalence rather than worm burden.[10] The rather obvious benefit of such a model is that it is based on human infection prevalence, which can be measured (or at least approximated from the currently best-available diagnostic tests), as opposed to worm burdens which cannot be directly measured in humans and are difficult to relate to egg count data. Under Barbour's model:

$$R_0 = \frac{1}{(1-y^*)(1-P^*)} \tag{0.9}$$

where, as before, y^* is the equilibrium infected snail prevalence and now P^* is the equilibrium prevalence of infection in humans. Real-world values of 3% snail infection and 50% human infection, immediately result in values of R_0 twice as high as the Macdonald model without accounting for any heterogeneities. Indeed, there is some acceptance of this modelling approach by those working in China and the Philippines on *S. japonicum*.[15,17]

Modelling Control Dynamics

As we alluded to earlier, regardless of our choice of model (e.g., Macdonald or Ross), the equilibrium analyses of the models only tell us the eventual equilibrium values of the system, but not how fast the system will arrive at those equilibria. In the real-world, there is a practical need to understand the comparative performance of competing control programs, which often depends upon what can be done in short time frames (e.g., 5-10 year control programs) and how large of an impact might be expected. Such issues ultimately lead to questions about the dynamic properties of the system.

We note from our work, that when various heterogeneities and delays are incorporated into the system, analytical solutions are essentially not possible. Thus, numerical simulations are required to understand the equilibrium and dynamics of the system. Moreover, often quite complex models are numerically simulated and calibrated to field data, which can be computationally intensive. Thus, in Chapter 7, considerable attention was paid towards optimizing simulation times via tricks, such as prefiltering the parameter space based on R_0. Indeed, in our work, we have spent considerable time to model the dynamic behavior of various combinations of disease controls (e.g., various combinations and levels of chemotherapy, snail control and sanitation).[14] We note also that the dynamics of these controls may act differently. For instance, in contrast to the Anderson-May method of modelling a chemotherapy program as described above,[8] we have modeled chemotherapy as a resetting of the worm burden state (i.e., an instantaneous drop in worm burden when chemotherapy is provided), which can lead to disease elimination when considered in terms of the break point phenomenon of the model system. In contrast, snail control may be implemented either as a instantaneous drop in snail populations, reduction in snail habitat, or additional mortality, which depends on how the snail control was implemented. Thus, some disease control strategies may be used to yield immediate effects on reducing disease morbidity (good public health benefit), while others may be aimed at long-term, sustainable disease control (moving towards disease elimination). This can be clearly seen in the time-series figures of that show sudden drops in worm burden due to chemotherapy, but gradual reductions in transmission over time due to combined control of snails and/or sanitation.[14]

Given the above dual objectives for disease control, it is interesting to comment upon the various stages of schistosomiasis control in the world. Global schistosomiasis control policy currently focuses strongly on the distribution of praziquantel. As the drug has become increasingly affordable, it has taken a leading role in disease control efforts.[18] In contrast, 15 years ago, the World Health Organization placed health education, access to safe water supplies and sanitation at the center of schistosomiasis control efforts.[19] Clearly, for some countries, drug treatment to control morbidity needs to remain the focus, particularly in environments where being able to reliably deliver the drug remains a challenge. However in China, where much of our work is focused, there is already good access to drugs, yet the disease persists. There is recent evidence of disease re-emergence and a rise in prevalence following a 10-year chemotherapy campaign in some endemic areas.[20] The situation in China and in other middle-income countries dealing with parasitic diseases is transitioning away from asking the dynamic questions of how to achieve short term reductions in morbidity, but rather how to effectively reduce and/or eliminate transmission over the long term. Moreover, a related question concerns the threat of disease (re)emergence and how might models be best-used to evaluate this threat in terms of its probability, likely time-frame and place of occurrence. Current efforts in our group are focused on exploring these latter issues, particularly in the context of using models to control and provide early warning for emerging schistosomiasis.

New Model Developments: Incorporating Population Heterogeneity and Connectivity

Quite early on it was recognized that one of the limitations of the simple schistosomiasis model was the assumption of population homogeneity, which led subsequently to adjustments such as those for worm aggregation and differences in exposure (e.g., different risk groups in our work). Still it was assumed that transmission occurred such that host populations were isolated from other hosts and parasite populations were isolated from other parasite populations. The real world, however, is made up of a complex mixture of populations that are defined by natural and social and political boundaries. Moreover, while parasites may be transported via hydrological or social networks, it is political boundaries that often govern how control is conducted in the real world. In China, for instance, each county organizes their own local control strategy against *S. japonicum*. In any given year, a county may decide to perform surveillance or mass control in some villages, but not in others.

Increasing evidence suggests that transmission actually occurs in groupings of heterogeneous populations that are in fact connected to one another in varying degrees. Connections may occur via numerous processes, most notably via hydrological channels that transport the larval stages of the parasite between environments and via socio-economic connections, such as host movement between environments. Generally, this understanding has led to the following model as described by Guarie and Seto[16] (modifying their notation slightly to match our's):

$$\begin{aligned} \frac{dM}{dt} &= \alpha\pi_C \hat{\Theta}\cdot\left(\Omega^T \cdot T_C^{-1}\right)\cdot \hat{N}\cdot Y - \gamma M \\ \frac{dY}{dt} &= \beta\pi_M\left(T_M^{-1} \cdot \Omega \cdot \hat{\Theta}' \cdot \hat{H}\right)\cdot(1-Y)M - \mu Y \end{aligned} \tag{0.10}$$

in which different heterogeneous populations are represented in matrix notation as M, a vector of village worm burdens and Y, a vector of infected snail prevalences. The model accounts for differential social contacts through the matrices Θ, Ω, per capita cercarial and miracidial outputs $\pi_{C,}$ π_M and spatially explicit transport for cercariae and miracidia that incorporate hydrological connectivity, survival and mortality in the matrices T_C,T_M and different vectors of human H and snail N populations.

Various elaborations of this system have been explored by our group for *S. japonicum*.[16,21] The more recent theoretical analysis of the system by Gurarie and Seto describes both the general characteristics of the system and a few surprising aspects that have practical disease control implications. First, we define the terms:

$$\begin{aligned} R_{SH} &= \hat{N}^{-1}\cdot\left(\Omega^T \cdot T_C^{-1}\right)\cdot \hat{N} \\ R_{HS} &= \left(T_M^{-1} \cdot \Omega\right) \end{aligned} \tag{0.11}$$

which represent the contribution of social and hydrological connectivities for snail-to-human and human-to-snail transmission, respectively. The product of these factors and the "internal potential" ρ results in the "Basic Reproduction Matrix":[16,22]

$$R = R_{HS}\cdot\hat{\rho}\cdot R_{SH} \tag{0.12}$$

Similar to the $R_0 > 1$ inequality for the simple Macdonald system (0.1), this matrix analog, R conditions sustained transmission for the distributed-connected system (0.10). Specifically, transmission is sustained when the largest eigenvalue of R satisfies the condition:[22]

$$\lambda_1(R) > 1 \tag{0.13}$$

We note that (0.12) and (0.13) precisely define the role that "internal potential" and "connectivity" play in transmission. Both are important in sustaining transmission. The underlying parameters that make up internal potential ρ relate to the local potential for transmission within each village in the absence of connectivity (i.e., R_0). In fact, in the absence of connectivity, the matrices simplify in such a manner that each individual village's R_0 conditions transmission.

The practical implications of the system on control are large. We can show that due to connectivity, infection may be high in certain villages. Yet focusing treatment on these high-infection villages may be inefficient. Instead, careful exploration of the network of connectivity and choosing key nodes that contribute to downstream infection may be much more useful in achieving region-wide control. Indeed, it can be shown that the structure and characteristic of various social and hydrological connectivities can determine whether region-wide schistosomiasis transmission will occur optimally, or not at all (Fig. 2).

We have observed this in our work in 20 villages in Xichang County located in Southwest China: the village with the highest infection prevalence (73% of residents were infected in 2000) was located downstream from an endemic village that pertained to a different county. High levels of cercariae were detected in the irrigation channels through which water entered

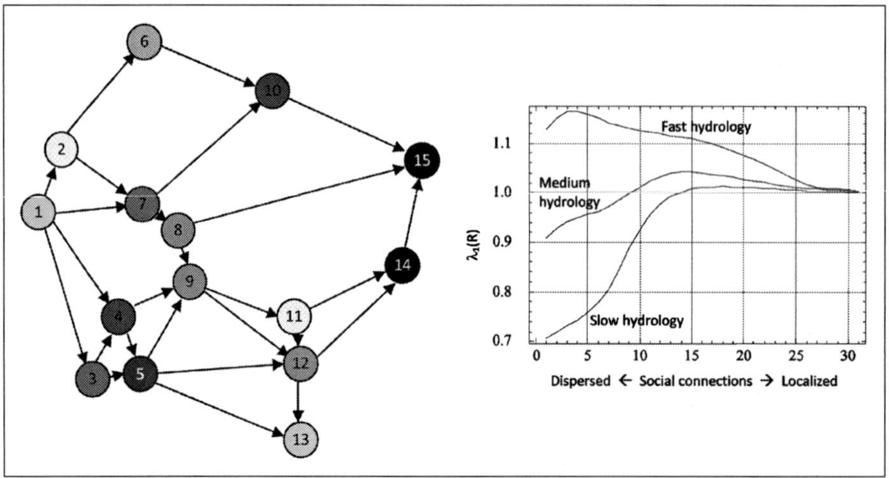

Figure 2. Network of connected villages with hydrological connections illustrated via arrows and each village's local R_0 shown in increasing shades of grey (left). Region-wide schistosomiasis transmission across this network can be shown to be a function of not only the local R_0, but also the nature of social and hydrological connectivities. For example, optimal levels of transmission (sustained transmission occurs when $\lambda_1(R) > 1$) depends upon hydrology and does not always coincide with the most dispersed social connectivity (right). See Gurarie and Seto (2008).

the village suggesting local infection may not only be dependent on village conditions but the influx of parasites from upstream.[7] Despite treatment of all infected residents in 2000, infection prevalence was 52% only two years later. We surmise that disease control efforts in this village that do not address the sources of parasite import will yield transient reductions in infection. The connected model confirms this hypothesis and suggests a more regional perspective to control activities may be useful, particularly in environments like those in China where rapid re-infection occurs and re-emergence is common. This understanding has led most recently to spatial analyses aimed at identifying connected watersheds that maybe used by local disease control agencies to better coordinate regional control strategies (Fig. 3).

While it has not yet been explored extensively for *S. japonicum*, we note that the dynamics of such a system maybe particularly interesting from the perspective of emerging disease and disease spread. Because population connectivities are explicitly modeled in the system, it would be possible to consider the impacts of increasing population connectivity that might occur via improved infrastructure, such as road and water resource development. In some cases, as we have shown in our work, increased social dispersion may actually reduce the potential for transmission by moving people away from hotspots in the network that sustain infection.[16] The practical analog to this is the increasing migration of labor away from rural environments to urban ones. Moreover, stochasticity may be incorporated into the model to consider the effects of chance perturbations to the network, including, for example, the occasional introduction of infected hosts into areas with low or no transmission. Such a model would allow for the identification of environments particularly favorable to the (re)initiation of transmission.

Conclusion

Given almost half a century of model development for schistosomiasis, surprisingly little use of models occurs in the planning and implementation of disease control programs. As described above, simple models and crude R_0-based analyses allow for easy calculations of the effect of chemotherapy and other control strategies, however, such calculations may lead to misleading results

Figure 3. Three-dimensional rendering of watersheds (bounded by white lines) in Xichang County, Sichuan Province, China, computed in a geographic information system and displayed in Google Earth. These visualizations were created to assist the county in better planning their regional control efforts by taking into account how villages may be hydrologically connected to one another.

because of the over-simplified assumptions of these models. Accounting for various heterogeneities in the transmission of the parasite has led to more complex models, which have suggested that the disease may be more difficult to control than suggested by simpler models. Indeed, there is a trend towards increasingly complex models because simple models are not able to mimic real-world field data. Another reason for increasing model complexity may also be less justified, simply because there exists computing power to consider such complex models via numerical simulation. Unfortunately, the trend towards more complex models and numerical simulation is pushing the field of schistosomiasis modelling outside of the realm of practical field use at the local or regional level. Indeed, after over a decade in our group of developing our *S. japonicum* model for China, there has not been the knowledge transfer necessary for our complex model to be run at a provincial-level Chinese research institute.

However, our modelling work has demonstrated some encouraging applications. The most important application has been the ability to use the models as an independent tool based on field data to assess whether current control policies seem reasonable. In this regard, we have been able to confirm regional and national-level policies promoting combinations of disease control measures as having the best chance at eliminating transmission. Much of this knowledge has come about from studying the more complex dynamics of the model, rather than the equilibria of the system, which is consistent with the practical question of how long it takes to reach appreciable levels of morbidity reduction from various combinations of controls. Even more encouraging are the recent developments in modelling heterogeneous, yet connected populations that are beginning to tackle some of the most challenging aspects of schistosomiasis control, such as how to reduce the potential for re-emergence. In our increasingly connected society, these connected models also have great potential to identify the impacts of changes in connectivity on transmission of parasitic diseases such as schistosomiasis.

Although many practical field decisions regarding schistosomiasis control continue to be based on experiential knowledge, there exists a new class of problems, for which there is little to no field experience. This is the case with disease emergence and re-emergence. In our group, the current work on developing connected models may be used to design surveillance systems and examine predictors of disease emergence. Such models may prove to be very useful in exploring the possible impacts of (re)emerging disease, making full use of the ability to run computer simulations for a variety of hypothetical scenarios when very little experiential knowledge is available from the field. Our group is also continuing our work on complex site-specific models to explore the effect of various mixtures of control strategies, which is very much needed to combat endemic transmission as well as to prevent disease re-emergence.

Acknowledgements

The authors are supported by the National Institutes of Health, National Institute for Allergy and Infectious Disease (NIH R01 AI68854). We thank the Sichuan Institute of Parasitic Disease for their collaboration on fieldwork that has led to the parameterization of our models.

References

1. Woolhouse ME. On the application of mathematical models of schistosome transmission dynamics. I. Natural transmission. Acta Trop 1991; 49(4):241-270.
2. Woolhouse ME. On the application of mathematical models of schistosome transmission dynamics. II. Control. Acta Trop 1992; 50(3):189-204.
3. Barbour AD. Schistosomiasis. In: Anderson RM, ed. Population dynamics of infectious diseases. London: Chapman and Hall, 1982:180-208.
4. Cohen JE. Mathematical models of Schistosomiasis. Ann Rev Ecol Syst 1977; 8:209-233.
5. Macdonald G. The dynamics of helminth infections, with special reference to schistosomes. Trans R Soc Trop Med Hyg 1965; 59(5):489-506.
6. Chitsulo L, Engels D, Montresor A et al. The global status of Schistosomiasis and its control. Acta Trop 2000; 77(1):41-51.
7. Spear RC, Seto E, Liang S et al. Factors influencing the transmission of Schistosoma japonicum in the mountains of Sichuan Province of China. Am J Trop Med Hyg 2004; 70(1):48-56.
8. Anderson RM, May RM. Helminth infections of humans: mathematical models, population dynamics and control. Adv Parasitol 1985; 24:1-101.
9. Barbour AD. Macdonald's model and the transmission of bilharzia. Trans R Soc Trop Med Hyg 1978; 72(1):6-15.
10. Barbour AD. Modeling the transmission of schistosomiasis: an introductory view. Am J Trop Med Hyg 1996; 55(5 Suppl):135-143.
11. Pesigan TP, Hairston NG, Jauregui JJ et al. Studies on Schistosoma japonicum infection in the Philippines. 2. The molluscan host. Bull World Health Organ 1958; 18(4):481-578.
12. Seto EYW, Lee YJ, Liang S et al. Individual and village-level study of water contact patterns and Schistosoma japonicum infection in mountainous rural China. Trop Med and Int Health 2007; 12:1199-1209.
13. Chen MG. Schistosoma japonicum and S. japonicum-like infections: epidemiology, clinical and pathological aspects. In: Jordan P, Webbe G, Sturrock RF, eds. Human Schistosomiasis. Wallingford: CAB International, 1993:237-270.
14. Liang S, Seto EYW, Remais JV et al. Environmental effects on parasitic disease transmission exemplified by schistosomiasis in western China. P Natl Acad Sci USA 2007; 104(17):7110-7115.
15. Riley S, Carabin H, Marshall C et al. Estimating and modeling the dynamics of the intensity of infection with Schistosoma japonicum in villagers of leyte, Philippines. Part II: Intensity-specific transmission of S. japonicum. The schistosomiasis transmission and ecology project. Am J Trop Med Hyg 2005; 72(6):754-761.
16. Gurarie D, Seto EY. Connectivity sustains disease transmission in environments with low potential for endemicity: modelling schistosomiasis with hydrologic and social connectivities. J R Soc Interface 2008; Epub ahead of print.
17. Williams GM, Sleigh AC, Li Y et al. Mathematical modelling of Schistosomiasis japonica: comparison of control strategies in the People's Republic of China. Acta Trop 2002; 82(2):253-262.
18. WHO. Preventive chemotherapy in human helminthiasis: coordinated use of anthelminthic drugs in control interventions: a manual for health professionals and programme managers. Geneva: WHO, 2006.
19. WHO. The control of schistosomiasis: Second report of the WHO expert committee. Geneva: WHO, 1993.

20. Zhou XN, Guo JG, Wu XH et al. Epidemiology of Schistosomiasis in the People's Republic of China, 2004. Emerg Infect Dis 2007; 12(10):1470-1476.
21. Xu B, Gong P, Seto E et al. A spatial-temporal model for assessing the effects of intervillage connectivity in schistosomiasis transmission. Annals of the Association of American Geographers 2006; 96(1):31-46.
22. Gurarie D, King CH. Heterogeneous model of schistosomiasis transmission and long-term control: the combined influence of spatial variation and age-dependent factors on optimal allocation of drug therapy. Parasitology 2005; 130:49-65.

Chapter 13

Modelling Climate Change and Malaria Transmission

Paul E. Parham* and Edwin Michael

Abstract

The impact of climate change on human health has received increasing attention in recent years, with potential impacts due to vector-borne diseases only now beginning to be understood. As the most severe vector-borne disease, with one million deaths globally in 2006, malaria is thought most likely to be affected by changes in climate variables due to the sensitivity of its transmission dynamics to environmental conditions. While considerable research has been carried out using statistical models to better assess the relationship between changes in environmental variables and malaria incidence, less progress has been made on developing process-based climate-driven mathematical models with greater explanatory power. Here, we develop a simple model of malaria transmission linked to climate which permits useful insights into the sensitivity of disease transmission to changes in rainfall and temperature variables. Both the impact of changes in the mean values of these key external variables and importantly temporal variation in these values are explored. We show that the development and analysis of such dynamic climate-driven transmission models will be crucial to understanding the rate at which *P. falciparum* and *P. vivax* may either infect, expand into or go extinct in populations as local environmental conditions change. Malaria becomes endemic in a population when the basic reproduction number R_0 is greater than unity and we identify an optimum climate-driven transmission window for the disease, thus providing a useful indicator for determing how transmission risk may change as climate changes. Overall, our results indicate that considerable work is required to better understand ways in which global malaria incidence and distribution may alter with climate change. In particular, we show that the roles of seasonality, stochasticity and variability in environmental variables, as well as ultimately anthropogenic effects, require further study. The work presented here offers a theoretical framework upon which this future research may be developed.

Introduction

The potential effects of global climate change and ozone depletion on human health have received increasing attention in recent years, with those due to changes in vector-borne disease (VBD) incidence and distribution thought to represent one of a range of major direct and indirect effects.[1-6] Malaria has arguably attracted the most attention of all VBDs,[4,7-9] due to both the sensitivity of its transmission dynamics to changes in environmental variables and its status as one of the biggest causes of worldwide mortality due to infectious disease, with an estimated 247 million cases in 2006.[10,11]

*Corresponding Author: Paul Parham—Grantham Institute for Climate Change, Department of Infectious Disease Epidemiology, Imperial College London, St. Mary's Campus, Praed Street, London W2 1PG, UK. Email: paul.parham@imperial.ac.uk

Modelling Parasite Transmission and Control, edited by Edwin Michael and Robert C. Spear.

Climate change may affect malaria transmission via changes in the ecology and behaviour of humans, *Anopheles* mosquitoes and *Plasmodium* parasites.[11] Perhaps the most important effect will be on the vector population itself. Immature stages of the lifecycle are aquatic and increasing rainfall, leading to increased availability of breeding sites, is a strong driver of abundance. The use of precipitation as an empirical predictor of incidence,[12-16] besides its direct impact on vector abundance,[17-22] is well-established. The role of temperature has also been considered in terms of its effects on vector abundance (as a physiological factor affecting the development rate of larvae, adult daily survival probability and biting rates due to increased processing rate of blood meals), local dispersal, parasite dynamics (such as reproduction rates as a function of temperature) and human behaviour.[4,23-25]

Given the links between malaria incidence and variation in environmental variables, mathematical models incorporating climate variables have been developed in recent years.[7,20,26-28] The growing importance of these models derives from the fact that while statistical models have been useful in elucidating relationships between environmental variables and transmission intensity,[9] process-based mathematical models permit a more explanatory understanding of the role of internal (due to biological processes) versus external (such as those due to changes in environmental variables) drivers of transmission. Dynamical models are fundamentally important, since the biological processes driving malaria transmission are embedded within a dynamically changing environment on a range of timescales. Moreover, to meaningfully capture the emergence of new outbreaks (e.g., due to changing climate making areas previously unsuitable for transmission more favourable, human movement or mosquito dispersal to previously disease-free regions), a dynamic model is required to capture invasion dynamics.

The modelling work to date incorporating environmental variables into dynamic transmission models has almost entirely focussed on the impact of changes in temperature (although see ref. 20). Most of the attention of previous modelling work has also been on capturing equilibrium dynamics based on derivation and analysis of static quantities such as R_0, the basic reproduction number of the disease. In this chapter, we firstly consider how rainfall may be incorporated within a dynamic transmission model and secondly highlight how important disease transmission issues not examined to date can only be addressed within such a dynamic modelling framework. More specifically, we examine the impact of climate change on mosquito (and hence malaria) extinction due to changes in environmental conditions, malaria invasion dynamics in previously disease-free regions and the effects of temporal variability in climate variables on mosquito population dynamics, invasion behaviour and endemic prevalence.

Mathematical Model Development

In terms of developing a dynamic framework for understanding the impact of climate change on malaria transmission, a deterministic or stochastic transmission model may be developed, embedded within an environment assumed static or fluctuating over the timescale of interest. Arguably, the most realistic framework is a stochastic transmission model within dynamic environmental conditions, in which the inclusion of environmental forcing would be key to obtaining a better understanding of the effects of climate change on the spread and control of malaria. Despite this, little research to date has examined the effects of temporal forcing by climate (i.e., variability in environmental variables occurring at intra-annual (seasonal), inter-annual, decadal and longer timescales) on malaria transmission, despite receiving attention for other infectious diseases.[29,30] Understanding the dynamical impact of trends and variability in climate over the next 40 years, for instance, may affect a goal of malaria eradication by 2050. Thus, although a more thorough theoretical study of the effects of temporal forcing is beyond the scope of this chapter, a key objective of the work presented here is to begin considering how seasonality effects may be incorporated into models and may impact vector population dynamics, invasion behaviour and changes in the R_0 of malaria.

We assume a deterministic model, which despite making simple biological assumptions, is expected to be sufficiently realistic to permit valuable insights into climate-driven malaria transmission dynamics. Let S_M, E_M and I_M represent the number of susceptible, exposed (but

Table 1. Malaria model parameters

Parameter	Definition (and Units)
$b(R,T)$	Adult mosquito birth rate (per day)
$\mu(T)$	Adult mosquito per capita death rate (per day)
$a(T)$	Mosquito biting rate (per day)
b_1	Proportion of bites by susceptible mosquitoes on infected humans that produce infection
$\tau_M(T)$	Duration of the sporogonic cycle (days)
$l_M(T)$	Survival probability of infected mosquitoes over the incubation period of the parasite
b_2	Proportion of bites by infectious mosquitoes on susceptible humans that produce infection
τ_H	Latent period of infection within humans (days)
$1/\gamma$	Human average duration of infectiousness (days)
$M(t)$	Total number of mosquitoes ($S_M(t) + E_M(t) + I_M(t)$)
N	Total number of humans ($S_H(t) + I_H(t) + R_H(t)$)

not infectious) and infectious mosquitoes respectively, along with S_H and I_H representing the analogous categories in humans (where we assume a fixed latent period of duration τ_H in humans). All model parameters, functional forms and baseline values assumed in this analysis are summarised in Tables 1-3. The model is described by the ordinary differential equations

$$\begin{aligned}
\frac{dS_M}{dt} &= b - ab_1\left(\frac{I_H}{N}\right)S_M - \mu S_M, \\
\frac{dE_M}{dt} &= ab_1\left(\frac{I_H}{N}\right)S_M - \mu E_M - ab_1\left(\frac{I_H(t-\tau_M)}{N}\right)S_M(t-\tau_M)l(\tau_M), \\
\frac{dI_M}{dt} &= ab_1\left(\frac{I_H(t-\tau_M)}{N}\right)S_M(t-\tau_M)l(\tau_M) - \mu I_M, \\
\frac{dS_H}{dt} &= -ab_2\left(\frac{I_M}{N}\right)S_H, \\
\frac{dI_H}{dt} &= ab_2\left(\frac{I_M(t-\tau_H)}{N}\right)S_H(t-\tau_H) - \gamma I_H,
\end{aligned} \tag{1}$$

where the total mosquito population $M(t) = S_M(t) + E_M(t) + I_M(t)$ and human population $N(t) = S_H(t) + I_H(t) + R_H(t)$.

Functional Forms for Incorporating Temperature and Rainfall Effects and the Derivation of R_0

We assume that the mosquito birth rate b depends on rainfall (through the dependence on breeding site availability) and temperature. Let us write the birth rate as

$$b(R,T) = \frac{B_E p_E(R) p_L(R,T) p_P(R)}{\tau_E + \tau_L(T) + \tau_P}, \tag{2}$$

Table 2. Functional forms for parameters appearing in Equation (1)

Parameter	Functional Form	Units	Reference
$a(T)$	$\frac{T-T_1}{D_1}$	Per day	34
$p(T)$	$e^{\frac{-1}{AT^2+BT+C}}$	Dimensionless	4
$\tau_M(T)$	$\frac{DD}{T-T_{min}}$	Days	35

where B_E is the number of eggs laid per adult per oviposition, $p_E(R)$, $p_L(R,T)$ and $p_p(R)$ are the daily survival probabilities of eggs, larvae and pupae, τ_E, τ_L (T) and τ_p are the duration of each development stage, $1/\gamma$ is the average duration of infectiousness in humans and we highlight the parameter dependence on daily temperature T (in °C) and rainfall R (in mm). The average larval duration depends on temperature as $\tau_L(T) = 1/(c_1T + c_2)$,[28,31] corresponding to larval daily survival probability $p_L(T) = e^{-(c_1T+c_2)}$, while the development rate of eggs and pupae is relatively independent of temperature.[32] While rainfall has been shown to positively correlate with malaria incidence, it has also been suggested that excessive rainfall may flush out larvae and breeding sites[31,33] and we assume a quadratic relationship between the daily survival probabilities of eggs, larvae and pupae and rainfall. For larvae, we assume

$$p_L(R) = \left(\frac{4p_{ML}}{R_L^2}\right)R(R_L - R) \tag{3}$$

where p_{ML} is the maximum daily survival probability (i.e., when there is optimum rainfall for mosquito breeding) and R_L is the rainfall limit beyond which breeding sites get flushed out and no immature stages survive. We assume an analogous expression for eggs and pupae. We also assume that temperature and rainfall act independently on the survival probability of larvae such that $p_L(R,T) = p_L(R)\, p_L(T)$, although this is likely to be an approximation in reality. Increased evaporation of breeding sites or the melting of snow packs as temperatures increase provide two examples of this. Substituting these expressions into (2) gives the birth rate as a function of temperature and rainfall and this functional form is plotted in Figure 1.

As well as $b(R,T)$, the biting rate $a(T)$, mosquito mortality hazard $\mu(T)$ and probability $l(\tau_M)(T)$ of a mosquito surviving the duration of the sporogonic cycle also depend on temperature.[4] These are summarised in Table 2 and plotted in Figure 2, where we additionally note that $\mu(T) = -ln\, p(T)$ and $l(\tau_M)(T) = p(T)^{\tau_M(T)}$. Note that the expression in ref. 4 for $p(T)$ is true only at favourable humidities for mosquito development. Survival drops off rapidly at relative humidities below 50-60%,[36] highlighting that $p(T)$ is also indirectly dependent on rainfall through its impact on relative humidity and illustrates again the interdependence of temperature and rainfall. We assume that the proportions of bites by susceptible mosquitoes on infectious humans (b_1) and infectious mosquitoes on susceptible humans (b_2) that produce infection are independent of environmental conditions.

Arguably, one of the most important concepts in infectious disease epidemiology is that of the basic reproduction number R_0, defined as the average number of secondary cases generated per infectious individual over their duration of infectiousness in an entirely susceptible population.[37] This may be similarly defined for VBDs as the product of the number of vectors infected per person and the number of people infected per vector (over their respective infectious periods). While R_0 may be calculated in a variety of ways for infectious disease models, calculation for models assuming homogeneous mixing is relatively straight forward. The transmission model (1) has two equilibrium

Table 3. Baseline parameter values assumed for model simulations

Parameter	Assumed Value	Units
B_E	200	Dimensionless
p_{ME}	0.9	Dimensionless
p_{ML}	0.25	Dimensionless
p_{MP}	0.75	Dimensionless
R_L	50	mm
τ_E	1	days
c_1	0.00554	(°C days)$^{-1}$
c_2	−0.06737	(days)$^{-1}$
τ_P	1	days
T_1	19.9	°C
D_1	36.5	°C days
b_1	0.04	Dimensionless
A	−0.03	(°C^2 days)$^{-1}$
B	1.31	(°C days)$^{-1}$
C	−4.4	days^{-1}
b_2	0.09	Dimensionless
τ_H	10	days
DD	111 (*P. falciparum*)	°C days
	105 (*P. vivax*)	
$T_{\min}$	16 (*P. falciparum*)	°C
	14.5 (*P. vivax*)	
γ	1/120	days^{-1}

states, namely the disease-free state and the endemic equilibrium. In the latter case, it is readily shown that the prevalence in mosquitoes is given by

$$I_M^* = \frac{M(R_0 - 1)}{\left(\frac{R_0}{l(\tau_M)}\right) + \left(\frac{ab_2 M}{\gamma N}\right)}, \tag{4}$$

while for the human population

$$I_H^* = \frac{N(R_0 - 1)}{R_0 + \left(\frac{ab_1}{\mu}\right)} \tag{5}$$

where (ab_1/μ) is Macdonald's index of stability.[35] Stability analysis about the endemic state demonstrates that malaria will persist when $R_0 > 1$ where

$$R_0 = \frac{Ma^2 b_1 b_2 l(\tau_M)}{\gamma \mu N}, \tag{6}$$

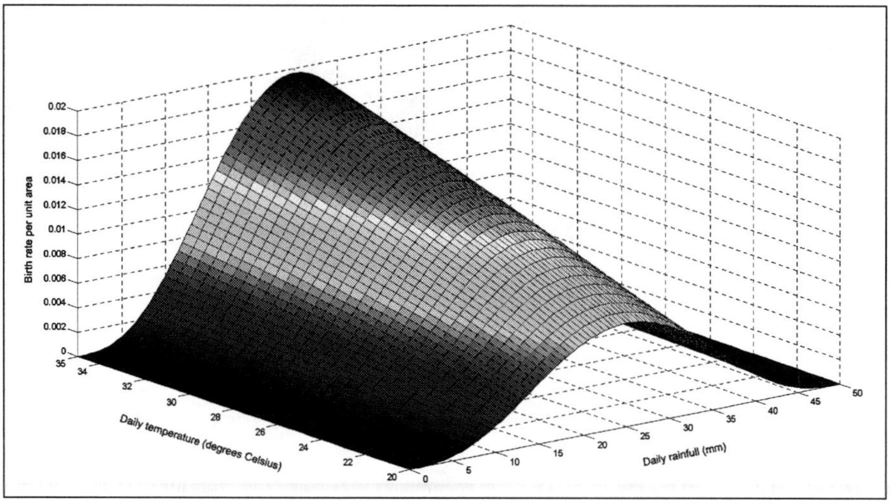

Figure 1. Mosquito birth rate b (R,T) as a function of daily rainfall R (in mm) and temperature T (in °C).

although note that this is conditional on assuming static environmental conditions. The expression for R_0 when we account for fluctuations in temperature and rainfall is considerably more complex[38-40] and we postpone a more theoretical discussion of the implications of environmental variability on R_0 (and transmission dynamics more generally) to future work.

Vector Population Dynamics

The dependence on mosquito abundance in the nonlinear vector-human transmission term of (1) results in considerable sensitivity of the model to vector population dynamics. Thus, understanding the dependence of $M(t)$ on temperature and rainfall (and seasonality therein) is key.

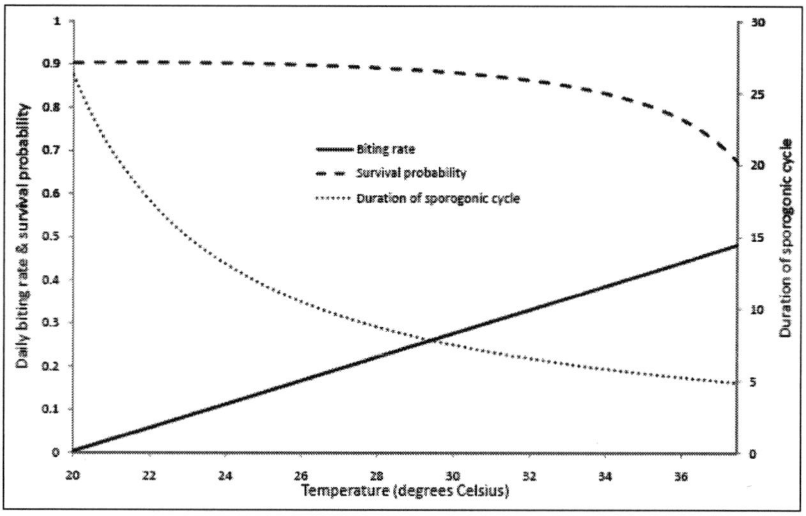

Figure 2. *Anopheles* biting rate and survival probability and *Plasmodium* incubation period as a function of temperature.

Adding equations in (1) related to the mosquito population gives the deterministic model for the vector population

$$\frac{\mathrm{d}M(t)}{\mathrm{d}t} = b(R,T) - \mu(T)M. \tag{7}$$

In reality, mosquito population dynamics behave stochastically and a full understanding of the impact of the mosquito population on transmission dynamics requires the stochastic equivalent of (7). If $p_M(t)$ denotes the probability of having $M(t)$ mosquitoes at time t, the master equation is

$$\frac{\mathrm{d}p_M(t)}{\mathrm{d}t} = b(t)p_{M-1}(t) + \mu(t)(M+1)p_{M+1}(t) - (b(t) + \mu(t)M)p_M(t) \tag{8}$$

for $M(t) = 0,1,2,...$ and where $p_{-1}(t) = 0$. If we define the probability generating function to be

$$G(z,t) = \sum_{i=0}^{\infty} p_i z^i ,$$

it is readily shown by successively multiplying (8) by z_i for each i and adding, we can convert (8) into a partial differential equation for $G(z,t)$ as

$$\frac{\partial G(z,t)}{\partial t} = b(t)(z-1)G(z,t) - \mu(t)(z-1)\frac{\partial G(z,t)}{\partial z}, \tag{9}$$

which can be solved using a variety of methods. We are interested in the dependence of the mosquito population on temperature and rainfall and their seasonal variability, a full understanding of which may be obtained by solving (9). The full solution of (9), however, for the case of general $b(t)$ and $\mu(t)$ is complex, so we simplify the analysis by considering only the dependence of the dynamics on the mean values of these environmental factors, not their variability.

We consider the mosquito population in Tanzania where malaria is highly-endemic, but the conclusions drawn about the effects of seasonality in environmental factors are general. Using temperature and rainfall data from WorldClim[41] (http://www.worldclim.org) and averaging across all regions in Tanzania, we fit the temporal profiles

$$T(t) = T_1(1 + T_2\cos(\omega_1 t - \Phi_1)) \tag{10}$$

and

$$R(t) = R_1(1 + R_2\cos(\omega_2 t - \Phi_2)), \tag{11}$$

giving the values in Table 4 (with the profiles plotted in Figure 3A. Substituting (10) and (11) into the earlier expressions for $b(R,T)$ and $\mu(T)$ and solving (7) gives the mean population dynamics plotted in Figure 3B, with the amplitude and frequency of vector oscillations strongly dependent on the seasonality factors T_2 and R_2. To better understand the sensitivity to variability in temperature and rainfall, Figure 3C plots the mean and standard deviation in mosquito numbers as a function of T_2 and R_2. Increasing seasonality in rainfall about a fixed mean (below the level at which flushing out occurs), broadly corresponding to more breeding sites, is found to always increase the mean size of the mosquito population, while increasing seasonality in temperature always reduces the vector population. Large amplitude variability in temperature is also found to drive mosquito populations to extinction, thus suggesting that extinction dynamics are more sensitive to changes in temperature than rainfall. This highlights that while understanding predicted changes in environmental variables from climate models is important, it is just as important to understand the impact of global warming on changes in the variability of climate variables.

Solution of (9) for general $b(t)$ and $\mu(t)$ would permit insight into the effects of climate variability on the stochastic population dynamics, but the complex and theoretical nature of the problem is beyond the scope of this chapter. Instead, we consider the extinction dynamics as a function of changes in mean temperature, fixing rainfall to simplify matters, but the analysis is

Table 4. Fitted temperature and rainfall seasonality parameters for Tanzania

Parameter	Definition	Fitted Value	Units
T_1	Mean temperature in the absence of seasonality	23.2	°C
T_2	Amplitude of seasonal variability in temperature	0.07	Dimensionless
ω_1	(Angular) frequency of seasonal oscillations in temperature	0.67	months^{-1}
Φ_1	Phase lag of temperature variability	1.53	Dimensionless
R_1	Mean monthly rainfall in the absence of seasonality	85.9	mm
R_2	Amplitude of seasonal variability in rainfall	0.98	Dimensionless
ω_2	(Angular) frequency of seasonal oscillations in rainfall	0.65	months^{-1}
Φ_2	Phase lag of rainfall variability	1.99	Dimensionless

readily extended to include changes in rainfall at fixed temperature. When $b(t) = b$ and $d(t) = d$, (9) can be solved using Laplace Transforms or characteristics.[42] If we assume M_0 initial mosquitoes, solution of (9) gives

$$G(z,t) = e^{\frac{b}{\mu}(z-1)(1-e^{-\mu t})}(1+(z-1)e^{-\mu t})^{M_0}, \tag{12}$$

whereupon substituting $z = 0$ gives, as a function of temperature and rainfall, the probability that the mosquito population fades out at or before time t as

$$p_0(t) = e^{-\frac{b(R,T)}{\mu(T)}(1-e^{-\mu(T)t})}(1-e^{-\mu(T)t})^{M_0}. \tag{13}$$

At extreme temperatures, mosquitoes are more likely to fade-out (e.g., due to thermal death at temperatures beyond around 40 °C),[31] although the risk decreases with increasing rainfall when $R < R_L$. Beyond these early stages of mosquito invasion, the probability that the population ultimately fades out is $e^{-\frac{b(R,T)}{\mu(T)}}$. Thus, reliable estimates of parameters included within the birth and death models (e.g., the rainfall threshold beyond which breeding sites are washed out) will enable robust conclusions not only about when and where mosquitoes may be driven to extinction, but also the implications for malaria transmission. While a full analysis of malaria fade-out requires a stochastic transmission model, the proportionality of vector abundance and R_0 means that the above analysis yields useful insight into the impact of changes in climate variables on malaria elimination.

Invasion Dynamics

As well as providing a useful framework for understanding how the stochastic dynamics of mosquitoes may lead to the establishment of malaria as environmental conditions change, dynamical models are vital for capturing the invasion dynamics in previously susceptible populations. In addition to the direct effects of climate change on transmission, indirect effects such as those on malnutrition, poverty or the more frequent occurrence of extreme weather events (e.g., floods or heat waves) may also contribute to changing distributions of incidence, as human and animal ecosystems and habitats become more susceptible. However, while malaria is, in general, most prevalent in the tropics and subtropics, predictions that increases in global temperatures may lead to emergence in currently temperate regions and at higher altitudes[27] are not universally accepted.[43] Nonetheless, it is well-accepted that changes in climate variables influential in malaria transmission are likely to cause the global distribution of incidence to change over the next 100 years. Thus, understanding the invasion behaviour into new regions represents a key challenge and one which has considerable public health implications.

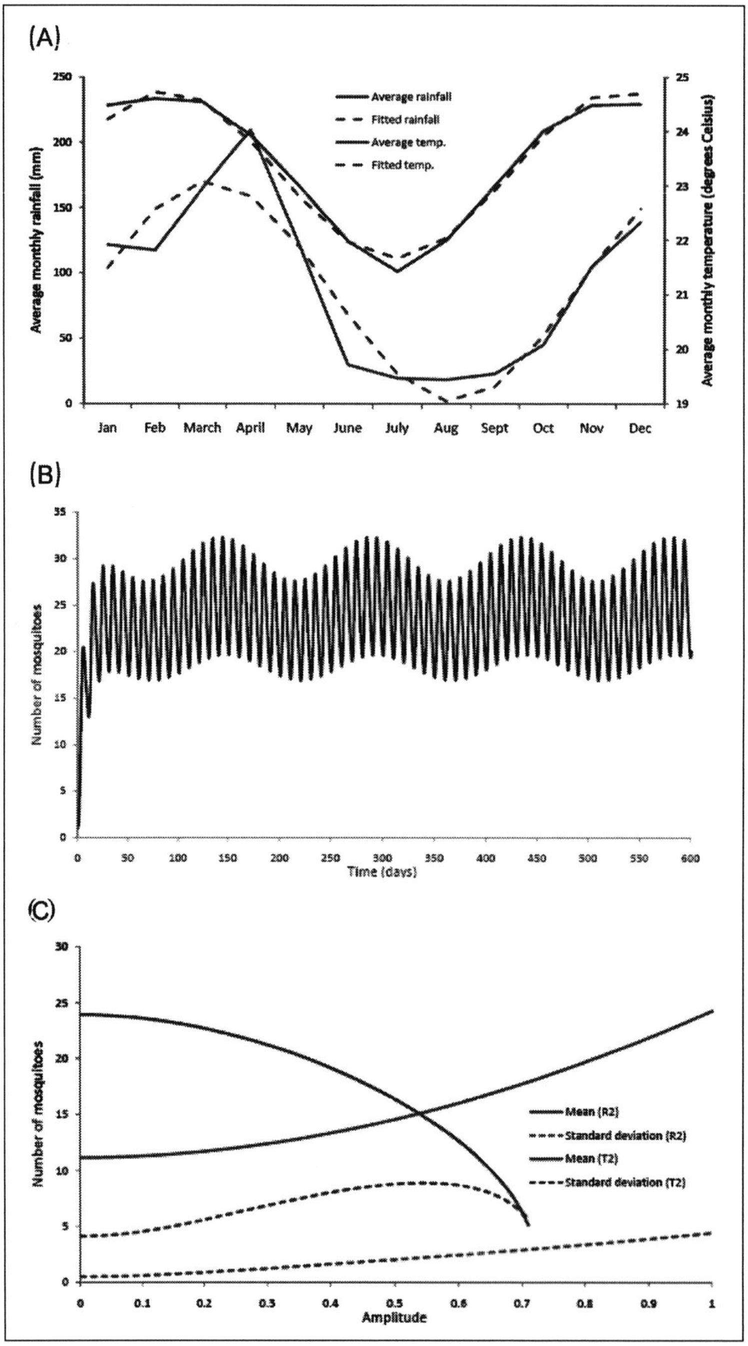

Figure 3. A) Annual temperature and rainfall profiles (10) and (11) fitted to WorldClim data[41] (http://www.worldclim.org) averaged across Tanzania, (B) predicted mosquito population dynamics for Tanzania from the solution to (7) (with parameters from Tables 2 and 3) and (C) the solution to (7) as a function of increasing seasonality in temperature (T_2) and rainfall (R_2).

Again, a more detailed stochastic transmission model, incorporating seasonally-varying environmental variables, is required for a more thorough understanding of invasion dynamics, but considerable progress about the expected behaviour can be obtained from (1). When most of the population are initially susceptible, $S_H(t-\tau_H)\approx N$ and $S_M(t-\tau_M)\approx M$, whereupon substituting into (1), the invasion dynamics are described by

$$\frac{dI_M}{dt}\approx ab_1\left(\frac{I_H(t-\tau_M)M}{N}\right)l(\tau_M)-\mu I_M, \tag{14}$$

$$\frac{dI_H}{dt}\approx ab_2 I_M(t-\tau_H)-\gamma I_H \tag{15}$$

and we look to solve for the growth rate of $I_H(t)$. We approach this problem in a way that allows illustration of the method and thus how to apply the technique to more complex transmission models.

In essence, we look to rewrite the system as a single equation in $I_H(t)$ and its derivatives, which we note currently depend on factors proportional to 1, $I_M(t)$, $I_M(t-\tau_H)$ and $dI_H(t)/dt$. Here, we have a dependence on four factors, but only two independent equations. However, a third independent equation may be obtained by differentiating (14), since this will depend on some combination of the same four factors. Considering the resultant equation, together with (14), at time $t+\tau_H$ gives the three independent equations

$$\begin{aligned}
\frac{dI_H(t+\tau_H)}{dt}&=ab_2 I_M(t)-\gamma I_H(t+\tau_H),\\
\frac{dI_M(t)}{dt}&=\left(\frac{ab_1M}{N}\right)I_H(t-\tau_M)l_{\tau_M}-\mu I_M,\\
\frac{d^2I_H(t+\tau_H)}{dt^2}&=\left(\frac{a^2b_1b_2M}{N}\right)I_H(t-\tau_M)l_{\tau_M}-ab_2\mu I_M(t)-\gamma \dot{I}_H(t+\tau_H),
\end{aligned}\tag{16}$$

so that $I_H(t)$ and its derivatives now depend on three factors and we have three independent equations. If $\mathbf{J}$ is the 3×3 matrix with elements containing factors proportional to $I_H(t)$ and its derivatives, we can rewrite (16) as the matrix equation

$$\mathbf{J}\begin{pmatrix}1\\ I_M(t)\\ \dfrac{dI_M(t)}{dt}\end{pmatrix}=0, \tag{17}$$

whereupon solving $\det \mathbf{J}=0$ leads to

$$\ddot{I}_H(t+\tau_H)+(\mu+\gamma)\dot{I}_H(t+\tau_H)+\mu\gamma(I(t+\tau_H)-R_0I(t-\tau_M))=0 \tag{18}$$

for the number of infectious humans. Substituting the trial solution $I_H(t)=e^{rt}$ gives

$$r^2e^{r\tau_H}+(\mu+\gamma)re^{r\tau_H}+\mu\gamma(e^{r\tau_H}-R_0e^{-r\tau_M})=0, \tag{19}$$

where r is the real-time growth rate and the solution reduces to the standard growth rate equation for an SEIR model (with analytical solutions) when $\tau_H=\tau_M=0$. Substituting parameters and functional forms from earlier, Tables 2 and 3 allow numerical solution of (19) and Figure 4A plots the outbreak doubling time $T_D=\ln(2)/r$ as a function of temperature at different mosquito densities. The rate of spread into a susceptible population is found to be extremely sensitive to temperature, with a clear window around 32-33°C where the doubling time is shortest. At lower temperatures, fewer mosquitoes survive long enough for the sporogonic cycle to complete, while at higher temperatures, mosquito survival probability drops off rapidly. Invasion dynamics are also found to be strongly dependent on vector abundance, itself driven by rainfall, with doubling times around 4-6 weeks at vector densities typical of rainy seasons and increasing to the order of months at lower abundances typical of dry seasons.

A more theoretical analysis of the effects of seasonality in the vector population on the growth rate has been considered elsewhere[40] and the applicability of such methods to more complex

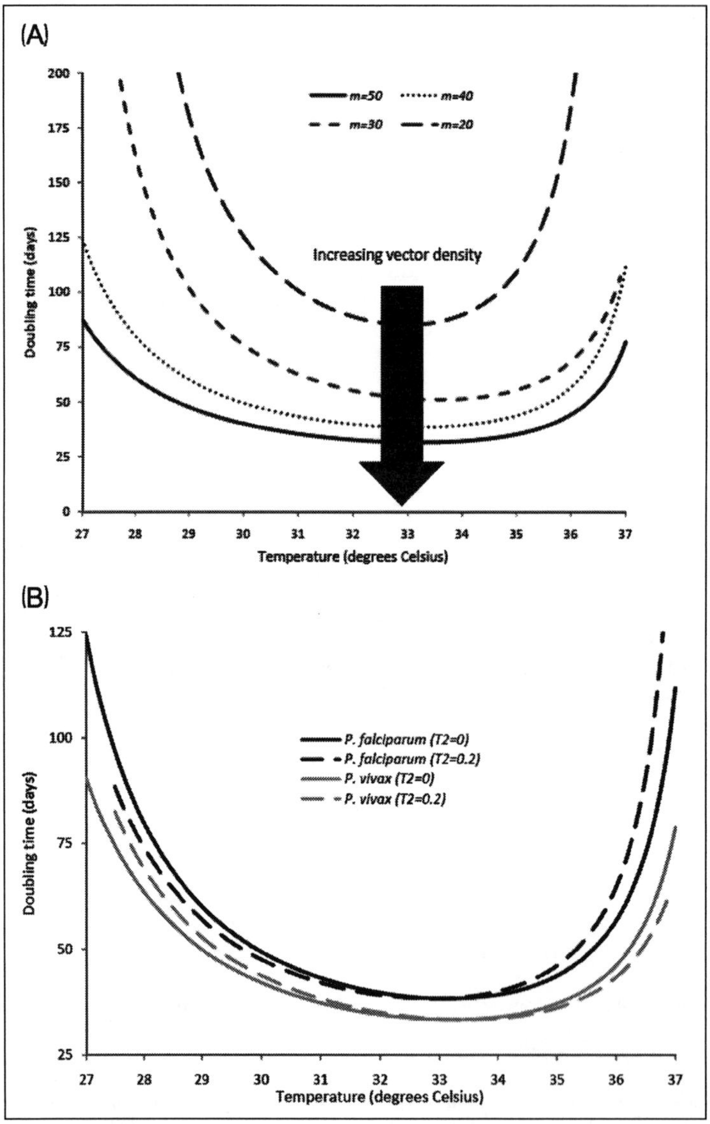

Figure 4. A) Doubling times of a *P. falciparum* outbreak as a function of temperature and mosquito density at fixed rainfall (and in the absence of seasonality). B) Predicted doubling times for *P. falciparum* and *P. vivax* with and without seasonality in temperature (where m = 40 and all other parameters as per (A)).

systems is beyond the scope of this chapter. However, approximate preliminary insight may be obtained by substituting (10), with parameter values given in Table 4, into (19) for parameters that depend on temperature and the results are shown in Figure 4B for *P. falciparum* and *P. vivax*. In the absence of seasonality, there is little difference between *P. falciparum* and *P. vivax*, with the latter spreading marginally quicker and potentially resulting in a greater outbreak threat at higher altitudes given the lower critical temperature for transmission (around 14.5°C and 16°C for *P. vivax* and *P. falciparum* respectively). Figures 4A,B combined also highlight that malaria invasion is considerably more sensitive to vector density than the strain of *Plasmodium* parasite. The crude inclusion of moderate seasonality in temperature, around three times greater than seasonality observed across Tanzania, is found to have little effect on the doubling times, although a more thorough analysis is required to fully understand the effects of variability in environmental variables on malarial invasion.

Implications for R_0 and Mapping Risk

It is readily shown from (1) that successful establishment and invasion of malaria into a previously unaffected population will lead to endemicity when $R_0 > 1$. Given the dependence of the vector population dynamics on temperature and rainfall, together with the functional forms contributing to R_0 in Table 2, an analytical expression for R_0 as a function of mean temperature and rainfall may be derived by substituting into (6). For the purpose of this analysis, we assume that the vector population is in equilibrium, so that $M(R,T) = b(R,T)/\mu(T)$ and there is no seasonality in rainfall nor temperature (so $R_2 = T_2 = 0$). For brevity, we do not write down the full expression, but R_0 as a function of these two variables is plotted in Figure 5.

Parameter uncertainty in the vector population model means that while precise quantitative conclusions should not be drawn from Figure 5, the qualitative dependence of R_0, capturing both the population risk of malaria becoming endemic and individual infection risk, on temperature and rainfall remains robust to these uncertainties. As with the invasion dynamics, we observe a clear window for optimum malaria transmission around 32-33°C, corresponding to where the balance between mosquitoes surviving long enough for completion of the sporogonic cycle and the rapid decline in mosquito survival at high temperatures is optimised. The approximate quadratic dependence of R_0 on temperature also offers insight into the question of how changes in local transmission risk may shift due to increases in temperature. Consider an endemic region currently at mean temperature T experiencing an increase in temperature by an amount ΔT. If

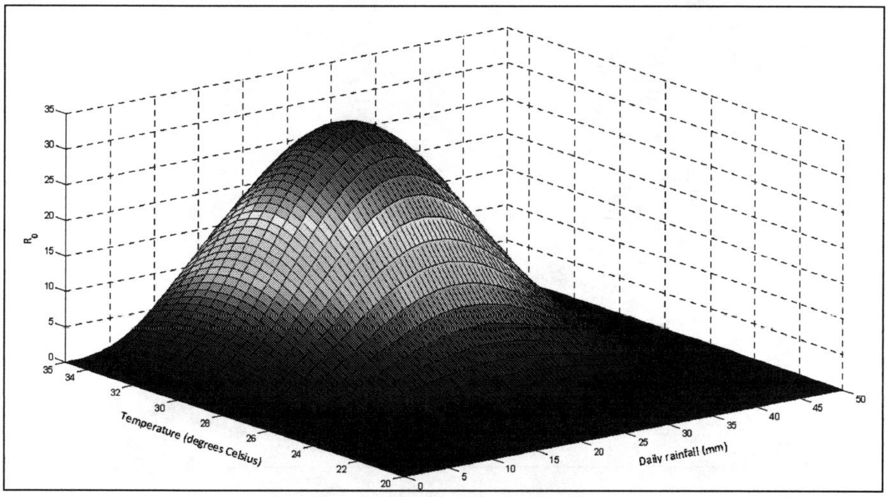

Figure 5. The dependence of R_0 on temperature and rainfall.

T_{max} represents the temperature at which the R_0 (T) profile peaks, it is clear that if $T + \Delta T < T_{max}$, malaria prevalence will get significantly worse in that region, while if $T > T_{max}$, global warming will lead to a decline in mosquito survival and a reduction in transmission.

For the case of Tanzania, Figure 6 plots the predicted mean temperature distribution for April (when rainfall peaks) under the A2a emission scenario[6] (data from WorldClim and where the calculation of mean temperature assumes uniformly distributed temperatures between monthly maxima and minima). Analysis of the expression quantifying the relationship between R_0 and temperature, and substitution of parameter values in Table 3, shows that T_{max} is around 32.9°C for *P. falciparum* and *P. vivax*. Temperature data presented in Figure 6 shows that the maximum mean temperature is predicted to be around 32.6°C for Tanzania and thus, given the monotonic relationship between R_0 and prevalence from (5) and the fact that we are in the regime $T + \Delta T < T_{max}$, we predict malaria prevalence to increase across Tanzania around the peak rainy season. The severity of the increase, however, will be regionally dependent and this highlights the need for implementation of control and mitigation strategies at a local level. Improvements in parameterisation of the model will lead to improved ability to make quantitative predictions about spatiotemporal dynamics and control, as well as addressing how changes in rainfall distribution, as well as temperature, may lead to changes in transmission risk. The impact of excessive rainfall, for instance, on vector abundance and transmission is only qualitatively captured here. Such an analysis also highlights the usefulness of assessing control strategies within a dynamic framework. Questions such as the relative magnitude and timing of interventions, as well as the impact of parameter uncertainty, variability and heterogeneity, be they human, mosquito or parasite, may also be considered, along with evaluating such questions within the context of limited resources. The direct impact of seasonality in vector dynamics on R_0 has been considered from a theoretical perspective elsewhere and we postpone a more thorough analysis of the dependence of control on variability in environmental conditions to future work.

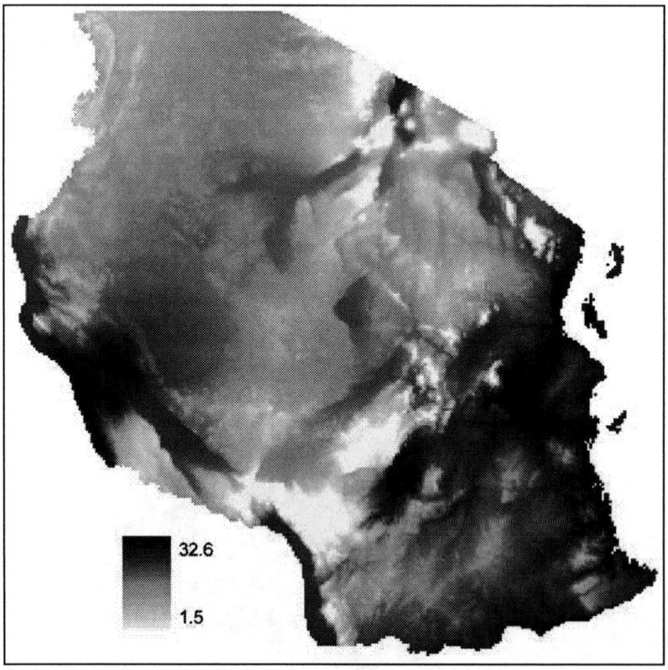

Figure 6. Predicted mean temperature distribution across Tanzania in April under the A2a emission scenario (data from WorldClim).

Conclusion

The complex transmission dynamics of malaria are strongly influenced by environmental conditions, with temperature, rainfall, humidity, wind speed and altitude, among others, shown to affect incidence. However, despite the opinion that malaria may represent one of the most sensitive VBDs to climate change over the next 100 years, most previous work on interactions between climate and malaria has been largely based on the use of statistical modelling approaches with little corresponding work carried out on using mathematical models to better understand the dynamical effects of this influence on disease transmission dynamics. Such process-based models may not only represent powerful tools for understanding how future trends and variability in key climate variables may affect global changes in malaria risk via their capacity to deal with the likely nonlinear and complex feedbacks in climate-pathogen relationships, but when combined with the ability to simulate 'what-if' scenarios, they may also represent valuable tools for policy-makers and future public health planning.[1,7,11]

In this chapter, we have shown the crucial need for the development of dynamic modelling frameworks to better understand the dependence of malaria transmission on changes in temperature and rainfall. The most general and probably most realistic modelling framework one could develop here would be a stochastic transmission model incorporating key environmental variables that are continuously changing. While we expect the field to undoubtedly progress towards this end, it is invaluable at the early stages of our understanding to construct simpler climate-linked transmission models that permit an understanding of how global warming may affect the burden of disease. Thus, using a relatively simple deterministic transmission model, we have been able to demonstrate the importance of accounting for the dependence of mosquito abundance on temperature and rainfall, which by entering the dynamical model through the nonlinear transmission term, we show can strongly influence the establishment probability of malaria in previously disease-free regions. The potential for mosquito extinction with large seasonality in temperature highlights a second key result, namely that it is important not only to examine the effects of future average trends in climate variables, but also the variability about these trends.

In reality, this variability may represent genuine fluctuations of climate variables or uncertainty from climate model predictions or human behaviour (corresponding to a range of possible emission scenarios). Indeed, it is clear from this chapter that seasonality and variability in climate on longer timescales can have profound effects on establishment, invasion and persistence and this remains a priority area for future research. Changes in patterns of malaria incidence due to changes in environmental conditions on longer timescales, e.g., due to the effects of El Niño-Southern Oscillation (ENSO), have been considered in the literature[44-47] and this continues to merit further research. Similarly, it is clear that when these studies are combined with corresponding concepts from infection system dynamics, such as the dependence of R_0 on temperature and rainfall and the identification of an optimal transmission window, such integrated analyses will prove crucial to improving understanding about how long-term global climate change will affect local environmental conditions and, in turn, whether a region is likely to experience a worsening or improvement in prevalence as global warming becomes more severe.

The results also highlight that considerable work, experimental, theoretical, modelling and policy-based, still remains to be statisfactorily carried out and modelled in an integrated fashion if we are to more realistically capture the impact of climate change on disease transmission. Although certain aspects of transmission dynamics are physiological (and therefore deterministic) drivers of transmission, it is clear that heterogeneities across the human, mosquito and parasite populations introduce considerable uncertainty into the system. It should also be borne in mind that a more realistic modelling approach should take spatial heterogeneities into account and thus realistic transmission models need to be spatial if they are to better predict spatiotemporal disease persistence and spread. A key challenge in incorporating environmental variables into malaria models is also selecting the appropriate scale at which to model, not only spatially, but over the most appropriate timescale and hierarchical level.[48] This is driven not only by the resolution of available climate data (either from remote sensing or output from climate models), but also from the knowledge that modelling at too fine a scale may translate poorly into global observables, while oversimplifying local heterogeneities may neglect key environmental or biological processes influencing observations. Another challenge is how best to integrate ecological drivers with sociological processes of disease transmission in vulnerable communities, an area only now beginning to be examined.[1,2,23,25,26,43]

The quality of incidence data, as well as the reliability and depth of experimental knowledge on the effects of environmental variables on the transmission cycle, also requires considerable research and this remains an important priority. The impact of excessive rainfall, for instance, on vector abundance and transmission is only qualitatively captured here. Mathematical models can thus act as a useful guide for data collection by identifying areas where improvements in data quality may lead to substantial improvements in their explanatory power.

Finally, our results also underscore the usefulness of assessing control strategies within a dynamic framework. In particular, we suggest that only such frameworks will allow fuller exploration of key questions such as the relative magnitude and timing of interventions, as well as the impact of parameter uncertainty, variability and heterogeneity, be they human, mosquito or parasite, along with reliable evaluation of these questions within the context of limited resources. The dynamical modelling approach described here may thus provide a useful framework not only for obtaining a better understanding of the integrated impact of climate change on disease transmission dynamics, but also to serve as a tool for policy-makers developing mitigation and intervention strategies that may be used to tackle the potentially tough challenges that lie ahead.

Acknowledgements

Both authors would like to acknowledge and thank the Grantham Institute for Climate Change at Imperial College London for support and funding of this research. This work forms part of the Institute's Health Theme-associated objective in developing and testing robust integrated assessment tools for evaluating, predicting and developing policy options to combat the impact of global climate change on human health.

References

1. Martens P. Global atmospheric change and human health: an integrated modelling approach. Clim Res 1996; 6(2):107-112.
2. Patz JA, Balbus JM. Methods for assessing public health vulnerability to global climate change. Clim Res 1996; 6:113-125.
3. Patz JA, Epstein PR, Burke TA et al. Global climate change and emerging infectious diseases. JAMA 1996; 275(3):217-223.
4. Martens P. Health and Climate Change: Modelling the Impacts of Global Warming and Ozone Depletion). London:Earthscan Publications Ltd:1998.
5. Department of Health. Health effects of climate change in the UK. Department of Health 2001.
6. IPCC. IPCC Fourth Assessment Report, Cambridge University Press 2007.
7. Martens P, Niessen LW, Rotmans J et al. Potential impact of global climate change on malaria risk. Environ Health Perspect 1995; 103(5):458-464.
8. Martens P, Kovats RS, Nijhof S et al. Climate change and future populations at risk of malaria. Glob Environ Change 1999; S9:89-107.
9. Rogers DJ, Randolph SE. The Global Spread of Malaria in a Future, Warmer World. Science 2000; 289:1763-1766.
10. World Health Organization. World Malaria Report 2008. WHO 2008.
11. Martens WJM, Jetten TH, Focks DA. Sensitivity of malaria, schistosomiasis and dengue to global warming. Clim Change 1997; 35(2):145-156.
12. Bi P, Tong S, Donald K et al. Climate Variables and Transmission of Malaria: A 12-Year Data Analysis in Shuchen County, China. Public Health Reports 2003; 118:65-71.
13. Koenraadt CJM, Paaijmans KP, Githeko AK et al. Egg hatching, larval movement and larval survival of the malaria vector Anopheles gambiae in desiccating habitats. Malar J 2003; 2:20.
14. Thomson MC, Mason SJ, Phindela T et al. Use of rainfall and sea surface temperature monitoring for malaria early warning in Botswana. Am J Trop Med Hyg 2005; 73(1):214-221.
15. Wiwanitkit V. Correlation between rainfall and the prevalence of malaria in Thailand. J Infect 2006; 52:227-230.
16. Briet OJT, Vounatsou P, Gunawardena DM et al. Temporal correlation between malaria and rainfall in Sri Lanka. Malar J 2008; 7:77.

17. Molineaux L, Storey J, Cohen JE et al. A longitudinal study of human malaria in the west African savanna in the absence of control measures: relationships between different Plasmodium species, in particular P. falciparum and P. malariae. Am J Trop Med Hyg 1980; 29:725-737.
18. Muir DA. Anopheline mosquitos: vector reproduction, life-cycle and biotope. In: Wernsdorfer WH, McGregor I, eds. Malaria: Principles and Practice of Malariology. London: Churchill Livingstone; 1988. 431–451.
19. Charlwood JD, Kihonda J, Sama S et al. The rise and fall of Anopheles arabiensis (Diptera: Culicidae) in a Tanzanian village. Bull Entomol Res 1995; 85:37-44.
20. Hoshen MB, Morse AP. A weather-driven model of malaria transmission. Malar J 2004; 3:32.
21. Koenraadt CJM, Githeko AK, Takken W. The effects of rainfall and evapotranspiration on the temporal dynamics of Anopheles gambiae s.s. and Anopheles arabiensis in a Kenyan village. Acta Tropica 2004; 90:141-153.
22. Ndiaye PI, Bicout DJ, Mondet B et al. Rainfall triggered dynamics of Aedes mosquito aggressiveness. J Theor Biol 2006; 243:222-229.
23. Martens P, Hall L. Malaria on the move: Human population movement and its impact on malaria transmission. Emerg Infect Dis 2000; 6(2):7-13.
24. Gubler DJ, Reiter P, Ebi KL et al. Climate Variability and Change in the United States: Potential Impacts on Vector- and Rodent-Borne Diseases. Environmental Health Perspectives 2001; 109(2):223-233.
25. Tatem AJ, Hay SI, Rogers DJ. Global traffic and disease vector dispersal. PNAS 2006; 203(16):6242-6247.
26. Lindsay SW, Birley MH. Climate change and malaria transmission. Ann Trop Med Parasitol 1996; 90:573-588.
27. Lindsay SW, Martens P. Malaria in the African highlands: past, present and future. Bull World Health Organ 1998; 76(1):33-45.
28. Craig MH, Snow RW, le Sueur D. A Climate-based distribution model of malaria transmission in Sub-Saharan Africa. Parasitol Today 1999; 15(3):105-111.
29. Altizer S, Dobson A, Hosseini P et al. Seasonality and population dynamics: infectious diseases as case studies. Ecol Letters 2006; 9:467-484.
30. Grassly NC, Fraser C. Seasonal infectious disease epidemiology. Proc R Soc B 2006; 273:2541-2550.
31. Jepson WT, Moutia A, Courtois C. The malaria problem in Mauritius: the bionomics of Mauritian anophelines. Bull Entomol Res 1947; 38:177-208.
32. Depinay JMO, Mbogo CM, Killeen G et al. A simulation model of African Anopheles ecology and population dynamics for the analysis of malaria transmission. Malar J 2004; 3(29).
33. Paaijmans KP, Wandago MO, Githeko AK et al. Unexpected high losses of Anopheles gambiae larvae due to rainfall. PLoS ONE 2007; 2(11):e1146.
34. Detinova TS. Age-grouping methods in diptera of medical importance. WHO Monograph 1962; 47: World Health Organisation, Geneva.
35. Macdonald G. The epidemiology and control of malaria. London:Oxford University Press:1957.
36. Warrell DA. Essential malariology, Arnold 2002.
37. Anderson RM, May RM. Infectious Diseases of Humans: Dynamics and Control. London:Oxford University Press:1991.
38. Bacaër N, Guernaoui S. The epidemic threshold of vector-borne diseases with seasonality. J Math Biol 2006; 53:421-436.
39. Bacaër N. Approximation of the basic reproduction number R_0 for vector-borne diseases with a periodic vector population. Bull Math Biol 2007; 69:1067-1091.
40. Bacaër N, Ouifki R. Growth rate and basic reproduction number for population models with a simple periodic factor. Math Biosci 2007; 210:647-658.
41. Hijmans RJ, Cameron SE, Parra JL et al. Very high resolution interpolated climate surfaces for global land areas. Int J Climatol 2005; 25:1965-1978.
42. Debnath L. Nonlinear Partial Differential Equations for Scientists and Engineers. Berlin:Springer 1997.
43. Reiter P. Climate change and mosquito-borne disease: knowing the horse before hitching the cart. Rev Sci Tech 2008; 27(2):383-398.
44. Bouma MJ, van der Kaay HJ. The El Nino Southern Oscillation and the historic malaria epidemics on the Indian subcontinent and Sri Lanka: an early warning system for future epidemics? Trop Med Int Health1996; 1(1):86-96.
45. Bouma MJ, Poveda G, Rojas W et al. Predicting high-risk years for malaria in Colombia using parameters of El Nino Southern Oscillation. Trop Med Int Health 1997; 2(12):1122-1127.
46. Hay SI, Myers MF, Burke DS et al. Etiology of interepidemic periods of mosquito-borne disease. PNAS 2000; 97(16):9335-9339.
47. Bi P, Parton KA, Tong S. El-Nino-Southern Oscillation and Vector-Borne Diseases in Anhui, China. Vector Borne Zoonotic Dis 2005; 5(2):95-100.
48. Levin SA. The problem of pattern and scale in ecology. Ecology 1992; 73(6):1943-1967.

Chapter 14

Modelling the Transmission of *Trypanosoma cruzi*: The Need for an Integrated Genetic Epidemiological and Population Genomics Approach

Michel Tibayrenc*

Abstract

This chapter describes what should be an integrated approach to the genetic epidemiology and population genomics of Chagas disease. Many studies have been conducted on the genetic diversity of *Trypanosoma cruzi* and the various triatomine bug species able to transmit Chagas disease. Far less research has analyzed the role played by the host's genetic variability on the transmission and severity of the disease. An integrated genetic epidemiology/population genomics approach would analyze these three components of the transmission chain together as well as their possible interactions (co-evolution phenomena). This is facilitated by the recent impressive progress in mega biotechnologies and by the fact that Chagas disease is an ideal model for experimental evolution approaches.

Introduction

For several years, I have pleaded for an integrated approach to the epidemiology of infectious diseases. Most authors focus on only one component of the transmission chain of a given disease: either the host or the pathogenic agent, or (in the case of vector-borne diseases such as Chagas disease) the vector. It is clear that these three actors play a role in the same performance as part of a single biological phenomenon: the coevolution between the host, the pathogen and the vector. An integrated approach to this global phenomenon is therefore sorely needed.[1,2] Scientists have a natural tendency to specialize, even to overspecialize. For example, my experience tells me that people working on the population genetics of African trypanosomes are poorly informed about similar studies conducted on *T. cruzi*. This is all the more distressing since comparative approaches are extremely informative, delineating the general laws that govern the evolution of organisms, while underscoring the specificities of each case.[3,4]

When transmission and severity of infectious diseases are concerned, it is very hard to know whether different genotypes of the pathogen are able to cause different clinical forms without knowing what the role of the host's genetic diversity could also be. An integrated approach consists in analyzing both phenomena jointly and in including studies on the vector when researching vector-borne diseases. Even more specifically, it aims at dissecting the possible interactions between the three (co-evolution phenomena). The international congresses MEEGID (Molecular Epidemiology and Evolutionary Genetics of Infectious Diseases) and the journal *Infection, Genetics*

*Michel Tibayrenc—Genetics and Evolution of Infectious Diseases, IRD, BP 64501, 34394 Montpellier Cedex 5, France. Email: michel.tibayrenc@ird.fr

Modelling Parasite Transmission and Control, edited by Edwin Michael and Robert C. Spear.

and Evolution (http://www.elsevier.com/locate/meegid) both aim to broaden the scope of this research by encouraging and practicing such an integrated approach. Interestingly, Chagas disease is an exemplary case for this approach.[5] I will explain below why this is more than ever true today. I will start by describing today's situation for each of the three actors: the pathogen, the vector(s) and the host(s).

Trypanosoma cruzi, World Champion of Pathogens for Population Genetics

The scientific community that works on Chagas is a tiny one. In fact, almost all of us know each other personally. It is like a big family, in which the contribution of the Latin American element has been and still is, prominent. Considering the small size of this group, *T. cruzi* would not be expected to be the pathogenic agent whose genetic diversity is among the best known, possibly more than the heavy weights *Escherichia coli* and *Candida albicans*. This fact leads me to advocate setting up *T. cruzi* as one of the landmark models of modern biology, together with the legendary *E. coli*, *Drosophila melanogaster*, *Caenorhabditis elegans* and *Mus musculus*. A historical view of how this unexpected situation came about is useful.

The Isoenzyme Saga

In the late 1960s and the 1970s, isoenzyme markers became immensely popular.[6] Isoenzymes are electrophoretic variants of the same enzyme that reveal the sequence diversity of the genes that code for them (Fig. 1). For the first time, they made it possible to directly unravel the genetic diversity of organisms: population genetics stopped being a speculative affair and entered the enchanted world of direct observation. Thousands of papers based on what had quickly become the gold standard have been published, covering virtually the entire living reign. This made it possible to firmly establish the mendelian inheritance (codominant markers) of isoenzymes. Fortunately, bacteriologists and parasitologists did not miss the train and took advantage of the interesting properties of isoenzymes to clarify the subspecific variability and population structure of their pet bugs.[7,8] This has been especially true for *T. cruzi*, in particular through the pioneering studies of Miles and collaborators[9] and studies based on a population genetics approach by our group.[10] It is worth noting that all the results on the population structure and evolutionary pattern of *T. cruzi* based on isoenzymes have been fully confirmed by more fashionable molecular methods (see below).

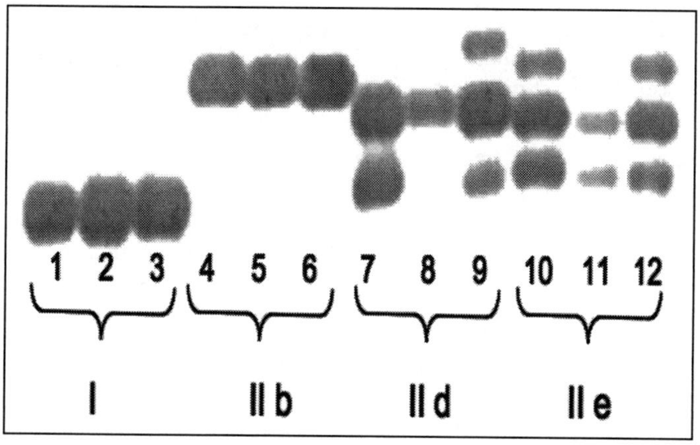

Figure 1. An isoenzyme gel for the genetic locus *Glucose phosphate isomerase* (*Gpi*) showing the genetic polymorphism of different genetic subdivisions (discrete typing units; see text) of *T. cruzi* (experiment and photograph by Jenny Telleria, IRD Montpellier, France).

The Molecular Biology Wave

In the 1980s, "the Maniatis"[11] became the laboratory bible and young researchers dangerously started spotting their fingers with ethidium bromide, fascinated by the orange fluorescent banding patterns visible on gels under UV light. In a flash, isoenzymers became nerdish. In the Chagas community, the first group to step on stage was Morel's group,[12] who showed off the very esthetic schizodeme profiles of strains (RFLP patterns of kinetoplast DNA). Many other molecular techniques were later added to the display of strain typers, including miniexon gene polymorphism,[13] microsatellites[14] and random primed polymorphic amplified DNA or RAPD,[15] among others. All these studies revealed a striking pattern: they showed a constantly converging picture of *T. cruzi* subspecific variability, also congruent with the picture drawn by the nerdy isoenzymes. This was all grist for the population geneticists' mill (see below).

The Sequencing, Genomic and Postgenomic Era

We are still in the midst of this revolution. *T. cruzi* was not left by the wayside in this new wave. Its genome is now fully sequenced.[16] Several other strains are currently being sequenced as well (El Sayed, personal communication). Microarrays and proteomic analyses are on the way. The challenge is now to channel and filter the coming flood of data so that it remains informative. Fortunately, in the case of *T. cruzi*, theory came before the data flood and we have a robust population model available that provides a relevant framework for all studies investigating the genetic, genomic and phenotypic diversity of this parasite. The story is briefly expounded below.

Is T. cruzi *a "Good" Species?*

This is the first question to raise from an epidemiological point of view, especially where molecular epidemiology is concerned. Defining a good species refers to the definition of a species itself. Here is not the place to rekindle the debate.[17] Briefly, species are generally defined as: (i) a mating community (the biological species concept) or (ii) a clade (a monophyletic line with only one ancestor; the phylogenetic species concept) or (iii) a set of organisms that share remarkable phenotypic traits (the phenotypic species concept). Undoubtedly, *T. cruzi* meets the criteria for (ii) and (iii). All phylogenetic studies have brought all *T. cruzi* strains into a unique clade that is distinguishable from closely related taxa (*T. cruzi marenkellei*, a close cousin of *T. cruzi* that parasitizes bats, is the best example of such an outgroup). Moreover, all *T. cruzi* strains share a set of specific phenotypic characters (morphological aspects, vectorial transmission by triatomine bugs, potential host range extended to all mammals, but restricted to them, geographical distribution limited to the new world, potential pathogenicity). Consequently, from the point of view of the phylogenetic and phenotypic concepts, *T. cruzi* is a good species. The fact that *T. cruzi* is a unique clade makes it possible to design many molecular markers that will be specifically shared by all strains of the taxon (in the cladistic jargon, synapomorphic characteristics; see also the concepts of DTU and tags below).

The Population Structure of T. cruzi*: Sex or No Sex?*

In the 1980s, *T. cruzi* found itself enrolled in the noisy clonality/sexuality debate that roused bacteriologists and parasitologists.[18] There is now consensus that this parasite is somewhat of a paradigm of the preponderant clonal evolution model.[19,20] This means that its multi-locus genotypes copy themselves like genetic photocopies and are extremely stable in space and time, even at an evolutionary scale. Two important features of the model, often neglected by scientists who read only the abstracts of the papers, are that: (i) sex (in the broad sense: any kind of genetic exchange between two different cells) is not supposed to be totally absent, but only rare and not frequent enough to break the prevalent pattern of clonality; (ii) clonality is taken here in its genetic meaning and refers to all situations where descendant multi-locus genotypes are virtually identical to parental genotypes, whatever the actual mating system. It could be mitotic clonality, or parthenogenesis, or an extreme selfing situation, or extreme homogamy.[3]

Scarcity or absence of genetic recombination has been established in *T. cruzi* by evidencing an extreme linkage disequilibrium (LD; nonrandom association of genotypes occurring at separated

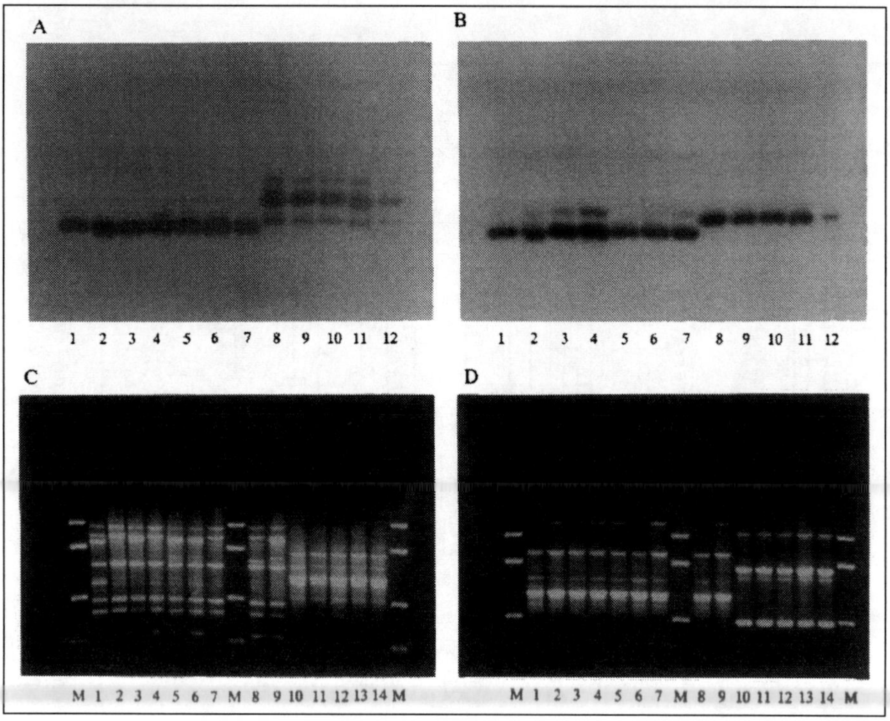

Figure 2. Four electrophoretic experiments, two isoenzyme gels with two different genetic loci (top), two random primed amplified polymorphic DNA (RAPD) surveying two different genomic regions (bottom). The same *T. cruzi* strains are surveyed on the four gels; however, RAPD experiments have two additional strains. On the four gels, only rwo main genotypes (DTU 1 and DTU 2; see text) are observed; lines 1-7 and 8-12 for isoenzymes, lines 1-9 and 10-14 for RAPD. M lines on RAPD gels are molecular weight ladders. Genotypes DTU 1 are constantly linked together and the same is observed for genotypes DTU 2. Crossed genotypes, which would be the result of genetic recombination (for example gel A line 1/gel B line 8 + gel C line 1 + gel D line 10), have never been observed on more than 600 *T. cruzi* strains characterized by our team to date. This kind of strong association between genotypes occurring at different loci is by definition a linkage disequilibrium (see text) and is a manifestation of preponderant clonal evolution in *T. cruzi*. Copyright 1993, Proc Natl Acad Sci USA. Tibayrenc M et al. 90:1335-39.[15]

loci), as shown in Figure 2. A striking manifestation of LD is that phylogenetic trees established from different genetic markers are very similar (Fig. 3). The contrary arises in those pathogens where recombination is abundant, such as the bacterium *Helicobacter pylori*.

The model has considerable implications in terms of applied research: (i) the stability of multi-locus genotypes makes them ideal targets for molecular epidemiology (strain typing and tracking); (ii) since genotypes are genetically separated from each other, their evolutionary fate is to accumulate more and more divergent mutations, including for those genes that govern medically relevant properties (pathogenicity, drug resistance). Clones that are genetically similar should tend to be also phenotypically similar and vice versa.

There is also a consensus on the number of genetic subdivisions that are observable within *T. cruzi*: there are two main clusters,[3,13] which have been named by a group of anonymous experts *T. cruzi* I and *T. cruzi* II (TC I and II).[21] TC II is itself subdivided into five lesser clusters (TC IIa-e).[22] There has been a debate on the actual evolutionary nature of these clusters. The presence of

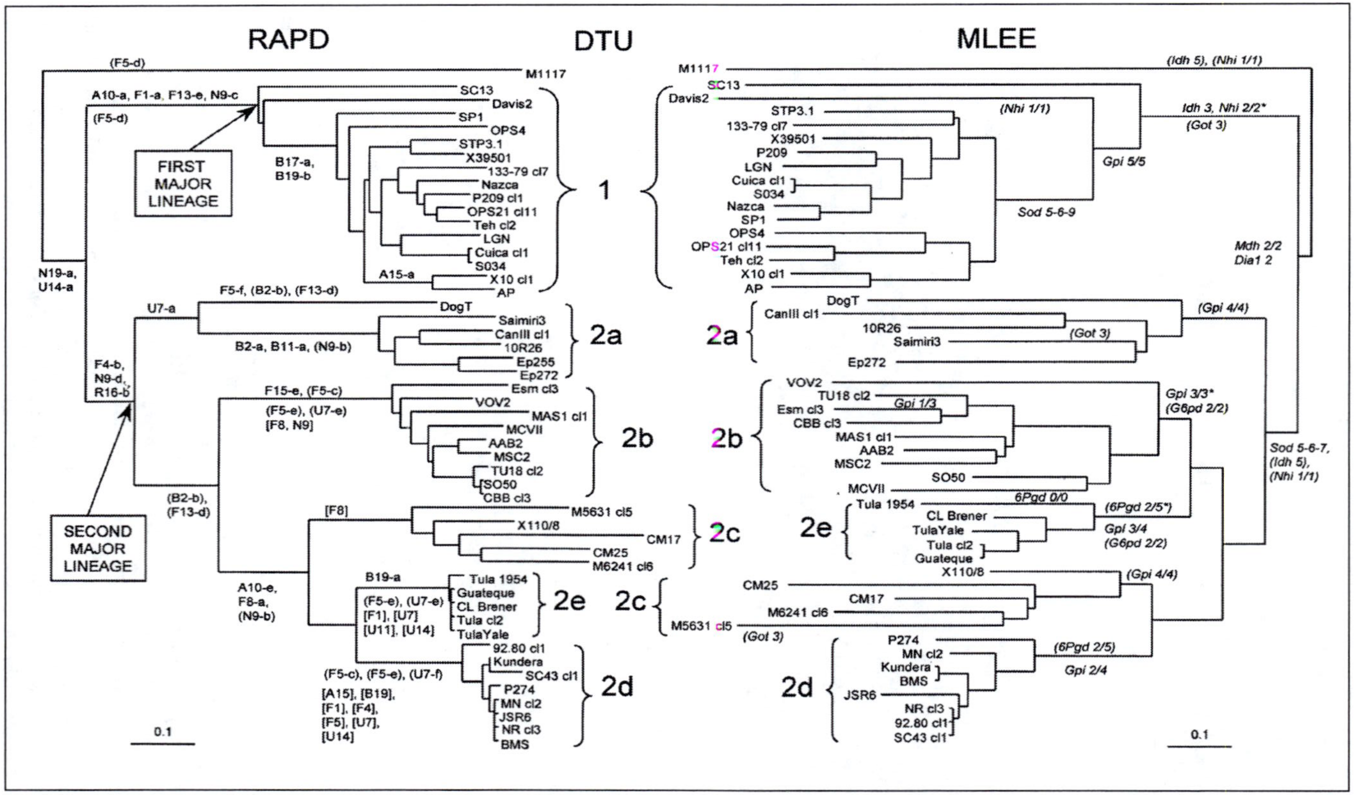

Figure 3. Two phylogenetic trees depicting the evolutionary relationships among *T. cruzi* genotypes: isoenzymes (left) and RAPD (right). The strong similarity between the two trees is an extreme manifestation of linkage disequilibrium (see Fig. 2). Copyright 2000, Int J parasitol. Brisse S et al. 30:35-44.[22]

some genetic recombination and hybrid genotypes prevents one from calling them clades. We have proposed the operational term of discrete typing units (DTU):[1] sets of strains that are genetically more similar to each other than to any other strains and that share common molecular or serological markers (tags). This terminology is widely accepted in the Chagas research community.[a] We will henceforth refer to DTU I, IIa, IIb, IIc, IId and IIe. Figure 1 shows that isoenzyme electrophoresis shows drastically distintc patterns between *T. cruzi* DTUs.

As already noted, the clonal model does not state that sex (recombination) is absent, but only that it is rare and is not a mandatory mechanism of reproduction, as it is in humans and fruitflies. Several natural *T. cruzi* genotypes appear to have a hybrid origin, subsequently stabilized by clonal propagation.[23-26] Moreover, the potential for genetic recombination in *T. cruzi* has been fully confirmed by experiments.[27]

The story therefore can be summarized as follows: *T. cruzi* undergoes predominant clonal evolution, which permits it to stabilize favorable genotypic combinations. Occasionally it mates, which makes it possible to rapidly generate new genotypes subsequently stabilized by clonal propagation. It is a typical case of reticulate evolution, a pattern also observed in many plant species.

The important facts to be kept in mind for the topic of this chapter: (i) the above-mentioned picture of subspecific variability could be refined; however, it is improbable that further studies will upset this picture. (ii) *T. cruzi* DTUs are robust units of analysis corroborated by many different studies, easy to specifically identify with appropriate tags. They actually behave like distinct taxa. They constitute convenient units of analysis for epidemiological tracking, applied research (vaccine and drug design) and experimental evolution studies.

However, a wide gap in our knowledge on *T. cruzi*'s genetic variability persists: the biological and epidemiological differences among *T. cruzi* clonal genotypes and DTUs remain imperfectly known. A relation between *T. cruzi* isoenzyme genotypes (zymodemes) and clinical forms of the disease has been long suspected,[28] but never fully confirmed. Long term experimental surveys in our laboratory have shown significant statistical associations between the genetic distances recorded among *T. cruzi* genotypes on one hand and biomedical differences on the other hand[29-34] (see also refs. 35-36). The biomedical properties that have been surveyed include growth speed in acellular and cell cultures, transmissibility through vectors, pathogenicity in mice and in vitro and in vivo susceptibility to antichagasic drugs (série). Only one study investigating in vitro drug sensitivity showed no statistical association.[37] There is clearly "something" there when the working hypothesis of a link between *T. cruzi* genotypic and biological diversity is tested. However, the picture is far from black and white and the statistical sets are limited for the moment. An interesting hypothesis states that different *T. cruzi* genotypes have different organ tropisms.[38] However, no firm conclusions have been reached on this point either. This long-term debate therefore remains partly unanswered.

The Second Actor: The Vector

It would be more accurate to say the vectors, for there are many of them. Triatomine bugs are true bugs (heteropterous). They make up a subfamily (Triatominae) within the family reduvidae, which are basically predator bugs. The triatominae turned out to be obligatory blood feeders, adults of both genders and larvae of all stages.

From a population genetics and evolutionary point of view, the harvest is not as plentiful as for *T. cruzi*. However, many studies have been conducted on triatomines, which makes them a rather well-known group.

It is now strongly suspected that the adaptative trait of strict hematophagy occurred several times in the evolutionary history of reduviidae. The triatominae are thought to be a polyphyletic group (Schofield, personal communication). Three main genera are recorded in the triatomines: *Rhodnius*,

a. In an expert meeting held in Buzios (Brazil), August 2009, in which the author participated, these DTUs have been validated. However, they have been renumbered I, IV, II, III, V, and VI, respectively.

Triatoma and *Panstrongylus*. It is being debated whether each of these genera considered alone is monophyletic. Within these genera, many different species are able to transmit *T. cruzi*.[39]

Triatomines exhibit various interesting phenomena of partial speciation and sibling speciation, which makes them informative models for scientists interested in speciation processes. These situations have been conveniently explored by population genetics approaches.

Since the first pioneering isoenzyme studies of the 1980s,[40] many population genetics and phylogenetics analyses have been conducted on various triatomine species with markers ranging from microsatellites[41,42] to RAPD,[43,44] mitochondrial genes[45] and gene sequences.[46-48] Interestingly, this rather classical molecular display has been completed by the complementary tools of cytogenetics[49] and morphometric analysis.[42,50,51]

The world of triatomine bugs is vast, with an extreme taxonomic, phylogenetic and ecological diversity, making this group a gold mine for population genetic and evolutionary studies.

Within the theme of the present chapter, an immense field of knowledge remains to be explored, with several unanswered key questions: (i) have all species of triatomines the same vectorial capacity (the answer is probably no); (ii) do species of the same genus tend to have comparable vectorial capacities; (iii) within the same species, do different populations have the same vectorial capacity; (iv) are different strains, genetic clones and DTUs of *T. cruzi* equally transmitted by triatomine bugs; (v) more generally, what complex phenomena of co-evolution and co-adaptation exist between the vector and the pathogen.

The Host

Again, it is preferable to say the hosts, since all mammalian species are potentially able to be contaminated by *T. cruzi*. Triatomine bugs feed on birds as well; however, these animals are resistant to Chagas disease. The reservoir of the disease is therefore virtually unlimited, since wild mammals as well as domestic animals are hosts. Chagas disease is a typical zoonosis. The transmission is enhanced by the fact that humans and domestic animals often live close together. For example, in Bolivia, farmers often have pet guinea pigs in their kitchen.

As for vector species, this wide range of mammalian host species provides abundant matter for co-evolution analyses, either in field surveys or in experimental studies. Dogs with their many breeds have been a choice model for experimental studies on *T. cruzi* pathogenicity.[52-55] Of course mice remain the easiest model to handle in experimental Chagas disease.[33,56,57] However, we still lack an overall picture of differential Chagas pathogenicity among different mammalian species and different populations and breeds of the same species. Moreover, the animal models do not clearly identify the candidate genes[58] that could be involved in the susceptibility to Chagas disease.

As for the human species, it is quite unexpected that our knowledge of the impact of human genetic diversity on the severity and clinical diversity of Chagas disease is poorly known. By comparison, in this field, we know much more about malaria,[59] tuberculosis,[60] leprosy,[61] schistosomiasis[62] and leishmaniosis.[63] This is all the more astonishing since Chagas disease should constitute a very favorable case to study this problem, for two reasons:

i. As for leprosy, the clinical phenotypes of the disease are quite clearly defined. It starts with an acute phase that can be discreet or on the contrary shows a severe septicemic syndrome. Approximately 10% of patients die at this stage. Those who survive enter the undetermined phase with no symptoms. Parasites hide in the cells. Unfortunately, after at few years, roughly 30% of patients begin showing symptomatic Chagas disease. The majority of them suffer from a cardiac form that leads to severe cardiac insufficiency. Digestive forms account for 3% of patent Chagas disease cases. They have the form of megacolon or megaesophagus. Some patients have a cardiac and digestive form. Some patients have no clinical symptoms, but have electrocardiogram abnormalities. Others have symptoms but negative serological tests.[64] Lastly, there are great differences in the way that antichagasic drugs act on different patients.[36] There is therefore a wide span of well-defined clinical forms to analyze in linkage studies.
ii. Chagas disease strikes a range of genetically diverse human populations, including different ethnic groups (mainly Amerindians, Causasians, Africans and people of mixed ancestry)

that could exhibit different susceptibilities to the disease. For example, Aymara Indians in Bolivia and Peru have lived only on the Altiplano for several thousand years. They have therefore long been protected from Chagas disease, since triatomine bugs do not survive at such altitudes. Their immunity against the disease could be lesser than that of other Amerindian populations who have coexisted with triatomine bugs for a long time.

A debate of importance for studying genetic susceptibility to Chagas disease is to know whether chronic Chagas disease is actually an autoimmune disease[65] or if the parasite is still present in the host's cells and causes a chronic inflammatory response. Both PCR experiments and the classical xenodiagnosis strongly suggest that the second hypothesis holds true.

The meager results obtained on human genetic susceptibility to Chagas diseases and its various clinical forms are now summarized. A familial component in the cardiac form of chronic Chagas disease has been suggested from a Brazilian study.[66] These results clashed with the results obtained from an Argentinian survey.[67] HLA polymorphism associated with Chagas disease has been analyzed in many studies. Some of them reported a total lack of association,[68] while on the contrary other research observed associations.[69] Some serological parameters seem to exhibit a notable heritability, as evidenced by extended pedigrees. In Brazil, the heritability of Chagas seropositivity would be no less than 0.556.[70] More specifically, the heritability of the IgA and IgG levels would be 0.33.[71] I have myself observed that many subjects in Bolivia seem to be quite resistant to Chagas disease contamination. In some areas where 100% of the thousands of triatomine bugs I have collected harbored *T. cruzi*, only 50% of the children who lived in those areas were seropositive. Nevertheless, all these children had been bitten hundreds or thousands of times in their lifetime.

To make a long story short, our knowledge on the role of human genetic diversity on Chagas transmission, severity and clinical polymorphism is extremely patchy and contradictory. This is the weak spot in the array of knowledge needed to design an integrated genetic epidemiology approach of Chagas disease.

The Future

Chagas disease could become a paradigm academic case for an integrated genetic epidemiology approach to transmissible diseases.[5] However, we still are far from this goal. Our knowledge on the parasite's genetic diversity is fairly advanced. Thanks to phylogenetic/population genetic/morphometric studies, the complexity of the world of triatomine bugs is being deciphered little by little. Although the role of the host's genetics remains poorly known, it is only a matter of applying the necessary effort. A black hole remains in our knowledge on the interactions and reciprocal impact between these three components. Actually, this is the case for all transmissible diseases. Research is strongly compartmentalized and parasitologists, entomologists and human/mammal geneticists rarely interact.

In the case of Chagas disease, the gaps in our knowledge could be filled using two approaches:

Field Studies

New powerful technologies (high-throughput sequencing, genome-wide scanning, microarrays, real-time PCR, high-resolution morphometric analysis) should make it possible to considerably improve our knowledge on the parasite, the vector and the host.

When humans are considered, as it is the case for any disease, studies on Chagas disease could greatly benefit from the megaprojects presently running on human genetic diversity, mainly the HapMap project (http://www.hapmap.org/). Genomics is presently making exponential progress and the few scientists working on human genetic susceptibility to Chagas disease should climb aboard this high-speed train. As emphasized above, in Chagas genetic epidemiology, the human side is the weak link in the chain. Of course studies dealing with the parasite and the vector would also greatly benefit from the impressive progress reached in genomics, proteomics and bioinformatics analysis.

Although it is important that specialists working on the parasite, the vector and the host continue developing their specific fields, our knowledge on these actors sorely needs them to interact more, including in field studies. It is crucial that epidemiologists, mammalogists, entomologists, human geneticists and clinicians include parasite isolation and characterization in their programs, so that it is better known, for example, which *T. cruzi* genotypes are more frequently harbored by given populations of given triatomine bug species, or more frequently isolated from given clinical forms of the disease. Such goals should be attainable by setting up multidisciplinary research consortia. Networking organizations would be an efficient means to attract valuable sources of funding. The long-term goal is to build an integrated population genomic approach,[72] joining together genomic and postgenomic studies investigating the pathogen, the vector and the host at the population level. Obviously, such an ambitious approach can be accomplished by extended collaborations between many teams of diversified expertise only.

Experimental Evolution

Integrated genetic epidemiology in the field is hard and involves heavy research protocols. Experiments on the evolution of Chagas disease are easier to design, since Chagas disease is an extremely favorable model for this kind of approach[5] and a complete Chagas transmission cycle is easy to maintain in experiments. The parasite is easy (although harmful) to culture, either in acellular or in cell (Vero cells) cultures. Many triatomine bug species are easy to rear in the laboratory. Various mice breeds can be infected by *T. cruzi*. Other laboratory animals such as dogs can also be used. A complete transmission cycle can even be maintained without laboratory animals by using artificial feeding devices for triatomine bugs. It is therefore quite feasible to survey the interactions between the parasite's, the host's and the vector's genetic variability by varying only one parameter at a time. A promising avenue of research is to explore the interactions between mixtures of *T. cruzi* genotypes, a situation that is frequent in natural cycles, in vectors as well as in Chagas disease patients. Our working hypothesis is that there is some sort of cooperation between different clonal genotypes that infect a single host (Ann Rev gen), so that the whole is more than the sum of the parts. This seems to have been verified in a number of preliminary experiments involving such mixtures of clonal genotypes.[34]

Conclusion

Chagas disease potentially constitutes a paradigm model for the integrated genetic epidemiology and integrated population genomic approaches. However, the Chagas scientific community, although talented, is tiny, a handicap in reaching this goal. It is therefore indispensable to attract other scientists to the enterprise. This is why it is crucial to sell *T. cruzi* and Chagas disease as a reference model for basic biology and evolution, as it is the case for *Escherichia coli*, *Candida albicans*, *Caenorhabditis elegans*, *Mus musculus* and *Drosophila melanogaster*.

References

1. Tibayrenc M. Genetic epidemiology of parasitic protozoa and other infectious agents: the need for an integrated approach. Int J Parasitol 1998; 28:85-104.
2. Tibayrenc M. Towards an integrated genetic epidemiology of parasitic protozoa and other pathogens. Annual Review of Genetics 1999; 33:449-77.
3. Tibayrenc M. Population genetics of parasitic protozoa and other microorganisms. Advances in Parasitology (eds. Baker JR, Muller R, Rollinson D.): 1995; 36:47-115.
4. Tibayrenc M. Towards a unified evolutionary genetics of microorganisms. Ann Rev Microbiol 1996; 50:401-29.
5. Tibayrenc M. Integrated genetic epidemiology of infectious diseases: the Chagas model. Mem Inst Oswaldo Cruz 1998; 93:577-80.
6. Tibayrenc M. Molecular epidemiology and evolutionary genetics of pathogens. In: Tibayrenc, M. ed. Encyclopedia of Infectious Diseases: Modern Methodologies. Hoboken: Wiley & Sons 2007; 337-55.
7. Milkman R. Electrophoretic variation in Escherichia coli from natural sources. Science 1973; 182:1024-26.
8. Kilgour V, Godfrey DG. Species-characteristic isoenzymes of two aminotransferases in Trypanosomes. Nature New Biol 1973; 244:69-70.

9. Miles MA, Souza A, Povoa M et al. Isozymic heterogeneity of Trypanosoma cruzi in the first autochtonous patients with Chagas' disease in Amazonian Brazil. Nature 1978; 272:819-21.
10. Tibayrenc M, Cariou ML, Solignac M. Interprétation génétique des zymogrammes de flagellés des genres Trypanosoma et Leishmania. C R Acad Sci Paris 1981; 292:623-25.
11. Sambrook J, Fritsch EF, Maniatis T. Molecular cloning: A laboratory manual. Cold spring harbor laboratory. 1980.
12. Morel CM, Chiari E, Plessmann Camargo E et al. Strains and clones of Trypanosoma cruzi can be characterized by pattern of restriction endonuclease products of kinetoplast DNA minicircles. Proc Natl Acad Sci USA 1980; 77:6810-14.
13. Souto RP, Fernandes O, Macedo AM et al. DNA markers define two major phylogenetic lineages of Trypanosoma cruzi. Molecular and Biochemical Parasitology 1996; 83:141-52.
14. Oliveira RP, Broude NE, Macedo AM et al. Probing the genetic population structure of Trypanosoma cruzi with polymorphic microsatellites. Proc Natl Acad Sci USA 1998; 95:3776-80.
15. Tibayrenc M, Neubauer K, Barnabé C et al. Genetic characterization of six parasitic protozoa: parity of random-primer DNA typing and multilocus isoenzyme electrophoresis. Proc Natl Acad Sci USA 1993; 90:1335-39.
16. El-Sayed NM, Myler PJ, Bartholomeu DC et al. The genome sequence of Trypanosoma cruzi, etiologic agent of Chagas disease. Science 2005; 309:409-15.
17. Tibayrenc M. The species concept in parasites and other pathogens: a pragmatic approach? Trends Parasitol 2006; 22:66-70.
18. Ørskov F, Ørskov I. Summary of a workshop on the clone concept in the epidemiology, taxonomy and evolution of the Enterobacteriaceae and other Bacteria. J Infect Diseases 1983; 148:346-57.
19. Tibayrenc M, Cariou ML, Solignac M et al. Arguments génétiques contre l'existence d'une sexualité actuelle chez Trypanosoma cruzi; implications taxinomiques. C R Acad Sci Paris 1981; 293:207-9.
20. Tibayrenc M, Ward P, Moya A et al. Natural populations of Trypanosoma cruzi, the agent of Chagas' disease, have a complex multiclonal structure. Proc Natl Acad Sci USA 1986; 83:115-19.
21. Anonymous. Taxonomy of Trypanosoma cruzi: a commentary on characterization and nomenclature. Memorias Instituto Oswaldo Cruz 1999; 94(Suppl 1):181-84.
22. Brisse S, Barnabé C, Tibayrenc M. Identification of six Trypanosoma cruzi phylogenetic lineages by random amplified polymorphic DNA and multilocus enzyme electrophoresis. Int J parasitol 2000; 30:35-44.
23. Bogliolo AR, Lauriapires L, Gibson WC. Polymorphisms in Trypanosoma cruzi: Evidence of genetic recombination. Acta Tropica 1996; 61:31-40.
24. Carrasco HJ, Frame IA, Valente SA et al. Genetic exchange as a possible source of genomic diversity in sylvatic populations of Trypanosoma cruzi. Am J Trop Med Hyg 1996; 54:418-24.
25. Machado CA, Ayala FJ. Nucleotide sequences provide evidence of genetic exchange among distantly related lineages of Trypanosoma cruzi. Proc Natl Acad Sci USA 2001; 98:7396-401.
26. Brisse S, Henriksson J, Barnabé C et al. Evidence for genetic exchange and hybridization in Trypanosoma cruzi based on nucleotide sequences and molecular karyotype. Infection, Genetics and Evolution 2003; 2:173-83.
27. Gaunt MW, Yeo M, Frame IA et al. Mechanism of genetic exchange in American trypanosomes. Nature 2003; 421:936-939.
28. Miles MA, Povoa M, Prata A et al. Do radically dissimilar Trypanosoma cruzi strains (zymodemes) cause Venezuelan and Brazilian forms of Chagas' disease? Lancet 1981; 8234:1336-40.
29. Laurent JP, Barnabé C, Quesney V et al. Impact of clonal evolution on the biological diversity of Trypanosoma cruzi. Parasitology 1997; 114:213-18.
30. De Lana M, Pinto A Da S, Barnabé C et al. Trypanosoma cruzi: compared vectorial transmissibility of 3 major clonal genotypes by Triatoma infestans. Exp Parasitology 1998; 90:20-5.
31. Pinto A da S, de Lana M, Bastrenta B et al. Compared vectorial transmissibility of pure and mixed clonal genotypes of Trypanosoma cruzi in Triatoma infestans. Parasitol Res 1998; 84:348-53.
32. Revollo S, Oury B, Laurent JP et al. Trypanosoma cruzi: impact of clonal evolution of the parasite on its biological and medical properties. Exp Parasitol 1998; 89:30-9.
33. De Lana M, Pinto A, Bastrenta B et al. Trypanosoma cruzi: Infectivity of clonal genotypes infections in acute and chronic phases in mice. Exp Parasitol 2000; 96:61-6.
34. Pinto AS, de Lana M, Britto C et al. Experimental Trypanosoma cruzi biclonal infection in Triatoma infestans: Detection of distinct clonal genotypes using kinetoplast DNA probes. Int J Parasitol 2000; 30:843-848.
35. Toledo MJ de O, de Lana M, Carneiro CM et al. Impact of Trypanosoma cruzi clonal evolution on its biological properties in mice. Exp Parasitol 2002; 100:161-72.

36. Toledo MJ de O, Bahia MT, Carneiro CM et al. Chemotherapy with benznidazole and itraconazole for mice infected with different Trypanosoma cruzi clonal genotypes. Antimicrobial Agents and Chemotherapy 2003; 47:223-30.
37. Villarreal D, Barnabé C, Sereno D et al. Lack of correlation between in vitro susceptibility to benznidazole and phylogenetic diversity of Trypanosoma cruzi, the agent of Chagas disease. Exp Parasitol 2004; 108:24-31.
38. Vago AR, Andrade LO, Leite AA et al. Genetic characterization of Trypanosoma cruzi directly from tissues of patients with chronic Chagas disease: differential distribution of genetic types into diverse organs. Am J Pathology 2000; 156:1805-09.
39. Lent H, Wygodzinsky P. Revision of the triatominae (Hemiptera, Reduviidae) and their significance as vectors of Chagas disease. Bull Amer Mus Nat Hist 1979; 163:127-516.
40. Dujardin JP, Tibayrenc M. Etude de 11 enzymes et données de génétique formelle pour 19 loci isoenzymatiques chez Triatoma infestans (Hemiptera: Rediviidae). Ann Soc belge Méd Trop 1985; 65:271-80.
41. Anderson JM, Lai JE, Dotson EM et al. Identification and characterization of microsatellite markers in the Chagas disease vector Triatoma dimidiata. Infect Genet Evol 2002; 1:243-8.
42. Dumonteil E, Tripet F, Ramirez-Sierra MJ et al. Assessment of Triatoma dimidiata dispersal in the Yucatan Peninsula of Mexico by morphometry and microsatellite markers. Am J Trop Med Hyg 2007; 76:930-7.
43. Pacheco RS, Almeida CE, Costa J et al. RAPD analyses and rDNA intergenic-spacer sequences discriminate Brazilian populations of Triatoma rubrovaria (Reduviidae: Triatominae). Ann Trop Med Parasitol 2003; 97:757-68.
44. Garcia AL, Carrasco HJ, Schofield CJ et al. Random amplification of polymorphic DNA as a tool for taxonomic studies of Triatomine bugs (Hemiptera: Reduviidae). J Med Entomol 1998; 35:38-45.
45. Sainz AC, Mauro LV, Moriyama EN et al. Phylogeny of Triatomine vectors of Trypanosoma cruzi suggested by mitochondrial DNA sequences. Genetica 2004; 121:229-40.
46. Martínez FH, Villalobos GC, Cevallos AM et al. Taxonomic study of the Phyllosoma complex and other Triatomine (Insecta: Hemiptera: Reduviidae) species of epidemiological importance in the transmission of Chagas disease: using ITS-2 and mtCytB sequences. Mol Phylogenet Evol 2006; 41:279-87.
47. Bargues MD, Marcilla A, Ramsey JM et al. Nuclear rDNA-based molecular clock of the evolution of triatominae (Hemiptera: reduviidae), vectors of Chagas disease. Mem Inst Oswaldo Cruz 2000; 95:567-73.
48. Bargues MD, Klisiowicz DR, Gonzalez-Candelas F et al. Phylogeography and genetic variation of Triatoma dimidiata, the main chagas disease vector in central America and its position within the genus Triatoma. Plos Neglect Infec Dis 2008; 2:e233.
49. Perez R, Panzera Y, Scafiezzo S et al. Cytogenetics as a tool for Triatomine species distinction (Hemiptera-Reduviidae). Mem Inst Oswaldo Cruz 1992; 87:353-61.
50. Rodríguez Rodríguez J, Fuentes González O, Nodarse JF et al. Morphometric changes of Triatoma flavida Neiva, 1911 (Hemiptera: Triatominae) in the transition from sylvatic to laboratory conditions. Rev Inst Med Trop Sao Paulo 2007; 49:127-30.
51. Feliciangeli MD, Sanchez-Martin M, Marrero R et al. Morphometric evidence for a possible role of Rhodnius prolixus from palm trees in house re-infestation in the State of Barinas (Venezuela). Acta Trop 2007; 101:169-77.
52. Guedes PM, Veloso VM, Gollob KJ et al. IgG isotype profile is correlated with cardiomegaly in Beagle dogs infected with distinct Trypanosoma cruzi strains. Vet Immunol Immunopathol 2008; 124:163-8.
53. Guedes PM, Veloso VM, Caliari MV et al. Trypanosoma cruzi high infectivity in vitro is related to cardiac lesions during long-term infection in Beagle dogs. Mem Inst Oswaldo Cruz 2007; 102:141-7.
54. Barr SC, Warner KL, Kornreic BG et al. A cysteine protease inhibitor protects dogs from cardiac damage during infection by Trypanosoma cruzi. Antimicrob Agents Chemother 2005; 49:5160-1.
55. Barr SC, Pannabecker TL, Gilmour RF Jr et al. Upregulation of cardiac cell plasma membrane calcium pump in a canine model of Chagas disease. J Parasitol 2003; 89:381-4
56. Cuervo H, Pineda MA, Aoki MP et al. Inducible nitric oxide synthase and arginase expression in heart tissue during acute Trypanosoma cruzi infection in mice: arginase I is expressed in infiltrating CD68(+) Macrophages. J Infect Dis 2008; [Epub ahead of print].
57. Faúndez M, López-Muñoz R, Torres G et al. Buthionine sulfoximine has anti-trypanosoma cruzi activity in a murine model of acute Chagas' disease and enhances the efficacy of nifurtimox. Antimicrob Agents Chemother 2008; 52:1837-9.
58. Tibayrenc M. Human genetic diversity and epidemiology of parasitic and other transmissible diseases. Adv Parasitol 2007; 64:378-428.
59. Garcia A, Marquet S, Bucheton B et al. Linkage analysis of blood plasmodium falciparum levels: interest of the 5q31-q33 chromosome region. Am J Trop Med Hyg 1998; 58:705-9.

60. Bellamy R, Beyers N, McAdam KP et al. Genetic susceptibility to tuberculosis in Africans: a genome-wide scan. Proc Natl Acad Sci USA 2000; 97:8005-9.
61. Abel L, Sanchez FO, Oberti J et al. Susceptibility to leprosy is linked to the human NRAMP1 gene. J Infect Dis 1998; 177:133-45.
62. Dessein AJ, Marquet S, Henri S et al. Infection and disease in human schistosomiasis mansoni are under distinct major gene control. Microbes and Infection 1999; 1:561-7.
63. Bucheton B, Abel L, El-Safi S et al. A major susceptibility locus on chromosome 22q12 plays a critical role in the control of kala-azar. Am J Hum Genet 2003; 73:1052-60.
64. Brenière SF, Poch O, Selaes H et al. Specific humoral depression in chronic patients infected by Trypanosoma cruzi. Revista do Instuto de Medicina Tropical Sao Paulo 1984; 26:254-8.
65. Kierzenbaum F. Is there autoimmunity in chagas disease? Parasitology Today 1985; 1:4-6.
66. Zicker F, Slith PG, Netto JCA et al. Activity, opportunity for reinfection and sibling history of heart diseases as risk factors for Chagas' cardiopathy. Amer J Trop Med Hyg 1990; 43:498-505.
67. Morini JC, Berra H, Davila HO et al. Alteration among first degree relatives with serological evidence of Trypanosoma cruzi infection. A sibship study. Mem Inst Oswaldo Cruz 1994; 89:371-5.
68. Fae KC, Drigo SA, Cuha-Neto E et al. HLA and beta-myosin heavy chain do not influence susceptibility to Chagas disease cardiomyopathy. Microbes and Infection 2000; 2:745-51.
69. Fernandes-Mestre MT, Layrisse Z, Montagnani S et al. Influence of the HLA class II polymorphism in chronic Chagas disease. Parasite Immunol 1998; 20:197-203.
70. Williams-Blangero S, VandeBerg JL, Blangero J et al. Genetic epidemiology of seropositivity for T. cruzi infection in rural Goias, Brazil. Am J Trop Med Hyg 1997; 57:538-43.
71. Barbossa CAA, Morton NE, Pao DC et al. Biological an cultural determinants of immunoglobulin levels in a Brazilian population with Chagas' disease. Human Genetics 1981; 59:161-3.
72. Tibayrenc M. A hard lesson for Europeans: the Asean CDC. Trends Microbiol 2005; 13:266-8.

Index

R

S

T

V

W